AF731026

Association and Nonprofit Sponsors

Gold Level

Florida Golf Course Superintendents Association
Mississippi Turfgrass Association
New Jersey Turfgrass Association
Tri-State Turf Research Foundation
United States Golf Association
Western Canada Turfgrass Association
Wisconsin Sports Turf Managers Association

Silver Level

Arkansas Turfgrass Association
Canadian Turfgrass Research Foundation
Carolinas Golf Course Superintendents Association
Connecticut Association of Golf Course Superintendents
GCSANJ Foundation
Golf Course Superintendents Association of New England
Idaho Golf Course Superintendents Association
Kansas Turfgrass Foundation
Louisiana-Mississippi Golf Course Superintendents Association
New England Regional Turfgrass Foundation, Inc.
New Jersey Landscape Contractors Association
Northwest Turfgrass Association
Ontario Turfgrass Research Foundation
Peaks & Prairies Golf Course Superintendents Association
Pennsylvania Turfgrass Council
Turfgrass Producers International
Wisconsin Golf Course Superintendents Association

Bronze Level

Alabama Turfgrass Association
Golf Course Superintendents Association of America
Golf Course Superintendents Association of Cape Cod
Lone Star Golf Course Superintendents Association
New York State Turfgrass Association, Inc.
Ohio Turfgrass Foundation
Oregon Turfgrass Foundation
Rhode Island Golf Course Superintendents Association
Rocky Mountain Regional Turfgrass Association
San Diego Golf Course Superintendents Association
Tennessee Turfgrass Association

Compendium of Turfgrass Diseases

FOURTH EDITION

Edited by

Lane P. Tredway
Syngenta Professional Solutions
Greensboro, North Carolina

Maria Tomaso-Peterson
Department of Biochemistry, Molecular Biology, Entomology, and Plant Pathology
Mississippi State University
Mississippi State, Mississippi

James P. Kerns
Department of Entomology and Plant Pathology
North Carolina State University
Raleigh, North Carolina

Bruce B. Clarke
Department of Plant Biology
Rutgers, The State University of New Jersey
New Brunswick, New Jersey

APS PRESS
The American Phytopathological Society
St. Paul, Minnesota, U.S.A.

Cover photos: Front, Eastimages/Shutterstock.com. Back, Top-left, top-right, and bottom-left, Courtesy L. P. Tredway—© APS. Bottom-right, Courtesy J. A. Roberts, Clemson University—© APS.

Library of Congress Control Number: 2022936201
International Standard Book Numbers:
Print: 978-0-89054-687-1
Online: 978-0-89054-688-8
Epub: 978-0-89054-689-5

Printed in the United States of America on acid-free paper.

The American Phytopathological Society
3285 Northwood Circle, Suite 100
St. Paul, Minnesota 55121, U.S.A.

Preface

Since the publication of the first edition in 1983, nearly 50,000 copies of *Compendium of Turfgrass Diseases* have been distributed worldwide, to all continents except Antarctica. Previous editions were translated into Chinese, Japanese, and Spanish. This fourth edition incorporates the many advances in turfgrass pathology since the publication of the third edition in 2005.

The compendium is intended to serve as a practical and technical reference for all those involved in the culture of fine turf. It is written primarily for those with some training in turfgrass science and plant pathology. The editors hope it will be a valuable resource for turfgrass managers, turfgrass scientists, plant pathologists, diagnosticians, extension specialists, consultants, agribusiness personnel, regulatory officers, and instructors and students of turfgrass management, landscape architecture, plant pathology, and related fields.

This book blends descriptive terminology with the technical language of plant pathology to better serve the diverse levels of knowledge and skills of the intended audience. The text is complemented by numerous images of disease symptoms, photomicrographs and drawings of many pathogens, diagrams of disease cycles, a glossary of terms, and an index. The book includes sections that briefly describe the characteristics of turfgrasses, the damage caused by noninfectious agents in the turfgrass ecosystem, the ecology and taxonomy of fungi pathogenic to turfgrasses, integrated management approaches, and basic diagnostic procedures.

Turfgrass diseases are described as they occur in the most common grasses maintained as lawns, sports fields, golf courses, bowling greens, sod production fields, and utility areas along roadways and airport runways. The control measures described are presented as principles or strategies, because specific measures vary greatly with individual circumstances. Detailed information on currently recommended control methods should be obtained from university turfgrass disease specialists, advisory officers, or crop protection specialists. It is not possible in a book of this type to catalog all appropriate references, but the selected references cited serve as points of access to more-comprehensive literature.

The fourth edition contains 333 figures, more than 200 of which consist of new, high-quality images that highlight the diagnostic features of common turfgrass diseases. Approximately half the text is new or revised. In particular, sections on root diseases, *Rhizoctonia* diseases, and *Pythium* diseases received major revisions to reflect our current understanding of these systems. Additional content related to diseases of warm-season grasses was added throughout. New sections on common physiological disorders are also included.

With the advent of molecular phylogenetics, many changes in fungal taxonomy have been made since 2005 and are reflected in this edition. A list of common disease names and current and past Latin binomials of pathogens is provided for reference in the appendix.

This revision has benefited from the contributions and reviews of many scientists. Each section was reviewed by the four editors, and approximately half the sections were sent to specialists who reviewed them and in some cases provided complete revisions. More than 40 scientists contributed directly to the information contained in this book.

The contributions of Richard Smiley (Oregon State University) and Peter Dernoeden (University of Maryland) are deserving of special recognition. Dr. Smiley was the lead author of the first three editions of *Compendium of Turfgrass Diseases* and was instrumental in establishing it as the authoritative reference on turfgrass pathology. Dr. Dernoeden served as a coauthor of the second and third editions, along with Bruce Clarke, and his writings and images continue to be prominent throughout this edition. The editors thank Dr. Smiley and Dr. Dernoeden for their contributions to the compendium and the field of turfgrass pathology.

In his capacity as a senior editor for APS PRESS, Barry Pryor (The University of Arizona) was the liaison between the publisher and the editors. He reviewed the manuscript and coordinated the peer review, and the editors thank him and two anonymous reviewers for their careful review and suggestions to improve this publication. The editors also thank Amy Rossman (U.S Department of Agriculture–Agricultural Research Service) for reviewing the appendix and the staff at APS PRESS for facilitating the publication process. Comments regarding errors or omissions are always welcome. With your suggestions and assistance, future editions can become even more valuable.

Syngenta Professional Solutions, Mississippi State University, North Carolina State University, and Rutgers, The State University of New Jersey, generously contributed the editors' time and materials for this revision. We also recognize the following individuals who provided images for this compendium:

Michael L. Agnew, Syngenta Professional Solutions, Kennett Square, Pennsylvania
Keenan L. Amundsen, Department of Agronomy and Horticulture, University of Nebraska, Lincoln
Amy Baetsan-Young, Syngenta Crop Protection, Stanton, Minnesota
Sam Bauer, Bauer Turf, Audubon, Minnesota
Jesse J. Benelli, Bayer Environmental Science, Kitchener, Ontario, Canada
T. Karl Danneberger, Department of Horticulture and Crop Science, The Ohio State University, Dublin
Michael S. Dieckhoff, Whispering Pines Golf Club, Trinity, Texas
Callie P. Freeman, Parker BioLabs, Hampstead, North Carolina
Paul R. Giordano, Harrell's LLC, Canton, Michigan
J. Mike Goatley, School of Plant and Environmental Sciences, Virginia Tech University, Blacksburg
Pat Gradoville, Gradoville & Hertzing Management Group, Palos Verdes Estates, California
James W. Hempfling, Bayer Environmental Sciences, Murrieta, California
Geunwha Jung, Stockbridge School of Agriculture, University of Massachusetts, Amherst

Junmin Liang, Institute of Microbiology, Chinese Academy of Sciences, Beijing
Matthew C. Martin, Department of Crop and Soil Sciences, North Carolina State University, Castle Hayne
Alfredo Martinez-Espinosa, Department of Plant Pathology, University of Georgia, Griffin
Steve J. McDonald, Turfgrass Disease Solutions, Spring City, Pennsylvania
Dean K. Mosdell, Syngenta Professional Solutions, Nipomo, California
Maggie Reiter, Department of Horticultural Science, University of Minnesota, St. Paul
Robert A. Webb, Jr., GreenPoint Ag, Yazoo City, Mississippi
Alan Windham, Department of Entomology and Plant Pathology, University of Tennessee, Nashville
Michael W. Wolpoff, Arroyo Trabuco Golf Club, Cypress, California
Frank P. Wong, Bayer Environmental Science, Cary, North Carolina
Wu Zhang, Lignnan Normal University, Zhanjiang, China

Lane P. Tredway
Maria Tomaso-Peterson
James P. Kerns
Bruce B. Clarke

Contributing Authors

Lisa A. Beirn
Syngenta Crop Protection
Washington, New Jersey

Stacy A. Bonos
Department of Plant Biology
Rutgers, The State University of New Jersey
New Brunswick, New Jersey

E. Lee Butler
Department of Entomology and Plant Pathology
North Carolina State University
Raleigh, North Carolina

B. B. Clarke
Department of Plant Biology
Rutgers, The State University of New Jersey
New Brunswick, New Jersey

Jo Anne Crouch
Mycology and Nematology Genetic
Diversity and Biology Laboratory
U.S. Department of Agriculture–
Agricultural Research Service
Beltsville, Maryland

Lukas A. Dant
Syngenta Crop Protection
Simpsonville, South Carolina

Peter H. Dernoeden
Department of Plant Science and Landscape Architecture
University of Maryland
College Park, Maryland

Michael A. Fidanza
Division of Science
Pennsylvania State University–Berks
Reading, Pennsylvania

Philip F. Harmon
Department of Plant Pathology
University of Florida
Gainesville, Florida

David W. Held
Department of Entomology and Plant Pathology
Auburn University
Auburn, Alabama

Bradley I. Hillman
Department of Plant Biology
Rutgers, The State University of New Jersey
New Brunswick, New Jersey

John E. Kaminski
Department of Plant Science
The Pennsylvania State University
University Park, Pennsylvania

James P. Kerns
Department of Entomology and Plant Pathology
North Carolina State University
Raleigh, North Carolina

Donald Kobayashi
Department of Plant Biology
Rutgers, The State University of New Jersey
New Brunswick, New Jersey

Paul L. Koch
Department of Plant Pathology
University of Wisconsin
Madison, Wisconsin

S. Bruce Martin
Department of Plant and Environmental Sciences
Clemson University
Florence, South Carolina

Ben A. McGraw
Department of Plant Science
The Pennsylvania State University
University Park, Pennsylvania

William A. Meyer
Department of Plant Biology
Rutgers, The State University of New Jersey
New Brunswick, New Jersey

G. Lee Miller, Jr.
Department of Botany and Plant Pathology
Purdue University
West Lafayette, Indiana

James A. Murphy
Department of Plant Biology
Rutgers, The State University of New Jersey
New Brunswick, New Jersey

Eric H. Reasor
PBI-Gordon Corporation
Shawnee, Kansas

Joseph A. Roberts
Department of Plant and Environmental Sciences
Clemson University
Florence, South Carolina

Richard W. Smiley
Oregon State University
Columbia Basin Agricultural Research Center
Pendleton, Oregon

Maria Tomaso-Peterson
Department of Biochemistry, Molecular Biology, Entomology, and Plant Pathology
Mississippi State University
Mississippi State, Mississippi

Lane P. Tredway
Syngenta Professional Solutions
Greensboro, North Carolina

James F. White, Jr.
Department of Plant Biology
Rutgers, The State University of New Jersey
New Brunswick, New Jersey

Contents

Compendium of Turfgrass Diseases

Introduction

Turfgrasses have been recognized for their importance to the quality of life for more than 2,000 years. Historical records establish that turfgrasses grew in the vast palace gardens of the emperor of the Han dynasty in China (100 B.C.E.); in the Persian garden carpets in the kingdom of Assyria (500 C.E.); on the sports fields of Akbar the Great, emperor of Hindustan (1600 C.E.); and in the lawn gardens of Britain in the thirteenth century.

Today, turfgrasses are cultivated in nearly all inhabited regions of the world. Turf species and cultivars of the family Poaceae (Gramineae) are remarkably adaptable; some have been adapted to subarctic regions and others to equatorial regions. Turfgrasses serve us in many important ways. As ornamental plants, they add beauty to the environment and improve the aesthetic value of our communities. The functional benefits of turfgrasses are often less obvious but may be more important than aesthetics. By serving as playing fields for many sports, they provide recreational spaces and help to limit injuries common to vigorous sports. Turfgrass is widely used to control water flow, sediment deposition, and wind erosion. It provides a utilitarian cover around houses and public and commercial buildings and in parks, cemeteries, and other facilities. Turfgrass adds value to property and reduces air and noise pollution, heat buildup, visual pollution, and glare. Turf is maintained on and around airfields to reduce dust, thereby extending airplane engine life and increasing safety. Similar benefits are obtained from turf maintained along highways.

The turfgrass industry is one of the largest segments of agriculture and horticulture in the United States, but the size and value of the industry are difficult to estimate since turf is a perennial crop and once established can remain in place for generations. A 2004 survey of the turfgrass industry in New York State revealed that turfgrass covered 10% of the land area of the state and was maintained mostly around private residences (82% of the total area planted with turfgrass); golf courses (3%), parks (2%), and other amenity uses accounted for smaller portions of the total. According to a 1999 survey in North Carolina, turfgrasses were grown on over 2.1 million acres in that state, with 69% around single-family dwellings, 16% along roadsides, 6% surrounding commercial properties, and 3% on golf courses. In 2002, the U.S. turfgrass industry was estimated to generate total revenue impacts of US$57.9 billion, employment impacts of 822,849 jobs, and value-added impacts of US$35.1 billion. Although golf courses comprise a small portion of turfgrass acreage, they generated approximately 40% of these economic and employment impacts.

Turfgrass plants are affected by numerous pathogens, insects, and physiological disorders. Estimating annual losses caused by disease is difficult, given that yield is not quantified in most turfgrass systems. Nevertheless, fungal diseases are commonly listed among the most significant challenges to turfgrass maintenance in humid climates, and approximately US$150 million is spent annually on turfgrass fungicides in the United States alone. Pathologists and other turfgrass scientists often focus on disease control as a primary research objective, and pathologists serve as key members of research institutes dedicated to improving turfgrass culture. Their collective efforts have greatly advanced our ability to diagnose diseases accurately and manage them effectively with an integrated approach including the use of resistant cultivars, suppressive cultural practices, and fungicide protection.

The first monograph on turf diseases and their control was by Monteith and Dahl in 1932. The science of turfgrass pathology gained more formal recognition with the publication of several reference works, including those by Howard et al. in 1951, Smith in 1959, Couch in 1962, Beard in 1973, Vargas in 1981, Smiley in 1983, and Tani and Beard in 1997. Several of these texts have been revised, and numerous other books and publications on turfgrass pathology and disease management are now available. The most definitive texts on turfgrass pathology as of this writing are those of Smith et al., published in 1989, Couch in 1995, and Latin in 2021. These excellent resources continue to serve as standard technical references on turfgrass pathology and are referenced throughout this fourth edition of *Compendium of Turfgrass Diseases*.

Selected References

Beard, J. B. 1973. Turfgrass: Science and Culture. Prentice-Hall, Englewood Cliffs, NJ.

Beard, J. B. 2002. Turf Management for Golf Courses. 2nd ed. Ann Arbor Press, Chelsea, MI.

Couch, H. B. 1962. Diseases of Turfgrasses. Reinhold, New York.

Couch, H. B. 1995. Diseases of Turfgrasses. 3rd ed. Krieger Publishing, Malabar, FL.

Farr, D. F., and Rossman, A. Y. n.d. Fungal Databases, Systematic Mycology and Microbiology Laboratory. U.S. Department of Agriculture, Agricultural Research Service, Washington, DC. https://nt.ars-grin.gov/fungaldatabases/

Haydu, J. J., Hodges, A. W., and Hall, C. R. 2002. Economic impacts of the turfgrass and lawncare industry in the United States. Univ. Fla. Inst. Food Agric. Sci. Coop. Ext. Publ. FE632. https://edis.ifas.ufl.edu/pdf/FE/FE63200.pdf

Howard, F. L., Rowell, J. B., and Keil, H. L. 1951. Fungus diseases of turfgrasses. Univ. R.I. Agric. Exp. Stn. Bull. 308.

Kerns, J. P., and Tredway, L. P. 2013. Advances in turfgrass pathology since 1990. Pages 733-776 in: Turfgrass: Biology, Use, and Management. Agron. Monogr. 56. J. C. Stier, B. P. Horgan, and S. A. Bonos, eds. Crop Science Society of America, Madison, WI.

Larsen, P. O., and Joyner, B. G., eds. 1980. Advances in Turfgrass Pathology. Harcourt Brace Jovanovich, Duluth, MN.

Latin, R. 2021. A Practical Guide to Turfgrass Fungicides. 2nd ed. American Phytopathological Society, St. Paul, MN.

Monteith, J., Jr., and Dahl, A. S. 1932. Turf diseases and their control. Bull. U.S. Golf Assoc. Green Sect. 12:85-188.

New York Agricultural Statistics Service. 2004. New York Turfgrass Survey. New York State Turfgrass Association, Latham, NY.

Phipps, M. S. 1999. 1999 North Carolina Turfgrass Survey. N.C. Dep. Agric. Consum. Serv. Publ. 199. https://repository.lib.ncsu.edu/handle/1840.20/38509

Roberts, E. C., and Roberts, B. C. 1989. Lawn and Sports Turf Benefits. Lawn Institute, Pleasant Hill, TN.
Schumann, G. L., and MacDonald, J. D. 1997. Turfgrass Diseases: Diagnosis and Management. CD-ROM. American Phytopathological Society, St. Paul, MN.
Smiley, R. W. 1983. Compendium of Turfgrass Diseases. American Phytopathological Society, St. Paul, MN.
Smith, J. D. 1959. Fungal Diseases of Turfgrasses. Sports Turf Research Institute, Bingley, U.K.
Smith, J. D., Jackson, N., and Woolhouse, A. R. 1989. Fungal Diseases of Amenity Turf Grasses. E. & F. N. Spon, London.
Tani, T., and Beard, J. B. 1997. Color Atlas of Turfgrass Diseases. Ann Arbor Press, Chelsea, MI.
Turgeon, A. J., and Kaminski, J. E. 2019. Turfgrass Management. Turfpath, State College, PA.
Vargas, J. M., Jr. 1981. Management of Turfgrass Diseases. Burgess Publishing, Minneapolis, MN.
Vargas, J. M., Jr. 2005. Management of Turfgrass Diseases. 3rd ed. John Wiley & Sons, Hoboken, NJ.

(Prepared by R. W. Smiley, P. H. Dernoeden, and B. B. Clarke; revised by J. P. Kerns and L. P. Tredway)

Grasses Managed as Turfs

The family Poaceae (formerly Gramineae) is composed of more than 10,000 species in 700 genera. About 40 species in 19 genera, belonging to the subfamilies Pooideae, Panicoideae, and Chloridoideae, are commonly managed as turfgrasses (Table 1). Each subfamily is further subdivided into tribes, seven of which include turfgrasses. Commercial cultivars are

TABLE 1. Taxonomy and Names of Selected Turfgrasses in the Family Poaceae[a,b]

Subfamily, Tribe, and Subtribe	Genus and Species	Common Name
Subfamily Pooideae (cool-season grasses)		
Tribe Poeae		
Subtribe Loliinae	*Bromus inermis* Leyss. subsp. *inermis*	Smooth brome
	Festuca arundinacea (Schreb.)	Tall fescue
	Festuca brevipila Tracey	Hard fescue
	Festuca ovina L. (syn. *F. ovina* subsp. *hirtula* (Hack. ex Travis) M. J. Wilk.	Sheep fescue
	Festuca pratensis Huds.	Meadow fescue
	Festuca rubra L. subsp. *rubra* Gaudin	Strong creeping red fescue
	Festuca rubra subsp. *commutata* Gaudin (syn. *F. rubra* subsp. *fallax* (Thuill.) Nyman	Chewings fescue
	Festuca rubra subsp. *littoralis* (G. Mey.) Auquier	Slender creeping red fescue
	Lolium multiflorum Lam.	Italian (annual) ryegrass
	Lolium perenne L.	Perennial ryegrass
	Poa annua L.	Annual bluegrass
	Poa compressa L.	Canada bluegrass
	Poa pratensis L.	Kentucky bluegrass
	Poa trivialis L.	Roughstalk bluegrass
Subtribe Coleanthinae	*Puccinellia distans* (Jacq.) Parl.	Weeping alkaligrass
Subtribe Agrostidinae	*Agrostis canina* L.	Velvet bentgrass
	Agrostis capillaris L.	Colonial bentgrass
	Agrostis castellana Boiss. & Reut.	Dryland browntop, highland bentgrass
	Agrostis gigantea Roth	Redtop
	Agrostis stolonifera L.	Creeping bentgrass[c]
Subtribe Holcinae	*Deschampsia cespitosa* (L.) P. Beauv.	Tufted hairgrass
Subtribe Aveninae	*Koeleria macrantha* (Ledeb.) Schult.	Junegrass, crested hairgrass
Subtribe Poinae	*Phleum nodosum* L.	Turf timothy
	Phleum pratense L.	Timothy
Tribe Triticeae	*Agropyron cristatum* (L.) Gaertn.	Crested wheatgrass
Subfamily Chloridoideae (warm-season grasses)		
Tribe Cynodonteae	*Cynodon dactylon* (L.) Pers.	Bermudagrass, common couchgrass
	Cynodon incompletus Nees var. *hirsutus* (Stent) de Wet & J. R. Harlan	Bradley Bermudagrass
	Cynodon × *magennisii* Hurcombe	Magennis Bermudagrass
	Cynodon transvaalensis Burtt Davy	African Bermudagrass
	Cynodon dactylon (L.) Pers. × *C. transvaalensis* Burtt Davy	Hybrid Bermudagrass
	Bouteloua curtipendula (Michx.) Torr.	Sideoats grama
	Bouteloua gracilis (Kunth) Lag. ex Griffiths	Blue grama
	Bouteloua dactyloides (Nutt.) J. T. Columbus (syn. *Buchloë dactyloides* (Nutt.) Engelm.)	Buffalograss
Tribe Zoysiinae	*Zoysia japonica* Steud.	Japanese lawngrass
	Zoysia matrella (L.) Merr.	Manila grass, Japanese carpetgrass
	Zoysia tenuifolia Willd. ex Thiele	Mascarene grass
Subfamily Panicoideae (warm-season grasses)		
Tribe Paspaleae	*Axonopus compressus* (Sw.) P. Beauv.	Tropical carpetgrass, carpetgrass
	Axonopus fissifolius (Raddi) Kuhlm.	Common carpetgrass
	Paspalum notatum Flüggé	Bahiagrass
	Paspalum vaginatum Sw.	Seashore paspalum
Tribe Paniceae	*Pennisetum clandestinum* Hochst. ex Chiov.	Kikuyugrass
	Stenotaphrum secundatum (Walter) Kuntze	St. Augustine grass
Tribe Andropogoneae	*Eremochloa ophiuroides* (Munro) Hack.	Centipede grass

[a] Prepared by R. W. Smiley, P. H. Dernoeden, and B. B. Clarke; revised by S. A. Bonos—© APS.
[b] A key source of current scientific names of turfgrasses is the U.S. Department of Agriculture, Agricultural Research Service, National Genetic Resources Program, Germplasm Resources Information Network (GRIN). www.ars-grin.gov/
[c] Creeping bentgrass is recognized in many regions by the name *A. stolonifera* var. *palustris* (Huds.) Farw.

derived by natural selection, traditional breeding of parental plants of the same species, and molecular techniques.

As a group, turfgrasses have diverse characteristics (Fig. 1). They are grown around the world, often with widely divergent levels of management. Many important diseases occur in grasses managed as turfs. Diagnoses of turfgrass diseases are expedited by identification of the host species. The following descriptions of turfgrass genera and the accompanying Key to Some Genera of the Family Poaceae Used as Turfgrasses are provided as aids to identification of turfgrass species.

Turfgrasses in the subfamily Pooideae are often referred to as "cool-season" grasses because they are adapted for optimal growth in temperate and subarctic climates. Optimal growth occurs at 15–25°C (60–75°F). Their geographic range is restricted by their lack of adaptation to prolonged or severe high temperatures or drought. These grasses are classified as C3 grasses because carbon is fixed primarily via the Calvin (C3) cycle during photosynthesis. Most turfgrasses in cool, humid climates belong to the genera *Agrostis, Festuca, Lolium,* and *Poa*. Limited culture of grasses in the genera

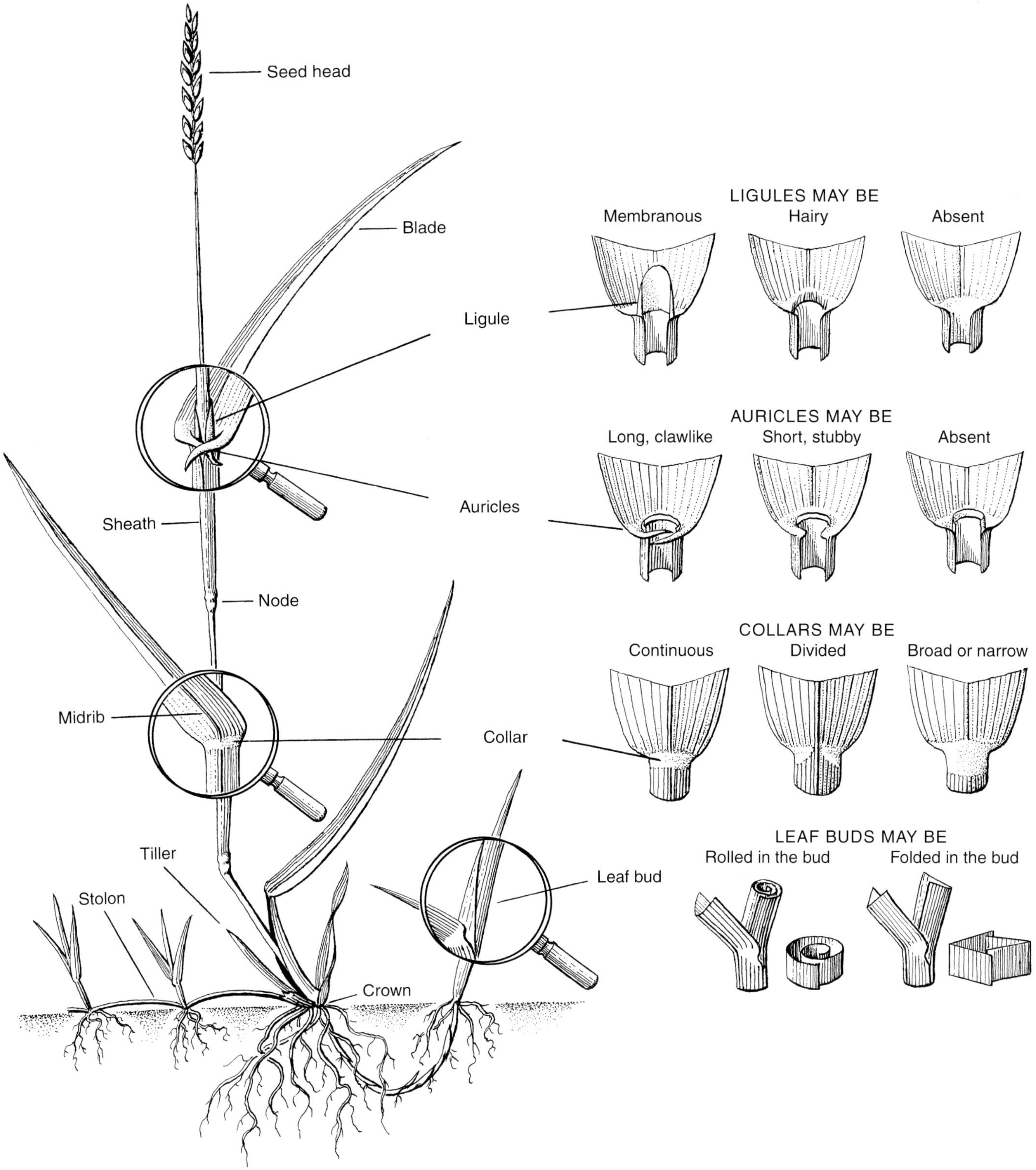

Fig. 1. Parts of a grass plant. (Reproduced, by permission, from Scotts Information Manual for Lawns, 1979, O. M. Scott & Sons, Marysville, OH.)

Key to Some Genera of the Family Poaceae Used as Turfgrasses[a]

I. Leaf blade rolled in the bud shoot (Fig. 1)
- A. Auricles absent or rudimentary
  - 1. Ligule a fringe of hairs
    - a. Rhizomes absent; stolons present .... *Buchloë*
    - b. Rhizomes and stolons present .... *Zoysia*
  - 2. Ligule membranous
    - a. Hairs absent on sheath, blade, and collar
      - 1) Sheath closed .... *Bromus*
      - 2) Sheath split (margins generally overlapping)
        - a) Ligule with prominent notch .... *Phleum*
        - b) Ligule without prominent notch
          - (i) Leaf blade and margins very rough; ligule 3–15 mm long .... *Deschampsia*
          - (ii) Leaf blade smooth; ligule 0.4–5 mm long, toothed, truncate, pointed, or round .... *Agrostis*
    - b. Hairs present on sheath, blade, or collar
      - 1) Stolons or rhizomes present .... *Paspalum*
      - 2) Stolons and rhizomes absent .... *Koeleria*
- B. Auricles present
  - 1. Blade glossy on undersurface; margin of ligule not ciliate; auricles blunt to clawlike
    - a. Auricles with short hairs .... *Festuca* (*F. arundinacea*)
    - b. Auricles without hairs .... *Lolium* (*L. multiflorum*)
  - 2. Blade not glossy on undersurface; margin of ligule ciliate or lacerate; auricles stiff, clawlike .... *Agropyron*

II. Leaf blade folded in bud shoot
- A. Auricles present, without hairs; lower sheath red at base; blade glossy on undersurface .... *Lolium* (*L. perenne*)
- B. Auricles absent
  - 1. Ligule a fringe of hairs
    - a. Rhizomes and stolons present
      - 1) Sheaths greatly overlapping .... *Cynodon*
      - 2) Sheaths not greatly overlapping .... *Pennisetum*
    - b. Stolons present; rhizomes absent
      - 1) Blade petiolate above ligule .... *Stenotaphrum*
      - 2) Blade not petiolate above ligule
        - a) Blade 1–2.5 mm wide, with sharp point, pubescent on upper surface near base .... *Bouteloua*
        - b) Blade 4–8 mm wide, obtuse .... *Axonopus*
  - 2. Ligule membranous
    - a. Ligule short membranous, short ciliate; collar continuous, broad, pubescent .... *Eremochloa*
    - b. Ligule membranous
      - 1) Blade prominently ridged on upper surface, narrow to bristle-like .... *Festuca* (*F. ovina, F. rubra*)
      - 2) Blade with boat-shaped tip, not prominently ridged or bristle-like .... *Poa*

[a] Prepared by R. W. Smiley, P. H. Dernoeden, and B. B. Clarke—© APS.

Agropyron, Ammophila, Bromus, Cynosurus, Dactylis, Deschampsia, Koeleria, Phleum, and *Puccinellia* also occurs in these regions.

Turfgrasses in the subfamilies Chloridoideae and Panicoideae are commonly called "warm-season" grasses because they are best adapted for growth in tropical and subtropical climates. Chloridoid grasses are also extensively grown in warm temperate climates. Optimal growth of warm-season grasses occurs at 25–35°C (75–95°F). These grasses are classified as C4 grasses because carbon fixation during photosynthesis occurs mainly via the dicarboxylic acid (C4) pathway. The geographic range of C4 grasses is restricted primarily by their lack of tolerance to prolonged or severe cold. Most turfgrasses in warm climates belong to the genera *Bouteloua, Buchloë, Cynodon,* and *Zoysia* in the subfamily Chloridoideae and the genera *Axonopus, Eremochloa, Paspalum, Pennisetum,* and *Stenotaphrum* in the subfamily Panicoideae.

Decumbent broadleaf plants are also used as well-managed amenity turfs in some regions. This compendium does not include diseases of *Cotula* spp. in New Zealand, *Dichondra* and *Trifolium* spp. in North America, or other turfs that are not members of Poaceae.

Turfgrass Genera

Agropyron. Wheatgrasses (*Agropyron* spp.) are coarse grasses with high tolerance for cold and drought. Several species are used for low-maintenance lawns, fairways, and roadside turfs in semiarid north temperate regions.

Agrostis. Bentgrasses (*Agrostis* spp.) are generally low-growing, fine-textured grasses. They can make high-quality greens for golf and bowling. They are susceptible to diseases, especially when managed with low cutting heights. Bentgrasses are adapted to moderately fertile, acidic soils that drain well. Four types (colonial, creeping, velvet, and dryland bentgrasses) are cultivated as turfgrasses. A fifth (redtop) is of limited use.

Axonopus. Carpetgrasses (*Axonopus* spp.) are coarse, light green, low-growing grasses adapted to soils of low fertility in subtropical regions. They are intolerant of low temperatures and drought. Two species of *Axonopus* are cultivated for use in lawns, along roads, and in other utility areas.

Bouteloua. The genus *Bouteloua* contains about 50 species, two of which are used as turfgrasses: *B. gracilis* (blue grama) and *B. curtipendula* (sideoats grama).

Blue grama is a fine-textured grass adapted to alkaline soils of low fertility in semiarid transitional climates. It has good heat and drought tolerance. It is used on roadsides and other low-maintenance sites and is sometimes used in secondary roughs on golf courses. Its range is similar to that of buffalograss (*B. dactyloides*).

Sideoats grama is occasionally planted as a turfgrass in semiarid regions. It is less tolerant of drought than blue grama. Unlike blue grama, it has a rolled vernation.

Bromus. Bromegrasses (*Bromus* spp.) are coarse grasses planted for soil stabilization on low-use sites, such as roadsides, in the central and western United States. They have good tolerance to drought, heat, and cold but are intolerant of close mowing or frequent traffic.

Buchloë. Buffalograss (*B. dactyloides*) is a fine-textured and strong sod-forming grass native to North America. It is tolerant of alkaline and droughty soils and extreme cold and heat. Buffalograss is used as a low-maintenance turf in semiarid temperate regions. Improved cultivars have been developed for lawns and golf course fairways and roughs.

Cynodon. Bermudagrasses, or couchgrasses (*Cynodon* spp.), are the most common warm-season turfgrasses grown throughout the warm, humid, and dry regions of the world. Their range extends from tropical regions to cool regions. Several species and hybrids are cultured for diverse uses, including golf greens and fairways, lawns, and athletic fields. Bermudagrasses are adapted to a wide range of soil conditions, but they are intolerant of dense shade, extremely cold climates, and very wet soils.

Deschampsia. Grasses in the genus *Deschampsia* vary from coarse to fine textured. Tufted hairgrass (*D. cespitosa*) performs well as a turf under shade and low-maintenance conditions. The primary disadvantage of *Deschampsia* spp. is their susceptibility to damage by insects, especially billbugs (*Sphenophorus* spp.). Information on disease susceptibility and stress tolerance in *Deschampsia* spp. is limited, but rust caused by *Puccinia* spp. has been reported to be severe in the northeastern United States.

Eremochloa. Centipedegrass, or Chinese lawngrass (*Eremochloa ophiuroides*), is widely used as a lawn and utility turf in the subtropics and tropics. It grows well in moist, acid, and infertile soils. It requires little maintenance, but it is intolerant of intensive wear.

Festuca. Fescues (*Festuca* spp.) vary from fine-leaf to coarse bunch types and sod-forming species. Fine-leaf fescues are grown in monoculture or planted in seed mixtures with *Agrostis, Poa,* and *Lolium* spp. and *F. arundinacea* (tall fescue). Fine-leaf fescues grow well at droughty or shaded sites and on less fertile and more acidic soils, but they do not tolerate low mowing or traffic during hot, dry weather. Tall fescue is a medium to coarse species that tolerates a wide range of soil conditions and moderate shade and has good wear tolerance. Five species of *Festuca* are used as turfgrasses.

Koeleria. Junegrass (*Koeleria macrantha*) is an important native prairie grass in parts of North America. As a turfgrass, it forms a dense, medium green turf. Junegrass performs well under low-maintenance conditions and prefers sandy soils.

Lolium. Perennial ryegrass (*Lolium perenne*) is extensively grown in monoculture or planted in seed mixtures with *Festuca* and *Poa* spp. It may be overseeded into *Cynodon* spp. to provide a green cover in winter. Perennial ryegrass tolerates a wide range of soil conditions but requires good soil fertility if it is to produce a high-quality turf. It may exhibit poor quality during seed head production and during periods of heat and drought stress, and it is intolerant of ice cover. Many diverse cultivars of perennial ryegrass have been developed, and they are susceptible to many diseases.

Italian ryegrass, or annual ryegrass (*L. multiflorum*), is used mostly in inexpensive seed mixtures to provide a temporary ground cover.

Paspalum. Bahiagrass (*Paspalum notatum*) is a coarse grass that grows well in sandy, infertile soils in subtropical regions. It is intolerant of low temperatures and extensive wear. This grass is used mostly as a turf for utility areas, such as roadsides.

Seashore paspalum (*P. vaginatum*) is a dense-growing, dark green, fine grass adapted to tropical and subtropical climates. It tolerates higher salt levels than many other turfgrass species. Improved, vegetatively propagated cultivars are used on golf greens, tees, and fairways where there are salinity problems.

Pennisetum. Kikuyugrass (*Pennisetum clandestinum*) is a medium to coarse grass adapted to fertile, moist soils. It has good heat, drought, and wear tolerance. Kikuyugrass is found in moist tropical regions of Mexico, Australia, and Africa. It is also found in coastal regions in southern California, where it is often considered a weed but can be managed as turf.

Phleum. Timothy grasses (*Phleum* spp.) are coarse and are added to seed mixtures for athletic fields in northern Europe and Asia and for roadsides in temperate regions of the United States. Two species of *Phleum* are used as turfgrasses.

Poa. Bluegrasses (*Poa* spp.) are the most common turfgrasses in humid temperate regions. Four species are used as turfgrasses.

Kentucky bluegrass, or smooth meadowgrass (*P. pratensis*), produces a dense, attractive turf that grows poorly on infertile and acid soils and readily becomes dormant during periods of drought. There is great variation in disease susceptibility among cultivars.

Annual bluegrass, or annual meadowgrass (*P. annua*), is widely distributed and often becomes dominant in moist and fertile soils. Annual bluegrass is generally not planted but is managed as a turf in many areas.

Roughstalk bluegrass (*P. trivialis*) grows well in wet and shaded sites. It is overseeded in winter-dormant Bermudagrass on golf courses. In cooler regions, it is generally considered a weed, especially when it is a contaminant in seed lots of other grasses.

Texas bluegrass (*P. arachnifera*) is native to parts of Texas and Oklahoma and is more tolerant of heat and drought stress than Kentucky bluegrass. This species is being bred with Kentucky bluegrass to produce hybrids incorporating the desirable characteristics of both species. When planted as turfgrass, these hybrids may expand the range of Kentucky bluegrass to the southern United States, where better heat tolerance and drought tolerance are needed.

Puccinellia. Alkaligrasses (*Puccinellia* spp.) are used as turfgrasses in saline and sodic soils of semiarid temperate regions. They are adapted for growth along roadways, especially where salt is spread for ice removal.

Stenotaphrum. St. Augustine grass (*Stenotaphrum secundatum*) is a coarse grass widely used in lawns in the subtropics and tropics. It is best adapted for growth in moist and fertile soils and shaded sites. Its zone of adaptation is restricted by its intolerance of cold.

Zoysia. Zoysiagrasses (*Zoysia* spp.) are medium, heat- and drought-tolerant grasses that can survive in most climatic regions from the tropics to the subarctic. There is considerable variation in cold tolerance and leaf texture among *Zoysia* spp. In general, growing regions for zoysiagrasses are limited to warm temperate and warm humid climates. Zoysiagrasses tolerate poor soils but are slow to become established. Three species, *Z. japonica* (Japanese lawngrass), *Z. matrella* (Manila grass), and *Z. tenuifolia* (Mascarene grass), and two hybrids are cultivated as turfgrasses.

Selected References

Beard, J. B. 2002. Turf Management for Golf Courses. 2nd ed. Ann Arbor Press, Chelsea, MI.

Bonos, S. A., and Huff, D. R. 2013. Cool-season turfgrass breeding. Pages 591-600 in: Turfgrass: Biology, Use, and Management. Agron. Monogr. 56. J. Stier, B. Horgan, and S. A. Bonos, eds. American Society of Agronomy, Madison, WI.

Braun, R. C., Patton, A. J., Watkins, E., Koch, P. L., Anderson, N. P., Bonos, S. A., and Brilman, L. A. 2020. Fine fescues: A review of the species, their improvement, production, establishment, and management. Crop Sci. 60:1142-1187.

Hanna, W., Raymer, P., and Schwartz, B. 2013. Warm-season grasses: Biology and breeding. Pages 543-590 in: Turfgrass: Biology, Use, and Management. Agron. Monogr. 56. J. Stier, B. Horgan, and S. A. Bonos, eds. American Society of Agronomy, Madison, WI.

McCarty, L. B. 2004. Best Golf Course Management Practices. 2nd ed. Prentice Hall, Upper Saddle River, NJ.
Meyer, W. A., and Funk, C. R. 1989. Progress and benefits to humanity from breeding cool-season grasses for turf. Pages 31-48 in: Contributions from Breeding Forage and Turf Grasses. K. H. Asay, J. F. Pedersen, and D. A. Sleper, eds. Crop Science Society of America, Madison, WI.
Turgeon, A. J., and Kaminski, J. E. 2019. Turfgrass Management. Turfpath, State College, PA.

(Prepared by R. W. Smiley, P. H. Dernoeden, and B. B. Clarke; revised by S. A. Bonos and W. A. Meyer)

Diseases and Their Causal Agents

When a pathogenic organism or an environmental factor causes an abnormality in the physiological processes and morphological development of a plant, the plant is considered to have a disease. Because plants are highly responsive to environmental extremes, a healthy plant must be fully adapted to the region where it is grown and to the use for which it is intended. Ideal conditions are seldom achieved, however, because many factors may alter the manufacture, translocation, and utilization of plant photosynthates, mineral nutrients, and water.

Agents of disease can be classified as either infectious or noninfectious. Infectious agents that cause turfgrass diseases include fungi, bacteria, nematodes, viruses, and mollicutes (Fig. 2). These biotic (living) entities invade a susceptible host plant, establish a parasitic relationship that causes disease symptoms, and then reproduce and spread to cause additional infections. Noninfectious agents include biotic or abiotic (nonliving) factors. Certain organisms such as algae, slime molds, or fairy ring fungi cause important turfgrass diseases, but they are classified as noninfectious since they do not invade the plant or establish a parasitic relationship. Abiotic factors, such as mechanical wear or extremes in temperature, light, or moisture, are also very common causes of turfgrass disease.

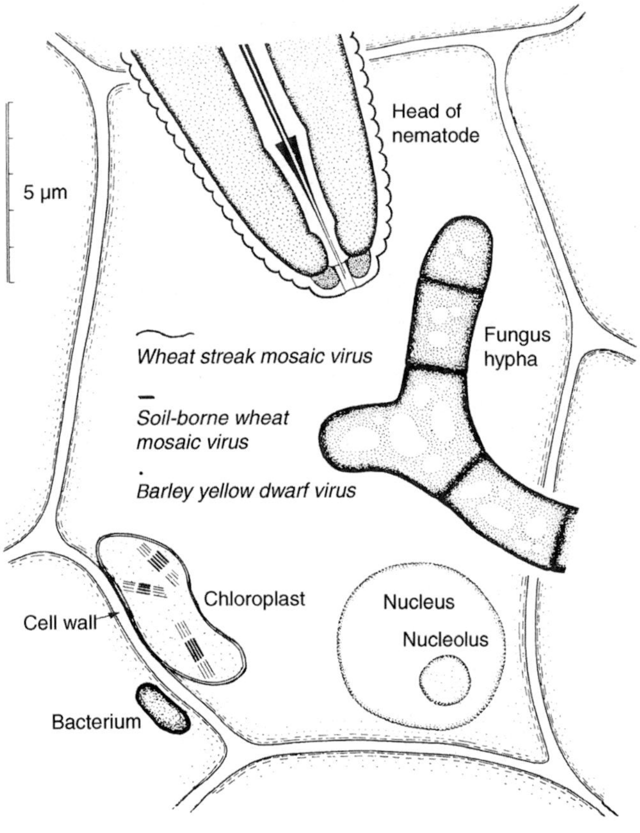

Fig. 2. Relative sizes of plant pathogens and a host plant cell. (Courtesy M. V. Wiese—© APS. Adapted from Agrios, G. N. 1969. Plant Pathology. Academic Press, New York. Reproduced, by permission, from Wiese, M. V. 1977. Compendium of Wheat Diseases. American Phytopathological Society, St. Paul, MN.)

Noninfectious agents are independently capable of causing disease, but they may also cause plants to become more susceptible to attack by infectious agents. Noninfectious and infectious agents can act in unison to cause complex, poorly defined diseases. In such instances, it may become necessary, if practical, to manage noninfectious agents before or at the same time that the infectious agents are managed. Management of noninfectious agents of disease often helps to prevent the occurrence of damaging levels of infectious diseases. For this reason, the control strategies discussed in this compendium emphasize the prevention of environmental and cultural conditions that predispose plants to infection by pathogens.

Infectious Agents

Fungi. Fungi cause the vast majority of diseases occurring in turfgrasses. Fungi were once considered to be close relatives of plants, but they are now recognized as belonging to a unique kingdom, Fungi, because their cell walls are composed of chitin and they contain no photosynthetic pigments. Fungi are heterotrophic organisms, and most are free living, serving important roles in the ecosystem such as decomposition of organic matter. Some fungi, however, can parasitize living plants and cause disease in the process.

Most fungi are multicellular and grow by producing microscopic, branching filaments called "hyphae." A mass of hyphae that is visible to the naked eye is called "mycelium." Many fungi also produce microscopic spores that aid in survival, dissemination, or infection. These spores may be produced directly from hyphae or within various spore-producing structures, many of which are visible to the naked eye or with a hand lens. Fungal spores are often referred to by more specific names based on how or where they are produced (e.g., conidia, ascospores, or basidiospores).

Many fungi undergo distinct phases of asexual and sexual reproduction that appear very different from one another. This hampered efforts to classify fungi for many years because the link between these different life forms was not always apparent, and separate nomenclatures were adopted for the asexual stages (anamorphs) and sexual stages (teleomorphs) of fungi. For example, the gray leaf spot pathogen could be referred to by the name of its anamorph, *Pyricularia oryzae,* or its teleomorph, *Magnaporthe oryzae.* With advances in DNA sequencing and other molecular techniques, it is now possible to definitively link the two life stages under one classification system. The International Code of Botanical Nomenclature now mandates a "one fungus, one name" approach, thus eliminating dual nomenclature for anamorphs and teleomorphs. The species epithets used in this volume are those currently recommended by the U.S. National Fungus Collections, maintained by the U.S. Department of Agriculture–Agricultural Research Service. Some common synonyms are provided in the appendix to this compendium, Diseases of Turfgrasses, and on the website of The American Phytopathological Society in the section "Common Names of Plant Diseases."

Chromists. Several important plant pathogens are currently placed in the kingdom Chromista, a collection of single-celled and multicelled eukaryotes that share a common ancestry unique from plants or fungi. Most chromists are capable of photosynthesis using chlorophyll c, whereas plants utilize chlorophyll a and b. The plant-pathogenic members of this group lack chlorophyll but share many other unique genetic and physiological characteristics; it is postulated that the plant pathogens lost the ability to photosynthesize and reverted to heterotrophy over evolutionary time.

Pythium and *Sclerophthora* spp. are often referred to as "water molds" because of their production of zoospores that are mobile in films of water. They are also commonly referred to as "oomycetes," a relic of when they were classified as fungi in the phylum Oomycota. While these organisms may appear similar to fungi in some ways, they are very different on a cellular level and are not sensitive to most fungicides used to control true fungi. Because of their unique adaptations to aqueous environments, these pathogens are most damaging under excessively wet conditions and can often be managed with proper irrigation practices and improvements in soil drainage. Classification of the oomycetes in the kingdom Chromista is somewhat controversial, with some researchers placing them in the kingdom Stramenopila.

Labyrinthula terrestris, the cause of rapid blight, belongs to a group referred to as "marine slime molds" that thrive in saline, aqueous environments. *L. terrestris* colonies form an ectoplasmic net, within which vegetative cells move and secrete digestive enzymes to garner nutrition from their substrate. Zoospores with flagella similar to those of *Pythium* spp. are released from sporangia to aid in dissemination. Rapid blight has been documented in cool- and warm-season turfgrasses managed in saline soils, especially during periods of prolonged drought. Planting salt-tolerant turfgrass species and reducing salinity levels in the soil are effective practices to manage *L. terrestris.*

Nematodes. Nematodes are tiny but complex animals (roundworms) that live primarily as saprophytes on dead plants or animals. Some parasitic species feed on the roots or shoots of turfgrasses, puncturing the plant cells and extracting nutrients with their needlelike mouthparts. Nematodes are abundant in turfgrass soils and cause various types of root dysfunction. They also cause wounds that enable fungi and other infectious agents to enter plants. These animals are important agents of disease in turfgrasses.

Bacteria. Bacteria are tiny, prokaryotic organisms that contain cytoplasm enclosed by cell walls but lack organized nuclei or other organelles. They do not require a living host for replication and growth. Most bacteria are saprophytes, but a few are infectious agents. The pathogenic species are short and rod shaped, about 0.5×2.0 µm. They have no means of penetration of plants and thus must enter a host through natural openings or through wounds created by insects, animals, or machines. Infected plants display a wide range of symptoms. Bacterial diseases of grasses are best controlled by planting resistant species or cultivars.

Most of the algal species associated with turfgrasses are also classified as bacteria. These cyanobacteria, or blue-green algae, are commonly observed on turfgrass surfaces during periods of wet or humid weather. In many cases, they are secondary colonizers occupying areas where turf density is low, but some are documented to produce herbicidal compounds that may directly injure turfgrasses.

Mollicutes are a group of bacteria that lack true cell walls. Phytoplasmas and spiroplasmas are two types of mollicutes that are documented to cause plant disease. These organisms are limited to the phloem of plants and are transmitted by the feeding activities of leafhoppers or other insects. Mollicutes are known to occur in grasses, but they are seldom detected as agents of diseases. Mollicutes can be controlled with tetracycline antibiotics or by limiting their transmission by insect vectors.

Viruses. A virus is a tiny, nonliving particle consisting of nucleic acid (RNA or DNA) in a protein coat. Viruses measure only a fraction of one micrometer and can be observed only through an electron microscope. All viruses are parasitic and multiply only in living cells. They cause disease by upsetting the metabolism of plant cells; the metabolic disorder may in turn cause cells to produce abnormal and injurious substances. Some viruses are transmitted from plant to plant primarily by the feeding activities of insect or nematode vectors, whereas others are readily spread through mechanical means.

Selected References

Agrios, G. N. 2005. Plant Pathology. 5th ed. Academic Press, San Diego, CA.

Cavalier-Smith, T. 2018. Kingdom Chromista and its eight phyla: A new synthesis emphasising periplastid protein targeting, cytoskeletal and periplastid evolution, and ancient divergences. Protoplasma 255:297-357.

Farr, D. F., and Rossman, A. Y. n.d. Fungal Databases, Systematic Mycology and Microbiology Laboratory. U.S. Department of Agriculture–Agricultural Research Service, Washington, DC. https://nt.ars-grin.gov/fungaldatabases/

Holliday, P. 1989. A Dictionary of Plant Pathology. Cambridge University Press, Cambridge.

Money, N. P. 1998. Why oomycetes have not stopped being fungi. Mycol. Res. 102:767-768.

Rossman, A. Y., and Palm, M. E. 2006. Why are *Phytophthora* and other Oomycota not true fungi? Outlooks Pest Manage. 17:217-219.

Schumann, G. L., and D'Arcy, C. J. 2009. Essential Plant Pathology. 2nd ed. American Phytopathological Society, St. Paul, MN.

Ulloa, M., and Hanlin, R. T. 2012. Illustrated Dictionary of Mycology. 2nd ed. American Phytopathological Society, St. Paul, MN.

(Prepared by R. W. Smiley, P. H. Dernoeden, and B. B. Clarke; revised by L. P. Tredway)

Ecological Groupings of Pathogens and Parasites

Microbial interactions in the turfgrass ecosystem are complex. Most of the organisms in the turf ecosystem are not pathogenic, but their activities have direct and indirect effects on pathogens. Although a discussion of microbial ecology is beyond the scope of this compendium, it is important to briefly address the various means of survival that characterize the principal pathogens of turfgrasses. Broad ecological groupings of pathogenic fungi and funguslike organisms, in which all representatives of the same group are likely to exhibit reasonably similar modes of survival and pathogenesis, can be discerned.

Fungi are unable to synthesize carbohydrates and must, therefore, derive their nutrients from dead or living plants, animals, or other substrates. An organism living on or in another living organism from which it obtains its food is a parasite. An organism that uses dead organic matter for food is a saprophyte.

Any entity that causes disease is a pathogen. Some organisms that live as saprophytes cause disease by altering certain environmental factors so that conditions become intolerable to living plants. These saprophytes, including most fairy ring fungi, are therefore pathogens but not parasites. The terms "pathogen" and "parasite" are often, but not always, interchangeable. Parasitism in the realm of plant pathology typically revolved around a tripartite system of classification: biotrophy, hemibiotrophy, and necrotrophy. However, recent research indicates that this grouping system may not be sustainable, and more-descriptive categories have been proposed. The work conducted to refine the current classification was based on genome-derived analysis of carbohydrate-active enzymes. This system defines four major divisions of fungal species associated with living plants: monomertrophs, polymertrophs, mesotrophs, and vasculartrophs. Many of the fungal pathogens examined in this research were not associated with turfgrasses, and the work focused only on fungal pathogens. Therefore, for this compendium, the current tripartite system is used to describe plant–pathogen interactions. However, the new categories are mentioned within each section.

Biotrophs (Monomertrophs)

Biotrophic fungi are highly specialized for synchronous growth with their hosts. They have evolved in such a way that they cause minimal damage to the plants that support and protect them. They do not directly kill or extensively damage host tissues. The distribution of these parasites is determined solely by their host species. More specifically, these pathogens typically produce a specialized feeding structure called a "haustorium." Since these organisms develop a close relationship with their hosts, typically gene-for-gene interactions occur in terms of effector development and host resistance deployment. Furthermore, plants infected with biotrophs often produce salicylic acid to signal defense mechanisms.

Mycorrhizal fungi are the most highly developed of the biotrophs. They form a persistent, often harmonious, nonpathological or symbiotic (mutually beneficial) association with plant roots. Numerous fungi establish mycorrhizal associations with roots of grasses, but the term "mycorrhizal fungi" usually refers to organisms that produce coenocytic mycelia and form vesicular-arbuscular mycorrhizae, also known as arbuscular or Glomalean endomycorrhizae. Vesicles of these fungi serve as food storage organs, and arbuscules facilitate the exchange of fungal- and plant-derived nutrients or growth compounds. There are numerous genera of endomycorrhizal fungi that infect turfgrasses and are commonly observed in the roots. They are generally considered beneficial when the availability of soil phosphorus is very low.

The primitive root biotrophs (e.g., *Lagena, Olpidium, Polymyxa,* and *Rhizophydium* spp.) and the foliar or stem biotrophs that cause powdery mildew (e.g., *Blumeria graminis*), yellow tuft (*Sclerophthora macrospora*), and rusts (e.g., *Puccinia* spp.) are less highly specialized than the mycorrhizal fungi. The smut fungi (e.g., *Urocystis* and *Ustilago* spp.) also are slightly less specialized. The damage that these parasites cause is not the result of a gross physiological change in the host, but rather it is a consequence of the diversion of nutrients to the fungus for production of fungal biomass. When hosts are young, it is often impossible to casually determine whether or not they are infected. The life span of the host is shortened only slightly because the more-mature tissues are affected the most. Improved growing conditions and the production of new tissue often reduce the damage from these parasites on perennial plants. These functionally obligate parasites cannot grow through soil or thatch toward susceptible plant tissue. To become infected, plant tissue must come in contact with a pathogen in its parasitic growth stage or with its survival spores. Wind, water, and vectors such as mollicutes may serve to bring the host and parasite together. Nematodes are biotrophic as well. These parasites feed on plant roots, but they do not directly kill the cells they feed on. Therefore, as root systems decline from nematode feeding or other stresses, nematode populations may also decline dramatically.

Hemibiotrophs (Mesotrophs)

Parasites classified as hemibiotrophs initially begin feeding on plants as biotrophs and at some point in the disease cycle switch to necrotrophy. Previously, turfgrass pathogens within this group were classified as facultative parasites or facultative saprophytes. This group is rather large and encompasses most of the fungal pathogens of turfgrasses. These organisms are typified by producing haustoria, appressoria, hyphopodia, or some other infection structures and can have either very narrow host ranges or in some cases fairly broad host ranges. Many of these pathogens are able to survive in decaying organic tissue, but normally they are not considered decomposers of that organic tissue. When plants are infected with these pathogens, they can use salicylic acid or jasmonic acid to signal plant defense responses. Research outside turfgrass systems demonstrates that these types of pathogens may produce effectors that produce a gene-for-gene response, but many do not. Host resistance is normally qualitative and/or quantitative. The pathogens of some of the most common diseases of turfgrasses, such as dollar spot, anthracnose, gray leaf spot, and snow molds, are classified as hemibiotrophic.

Necrotrophs (Polymertrophs)

Necrotrophs are parasites that grow and reproduce on dead or dying host tissue. Typically, they do not produce specialized infection structures, because they secrete toxins or enzymes to kill host tissue prior to colonization. Their host ranges can be extremely broad, and breeding for resistance to these organisms is challenging. Given the lack of any specialized interspecies communication between a necrotroph and its host, resistance is usually partial and quantitative in nature. The only effectors these organisms produce are toxins or other phytotoxic materials that are extracellularly excreted from the pathogen. Limited work has been conducted on the plant–microbe interactions of turfgrass pathogens, but species of *Pythium* and certain *Rhizoctonia* species are considered necrotrophic.

Selected References

Bulwalda, J. G., and Goh, K. M. 1982. Host-fungus competition for carbon as a cause of growth depressions in vesicular-arbuscular mycorrhizal ryegrass. Soil Biol. Biochem. 14:103-106.

Gange, A. C., Lindsay, D. E., and Ellis, L. S. 1999. Can arbuscular mycorrhizal fungi be used to control the undesirable grass *Poa annua* on golf courses? J. Appl. Ecol. 36:909-919.

Garrett, S. D. 1970. Pathogenic Root-Infecting Fungi. Cambridge University Press, Cambridge.

Hane, J. K., Paxman, J., Jones, D. A. B, Oliver, R. P., and de Witt, P. 2020. "CATAStrophy," a genome-informed tropic classification of filamentous plant pathogens—How many different types of filamentous plant pathogens are there? Front. Microbiol. doi.org/10.3389/fmicb.2019.03088

Harley, J. L., and Smith, S. E. 1987. Mycorrhizal Symbiosis. Academic Press, London.

Kommedahl, T., and Windels, C. E. 1979. Fungi: Pathogen or host dominance in disease. Pages 1-103 in: Ecology of Root Pathogens. S. V. Krupa and Y. R. Dommergues, eds. Elsevier Scientific, New York.

Salt, G. A. 1979. The increasing interest in "minor" diseases. Pages 281-312 in: Soil-Borne Plant Pathogens. B. Schippers and W. Gams, eds. Academic Press, New York.

Schoenbeck, F. 1979. Endomycorrhiza in relation to plant diseases. Pages 271-280 in: Soil-Borne Plant Pathogens. B. Schippers and W. Gams, eds. Academic Press, New York.

(Prepared by R. W. Smiley, P. H. Dernoeden, and B. B. Clarke; revised by J. P. Kerns)

Pathogens and Their Environment

The activity and survival of pathogenic fungi and fungus-like organisms in turfgrasses are highly dependent on numerous factors such as nutrition, temperature, moisture, aeration, pesticides, and competing organisms as well as host resistance and the competitive saprophytic ability of the pathogen.

Competition among microorganisms is thought to be limited more by the availability of nutrient substrates than by water or physical space. Carbon and nitrogen are the elements most likely to be deficient. Hence, the carbon-to-nitrogen ratio is regarded as an important determinant of microbial activity. The organisms that can reach nutrient substrates most rapidly have the greatest advantage. Turfgrass thatch is probably most limited in the amount of readily available nitrogen, which is quickly leached or immobilized into unavailable forms by thatch-inhabiting microbes. The addition of nitrogen fertilizer in a slow-release form or in small, repeated applications is necessary for active thatch decomposition. The rate of thatch decomposition is an indicator of the relative activity of microorganisms responsible for organic decomposition.

Nutrients are also lost from foliage and roots. Foliar losses include (1) leakage from injured cells; (2) leakage from the membranes of cells in senescing leaves; (3) leaching of nutrients from mature leaves by water, especially under low light, at high temperatures, or with prolonged wetting; and (4) guttation fluids. Leaf clippings left in the aerial canopy also serve as an important source of nutrients. Similarly, roots exude or leak nutrients, especially if they have been injured by drought.

Temperature is an important factor determining the competitive advantages of organisms as parasites or saprophytes. Pathogens that attack belowground tissue usually infect these tissues in response to specific soil temperature ranges. Leaf-to-leaf spread of foliar pathogens depends directly on temperature, especially nighttime temperature. Temperature governs the rate of hyphal growth, the mass of mycelium produced, and the infection potential of these pathogens.

A high concentration of water vapor in the atmosphere or a thin film of water on leaf surfaces is necessary to prevent the desiccation of fungal hyphae. Prolonged periods of leaf wetness or high humidity favor the spread of disease by increasing the time available for fungal growth and by increasing the removal of leachates from plant cells. Prolonged periods of rainfall that allow for soils to remain at or around field capacity also promote infection of belowground turf tissue.

Aeration is affected by the water and temperature status of soil and thatch. Fungi such as *Rhizoctonia* spp. and some *Fusarium* spp. tolerate comparatively dry environmental conditions. *Pythium* spp. and some *Fusarium* spp. can tolerate high levels of carbon dioxide (e.g., in wet soil), whereas *Bipolaris, Clarireedia,* and *Gaeumannomyces* spp. are intolerant of carbon dioxide. These characteristics influence the ability of these organisms to survive and grow in a subterranean environment.

The effects of pesticides and host resistance on the activity and survival of certain pests are considered in more depth in Part III (see Disease-Resistant Species and Cultivars, Disease Management with Synthetic Chemistries, and Nontarget Effects of Pesticides).

Selected References

Agrios, G. N. 2005. Plant Pathology. 5th ed. Academic Press, San Diego, CA.

Alexopoulos, C. J., Mims, C. W., and Blackwell, M. 1996. Introductory Mycology. 4th ed. John Wiley & Sons, New York.

Barnett, H. L., and Hunter, B. B. 1998. Illustrated Genera of Imperfect Fungi. 4th ed. American Phytopathological Society, St. Paul, MN.

Endo, R. M. 1972. The turfgrass community as an environment for the development of facultative fungal parasites. Pages 171-220 in: The Biology and Utilization of Grasses. V. B. Youngner and C. M. McKell, eds. Academic Press, New York.

Kerns, J. P., and Tredway, L. P. 2013. Advances in turfgrass pathology since 1990. Pages 733-776 in: Turfgrass: Biology, Use, and Management. Agron. Monogr. 56. J. Stier, B. Horgan, and S. A. Bonos, eds. American Society of Agronomy, Madison, WI.

Kirk, P. M., Cannon, P. F., David, J. C., and Stalpers, J. A., eds. 2001. Ainsworth & Bisby's Dictionary of the Fungi. 9th ed. C.A.B. International, Wallingford, U.K.

Margulis, L., McKhann, H. I., and Olendzenski, L., eds. 1993. Illustrated Glossary of Protoctista: Vocabulary of the Algae, Apicomplexa, Ciliates, Foraminifera, Microspora, Water Molds, Slime Molds, and the Other Protoctists. Jones and Bartlett, Boston.

Margulis, L., and Schwartz, K. V. 1998. Five Kingdoms. An Illustrated Guide to the Phyla of Life on Earth. 3rd ed. W. H. Freeman, New York.

Money, N. P. 1998. Why oomycetes have not stopped being fungi. Mycol. Res. 102:767-768.

Rossman, A. Y., Palm, M. E., and Spielman, L. J. 1987. A Literature Guide for the Identification of Plant Pathogenic Fungi. American Phytopathological Society, St. Paul, MN.

Smith, J. D., Jackson, N., and Woolhouse, A. R. 1989. Fungal Diseases of Amenity Turf Grasses. E. & F. N. Spon, London.

Ulloa, M., and Hanlin, R. T. 2012. Illustrated Dictionary of Mycology. 2nd ed. American Phytopathological Society, St. Paul, MN.

van Regenmortel, M. H. V., Fauquet, C. M., Bishop, D. H. L., Carstens, E. B., Estes, M. K., Lemon, S. M., Maniloff, J., Mayo, M. A., McGeoch, D. J., Pringle, C. R., and Wickner, R. B., eds. 2000. Virus Taxonomy: Seventh Report of the International Committee on Taxonomy of Viruses. Academic Press, San Diego, CA.

Waid, J. S. 1974. Decomposition of roots. Pages 175-211 in: Biology of Plant Litter Decomposition, vol. 2. C. H. Dickinson and G. J. F. Pugh, eds. Academic Press, London.

(Prepared by R. W. Smiley, P. H. Dernoeden, and B. B. Clarke; revised by J. P. Kerns)

Part I. Infectious Diseases

Diseases of Foliage and Crowns

Diseases affecting the foliage and crowns of turfgrass plants often induce distinct symptoms that can be recognized visually, and signs of the causal agent such as fungal hyphae or reproductive structures are often visible with the naked eye, a hand lens, or a dissecting microscope. Foliar diseases such as red thread, frost scorch, and brown patch were among the first turfgrass diseases described in the late 1800s to early 1900s. During the formative years of turfgrass pathology, most research focused on these and other foliar diseases such as dollar spot and leaf spots. Many foliar- and crown-infecting turfgrass pathogens have since been documented worldwide.

Foliar and crown diseases such as anthracnose, dollar spot, brown patch, and leaf spots remain some of the most widespread and damaging turfgrass diseases. Much has been learned about these diseases through many years of research, and integrated management strategies have been developed to lessen their impact on turfgrass health. Other foliar diseases, such as those caused by bacteria, are less well understood and have been the subject of more-recent research efforts.

The taxonomy of fungal pathogens has changed dramatically in recent years, brought about by techniques such as DNA sequencing and whole-genome analysis. Some of the most well-known turfgrass pathogens have been reclassified or are now referred to by the names of their sexual stages. For example, the dollar spot pathogen, originally described as *Sclerotinia homoeocarpa,* is now recognized as a complex of at least six different species in the new genus *Clarireedia. Microdochium nivale* is now referred to by its sexual stage, *Monographella nivalis,* and *Rhizoctonia zeae* is now referred to by its sexual stage, *Waitea zeae.*

Epidemics of new or previously unrecognized diseases are a regular occurrence in turfgrass systems. Whether these outbreaks are the result of new introductions or are brought about by changing management practices or climatic conditions is often unknown. Recently described diseases such as rapid blight, brown ring patch, bacterial decline, and lethal viral necrosis are foliar diseases that are new additions to this fourth edition of *Compendium of Turfgrass Diseases.*

(Prepared by R. W. Smiley, P. H. Dernoeden, and B. B. Clarke; revised by L. P. Tredway)

Anthracnose Diseases of Cool-Season Turfgrasses

"Anthracnose" is the name given to foliar diseases caused by fungi that produce asexual spores in acervuli. These diseases occur in cool- and warm-season turfgrass species throughout the world. In cool-season turf, the causal fungus, *Colletotrichum cereale* (formerly known as *C. graminicola*), often colonizes naturally senescing leaves and tillers, but it can also attack younger plant tissues. In cool, humid climates, anthracnose is particularly damaging in *Poa annua. Agrostis stolonifera* is most susceptible when grown on putting greens in the transition zone or subtropical environments. Additional cool-season turfgrass hosts include *Festuca, Lolium,* and other *Poa* spp. Anthracnose also occurs on warm-season turfgrasses, including *Axonopus, Cynodon, Eremochloa, Paspalum,* and *Zoysia* spp., but it is typically less destructive than on cool-season turfgrasses and is caused by other species of *Colletotrichum* (see Anthracnose Diseases of Warm-Season Turfgrasses).

Symptoms and Signs

Anthracnose symptoms may include foliar blight or a rotting of basal tissues (crowns, stolons, and roots). The two phases of the disease, referred to as "anthracnose foliar blight" and "anthracnose basal rot," are more common at different times of the year depending on turfgrass species. In *P. annua,* basal rot typically begins to develop in winter to spring and may progress into summer, whereas anthracnose foliar blight occurs more frequently during high-temperature stress of midsummer. In *A. stolonifera,* anthracnose basal rot develops during hot, humid conditions in the summer, whereas foliar blight can occur throughout the season, either during periods of heat stress in midsummer or under cloudy, wet conditions in the spring through autumn.

Foliar blight. Symptoms of anthracnose foliar blight include yellow or reddish brown discoloration of leaves (Fig. 3) and eventually a loss of shoot density (Fig. 4). It occasionally

Fig. 3. Yellowing of *Poa annua* on a golf green caused by anthracnose. (Courtesy J. Hempfling—© APS)

causes oblong, reddish brown leaf lesions (Fig. 5). The distinctive acervuli (fruiting bodies) of the pathogen, with their protruding black setae (hairlike structures), can be observed on green, yellow, or tan tissues of leaves, sheaths, or stems (Fig. 6). Large numbers of acervuli are often observed on dead plant tissue and are not necessarily diagnostic of an anthracnose outbreak. Therefore, it is important to carefully look for appressoria (Fig. 7) and acervuli on green or discolored tissue of living plants. *Bipolaris sorokiniana* and *Curvularia lunata* may be found in association with *C. cereale,* causing a decline in *P. annua* in the summer in some regions.

Fig. 4. Anthracnose of *Poa annua* on a golf fairway. (Courtesy R. W. Smiley—© APS)

Fig. 5. Anthracnose foliar blight lesions on *Poa pratensis*. (Courtesy R. W. Smiley—© APS)

Fig. 6. Acervuli of *Colletotrichum cereale* on leaves and a stem of *Poa annua*. (Courtesy B. B. Clarke—© APS)

Basal rot. Anthracnose basal rot initially appears as orange or yellow spots about 6–12 mm in diameter. As crown tissues become infected, small patches of plants or scattered individual plants turn yellow and die (Figs. 8 and 9). Individual plants may have both green, healthy-appearing tillers and yellowish orange, infected tillers. The central, or youngest, leaf is often the last to change color. Removal of the sheath tissue to expose the infected stem base often reveals a water-soaked or black rot of crown tissues where roots and new buds are produced (Fig. 10). As the disease progresses, it develops into patches or

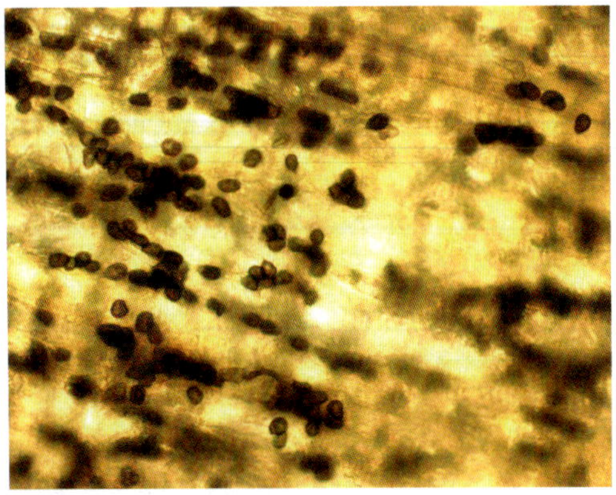

Fig. 7. Appressoria of *Colletotrichum cereale* on a leaf of *Poa annua*. (Courtesy L. Beirn)

Fig. 8. Small, yellow spot caused by anthracnose basal rot of *Poa annua* on a golf green in the spring. (Courtesy B. B. Clarke—© APS)

Fig. 9. Foliar bronzing caused by anthracnose basal rot of *Poa annua* on a golf green in the spring. (Courtesy J. E. Kaminski—© APS)

irregularly shaped patterns (Fig. 11) that are initially orange to reddish brown before turning yellow and dying. When large areas thin or die out completely (Fig. 12), the disease may be confused with melting-out or red leaf spot (see *Bipolaris* and *Exserohilum* Diseases and *Pyrenophora* (*Drechslera*) Diseases). Individual clones or genotypes of *A. stolonifera* may be attacked more severely than others, forming more-distinct patches in some cases.

In warm weather, anthracnose basal rot of *P. annua* and *A. stolonifera* appears as irregularly shaped patterns 1 m or more in diameter. Brown, circular patches similar to those caused by brown patch may also form. Plants often turn bright yellow or orangish yellow before dying; this symptom can be confused with symptoms of summer patch.

When discoloration or thinning is first observed, managers and diagnosticians are advised to carefully examine exposed stem bases or stolons to look for appressoria or hyphopodia, grayish black infection mats, or rotting tissues (Fig. 7). Appressoria and hyphopodia are small (64–316 μm in diameter), dark brown or black, and dome shaped. Infection mats first appear as small, pinhead-sized black specks (Fig. 13). The mats enlarge and turn black and may produce acervuli and setae (dark-colored bristle- or hairlike structures). Normally, there are no diagnostic lesions or signs of the pathogen on green leaf or sheath tissue during the early stages of basal rot. Turf managers often notice the discoloration only after stem tissues have begun to rot. In advanced stages, black aggregates of fungal mycelium, hyphopodia, infection mats, and acervuli are often present on infected stolons or stem bases. Acervuli and black mycelial aggregates on stem bases and stolons can be seen with a hand lens. In advanced stages of basal rot, acervuli develop on sheath or leaf tissue, at which point the plants generally die. Conclusive diagnosis of anthracnose basal rot depends on microscopic inspection of the conidia produced in acervuli. For unknown reasons, the pathogen seldom attacks both *P. annua* and *A. stolonifera* on the same green or even on the same golf course.

Causal Agent

C. cereale causes anthracnose of cool-season turfgrasses. This fungus is very common and forms small, dark, elongated acervuli within infected tissues. The presence of black setae arising from the acervuli is often used as a rapid diagnostic criterion for anthracnose. The setae are long, dark brown or black, septate spines (Fig. 6) measuring 32–120 × 6–8 μm; they are wider at the bases and taper toward the tips. Conidia are one celled, crescent shaped, guttulate, and 4–34 μm long (most are 19–30 μm) (Fig. 14). The sexual stage (*Glomerella graminicola*) rarely occurs in nature.

Research with isolates of *C. cereale* from *A. stolonifera* and *P. annua* indicates a degree of host specificity. However, the

Fig. 10. Water-soaked, rotted, blackened lower stems of *Poa annua* caused by anthracnose basal rot. (Courtesy P. J. Landschoot—© APS)

Fig. 11. Anthracnose basal rot on an *Agrostis stolonifera* putting green. (Courtesy M. Tomaso-Peterson—© APS)

Fig. 12. Anthracnose of *Poa annua* on a golf green in August. Note that the dark gray to black patches have coalesced into a large area colonized by algae. (Courtesy B. B. Clarke—© APS)

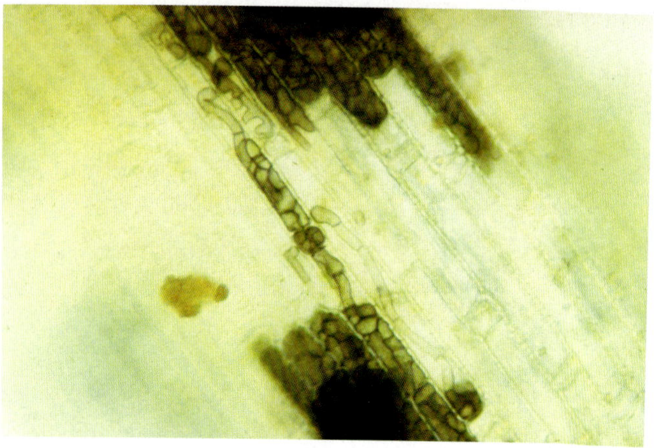

Fig. 13. Infection mat of *Colletotrichum cereale*. (Courtesy P. H. Dernoeden—© APS)

occurrence of races of the pathogen in turfgrasses is inconclusive. The apparent genetic similarities among isolates from the same host may account for the attack of *P. annua* in mixed stands of *P. annua* and *A. stolonifera* at one location and the attack of only *A. stolonifera* in similar mixtures elsewhere. Similarly, the preferential attack on *Festuca* spp. in turfs in which fescue is mixed with *P. pratensis* could also be explained by the presence of physiologic races.

Two other fungi have also been associated with turfs diagnosed with symptoms of anthracnose: *Microdochium bolleyi* and *Lectera colletotrichoides.* Their importance as potential pathogens of turfgrass is unknown.

M. bolleyi is common in some regions and appears similar to *C. cereale* upon initial inspection. The fruiting structure of *M. bolleyi* is a sporodochium, which may not be easily distinguished from an acervulus, except that *M. bolleyi* does not form setae. Conidia of *M. bolleyi* are one celled, crescent shaped, and 5–9 µm long and resemble conidia of *C. cereale.*

L. colletotrichoides has been isolated from *Festuca rubra* exhibiting symptoms and signs of anthracnose and was demonstrated to be pathogenic to *Festuca* spp. in greenhouse studies. It produces small, hemispherical sporodochia. Dark setae, about 100 µm long, surround the sporodochia and, like those of *C. cereale,* are clearly visible. Conidia are one celled, bean shaped, and about 5 µm long. Distinguishing between *Lectera* and *Colletotrichum* spp. requires diagnosis in a laboratory.

Disease Cycle

C. cereale is thought to survive unfavorable periods as mycelium in decaying plant debris or living host tissue. The fungus can penetrate foliage, stem, or root tissues of turfgrass plants growing under stressful conditions and can infect the plants when atmospheric humidity is high and leaves are wet. Acervuli form on blighted or necrotic tissue, and conidia are disseminated to uninfected plants by wind or water or mechanically by foot traffic or equipment. Conidia germinate within 2–6 hours and infect host tissue.

Epidemiology

Anthracnose is favored by overcast weather and other conditions that promote high humidity or allow excess water to remain in the leaf canopy. Conditions that weaken plants, including low mowing height, low fertility, drought stress, and extreme temperatures, can also stimulate anthracnose development. Symptoms may initially appear on mounds or other areas prone to moisture stress. Plants growing in soils deficient in phosphorus, potassium, or nitrogen are particularly susceptible to infection.

In *P. annua,* anthracnose can occur at almost any time of year. Anthracnose foliar blight generally occurs during periods of high-temperature stress in summer. Spring outbreaks of anthracnose basal rot in *P. annua* putting greens can be very destructive. *P. annua* weakened by environmental stress or low mowing during the previous summer or by winter injury is particularly susceptible. On putting greens, deep vertical cutting practices that extensively wound plant crowns may enable *C. cereale* to be more invasive; however, practices such as topdressing, other cultivation practices, and aerification do not increase the severity of this disease.

Fig. 14. *Agrostis stolonifera* leaf with numerous acervuli and one-celled, guttulate conidia of *Colletotrichum cereale.* (Courtesy M. Tomaso-Peterson—© APS)

Anthracnose is a common disease of *A. stolonifera* putting greens in the transition zone and subtropical climates and is likely encouraged by the physiological stresses imposed by these environments. Anthracnose basal rot is most common on *A. stolonifera* during hot, humid conditions and may be encouraged by excessive thatch accumulation and poor soil drainage. Basal rot often causes severe damage to *A. stolonifera* on putting surfaces and recovery is slow because the stolons and crowns are attacked. Anthracnose foliar blight is observed on *A. stolonifera* during prolonged periods of high heat or during periods of cloudy, wet weather.

Management

The effects of cultural practices on anthracnose in *P. annua* have been thoroughly investigated. Providing proper fertility, increasing mowing height, topdressing with sand, and avoiding drought stress are important measures for reducing disease severity. Turf should be fertilized to avoid deficiencies of nitrogen, phosphorus, and potassium. Nitrogen should not be applied at high rates during periods of drought or high-temperature stress. When applied frequently at low rates and in moderation, however, nitrogen and potassium help to reduce disease symptoms on putting greens and fairways in summer. A foliar nitrogen concentration of 3.4% (dry weight) is considered a minimum critical level for disease suppression on *P. annua* putting greens. Symptoms are more severe on *P. annua* grown at soil potassium levels below 50 ppm and less than 2% potassium in clipping tissue (dry weight). Disease symptoms are also more severe on *P. annua* that frequently experiences wilt stress. Water should be applied at levels sufficient to avoid both excessively wet and dry conditions. Light, frequent applications of topdressing sand throughout the growing season reduce disease symptoms and promote recovery of injured turf. On putting greens, midseason cultivation practices (e.g., solid tine aeration, vertical mowing, and grooming) do not enhance anthracnose severity, except for deep vertical cutting when the disease is active. The use of growth regulators that stress or weaken *P. annua* or delay recovery of injured turf should be avoided.

Observations of *A. stolonifera* support many of the same cultural recommendations for anthracnose management. Routine aerification and topdressing are important to minimize thatch accumulation and maintain soil drainage. During periods of summer stress, mowing height should be increased and mowing frequency reduced, and sufficient irrigation and fertility to support turf growth should be provided. Removal of trees and/or installation of fans can help to improve overall turf health and reduce anthracnose development.

In temperate climates, reducing *P. annua* populations in favor of *A. stolonifera* can reduce anthracnose incidence. Modern creeping bentgrass cultivars bred for improved heat tolerance are typically less susceptible to anthracnose in transition zone environments.

Preventive applications of fungicides are recommended where the disease is a chronic problem, but the use of best-management practices can substantially reduce fungicide inputs needed to control anthracnose. Strains of *C. cereale* that are resistant to one or more fungicides (e.g., benzimidazoles and quinone outside inhibitors) have been detected in the United States. Tank mixes can be used, or fungicides with different modes of action can be alternated. Sequential applications of active ingredients considered to have a high potential for fungicide resistance should be avoided.

Selected References

Avila-Adame, C., Olaya, G., and Köller, W. 2003. Characterization of *Colletotrichum graminicola* isolates resistant to strobilurin-related QoI fungicides. Plant Dis. 87:1426-1432.

Backman, P. A., Landschoot, P. J., and Huff, D. R. 1999. Variation in pathogenicity, morphology and RAPD marker profiles in *Colletotrichum graminicola* from turfgrasses. Crop Sci. 39:1129-1135.

Bailey, J. A., and Jeger, M. J., eds. 1992. *Colletotrichum:* Biology, Pathology and Control. CAB International, Wallingford, U.K.

Browning, M., Rowley, L. V., Zeng, P., Chandlee, J. M., and Jackson, N. 1999. Morphological, pathogenic, and genetic comparisons of *Colletotrichum graminicola* isolates from Poaceae. Plant Dis. 83:286-292.

Crouch, J. A., Clarke, B. B., and Hillman, B. I. 2006. Unraveling evolutionary relationships among the divergent lineages of *Colletotrichum* causing anthracnose disease in turfgrass and corn. Phytopathology 96:46-60.

Danneberger, T. K., Vargas, J. M., Jr., and Jones, A. L. 1984. A model for weather-based forecasting of anthracnose on annual bluegrass. Phytopathology 74:448-451.

Dernoeden, P. H. 2017. Creeping Bentgrass Management. 2nd ed. CRC Press, Boca Raton, FL.

Hempfling, J. W., Schmid, C. J., Wang, R., Clarke, B. B., and Murphy J. A. 2017. Best management practices effects on anthracnose disease of annual bluegrass. Crop Sci. 57:602-610.

Inguagiato, J. C., and Guillard K. 2016. Foliar N concentration and reflectance meters to guide N fertilization for anthracnose management of annual bluegrass putting green turf. Crop Sci. 56:3328-3337.

Inguagiato, J. C., Murphy, J. A., and Clarke, B. B. 2008. Anthracnose severity on annual bluegrass influenced by nitrogen fertilization, growth regulators, and verticutting. Crop Sci. 48:1595-1607.

Inguagiato, J. C., Murphy, J. A., and Clarke, B. B. 2009. Anthracnose disease and annual bluegrass putting green performance affected by mowing and lightweight rolling. Crop Sci. 49:1454-1462.

Inguagiato, J. C., Murphy, J. A., and Clarke, B. B. 2010. Anthracnose development on annual bluegrass affected by seedhead and vegetative growth regulators. Appl. Turfgrass Sci. 7. doi:10.1094/ATS-2010-0923-01-RS

Inguagiato, J. C., Murphy, J. A., and Clarke, B. B. 2012. Sand topdressing rate and interval effects on anthracnose severity of an annual bluegrass putting green. Crop Sci. 52:1406-1415.

Khan, A., and Hsiang, T. 2003. The infection process of *Colletotrichum graminicola* and relative aggressiveness on four turfgrass species. Can. J. Microbiol. 49:433-442.

Roberts, J. A., Inguagiato, J. C., Clarke, B. B., and Murphy, J. A. 2011. Irrigation quantity effects on anthracnose disease of annual bluegrass. Crop Sci. 51:1244-1252.

Roberts, J. A., and Murphy, J. A. 2014. Anthracnose disease on annual bluegrass as affected by foot traffic and sand topdressing. Plant Dis. 98:1321-1325.

Schmid, C. J., Clarke, B. B., and Murphy, J. A. 2017. Anthracnose severity and annual bluegrass quality as influenced by nitrogen source. Crop Sci. 57:S-285-S-292.

Schmid, C., Clarke, B. B., and Murphy, J. A. 2018. Potassium nutrition affects anthracnose on annual bluegrass. Agron. J. 110:2171-2179.

Smith, J. D. 1954. A disease of *Poa annua*. J. Sports Turf Res. Inst. 8:344-353.

Sutton, B. C. 1980. The Coelomycetes: Fungi Imperfecti with Pycnidia, Acervuli and Stromata. Commonwealth Mycological Institute, Kew, England.

(Prepared by R. W. Smiley, P. H. Dernoeden, and B. B. Clarke; revised by J. A. Murphy and B. B. Clarke)

Anthracnose Diseases of Warm-Season Turfgrasses

Tanso-byô (anthracnose) was described on *Eremochloa ophiuroides* in 2006 in Japan. Anthracnose of *E. ophiuroides* and *Axonopus compressus* was subsequently described in the United States and China, respectively. Three additional *Colletotrichum* species have been isolated from *Axonopus, Paspalum,* and *Zoysia* spp., but pathogenicity to these grasses has not been documented.

Symptoms

Anthracnose of *E. ophiuroides* affects the crown and leaf sheaths. The oldest leaf sheaths appear necrotic, while young leaf sheaths and leaves are chlorotic. Leaf blight progresses from the leaf tips to the bases of older leaves. Foliar lesions, as well as acervuli and setae, are absent on infected leaf sheaths. The stolons and roots are not affected. Irregular, diffuse patches of various sizes in the turf appear bright yellow or chlorotic. A loss of turf density may occur. Necrotic patches and subsequent bare areas develop and persist into the summer as plants die (Fig. 15).

Tanso-byô of *E. ophiuroides* is a foliar disease resulting in fusiform to ellipsoid necrotic lesions that develop a grayish white appearance as they mature. Acervuli and setae develop within lesions of infected tissues.

On *A. compressus,* conspicuous foliar lesions have pale brown centers surrounded by reddish brown borders. Lesions can coalesce to form large, irregular areas. Excessive moisture promotes infection of crowns and leaf sheaths, resulting in reddish brown areas within a turf stand (Fig. 16). Stolons and roots are not affected. Acervuli and setae develop in lesions and on associated infected leaves.

Causal Agents

Colletotrichum eremochloae infects *E. ophiuroides.* Light gray, septate mycelium may be observed colonizing the epidermal tissue of the inner leaf sheaths. Abundant, grayish brown to dark brown appressoria (12.5 × 12.5 µm) are produced in association with mycelium and serve as a diagnostic feature. In culture on potato dextrose agar (PDA), a 7-day-old colony produces mycelium that is gray on the inner half and white on the outer half. The perimeter of the colony has a distinct fluorescent yellow pigment when cultured in 12–16 hours of light per day. Conidia are hyaline, aseptate, falcate to fusiform, and average 23.4 × 5.6 µm.

C. caudatum, the causal agent of tanso-byô, produces acervuli and setae within necrotic lesions. The conidia are aseptate, hyaline, falcate to fusiform, and 20.5–38.5 × 2.8–4.2 µm. On PDA, the colony is grayish white to gray. The appressoria are grayish brown, ellipsoid to round, and 10.0–16.8 × 7.6–11.5 µm.

C. hainanense infects *A. compressus.* Falcate, aseptate, hyaline conidia (19.5–27.5 × 4–7 µm) are produced in acervuli. Appressoria are pale to brown, globose or lobate, and 10–16 × 8.5–11.5 µm.

C. axonopodi was isolated from *A. affinis* in Australia. In pure culture, mycelium is white, overlaid on brown mycelium with a pale tan orange pigment. Falcate to fusiform conidia average 24.2 × 4.2 µm. Appressoria are globose, ovoid or clavate, smooth or lobate to multilobate, and average 10.1 × 7.5 µm.

C. paspali was isolated from *Paspalum notatum* in Japan. The colony is whitish cream, and setae are absent. Conidia are falcate or fusiform and average 21.2 × 4.1 µm. Appressoria are variable in shape, averaging 12.3 × 9.9 µm.

C. zoysiae was isolated from leaves of *Zoysia tenuifolia* in Japan. On PDA, mycelium is gray to light gray and pale olivaceous gray at the margin. Caudate conidia are aseptate, hyaline, falcate, and 25–47.5 (including appendage) × 4.0–6.5 µm. Appressoria are elongate to round, lobate to multilobate, black, and 17.0–22.5 × 10.0–14.5 µm.

Colletotrichum spp. can be accurately identified by their morphological characteristics and by analyses of the nuclear rDNA internal transcribed spacer region, the 5' end of the DNA lyase gene (*Apn2*), the 3' end of *Apn2,* the 5' end of the mating-type gene *Mat1-2* (*Mat1/Apn2*), and the manganese superoxide dismutase (*Sod2*) gene.

Fig. 15. *Eremochloa ophiuroides* lawns exhibiting irregular, diffuse areas of yellow, thinning turf caused by anthracnose. (Courtesy M. Tomaso-Peterson—© APS)

Fig. 16. *Axonopus compressus* with anthracnose symptoms caused by *Colletotrichum hainanense*. (Courtesy W. Zhang—© APS)

Disease Cycle, Epidemiology, and Management

Anthracnose diseases of warm-season grasses are polycyclic and primarily occur when plants are transitioning out of winter dormancy, but they can persist into the summer months if disease pressure is high. Cool, springlike temperatures and abundant moisture favor disease outbreaks. Conidia and infected plant parts are disseminated by mowing and water movement. The turf recovers during the summer months when conditions are favorable for growth.

To date, little research has been conducted on control measures; however, fungicides that are effective on *C. cereale* may prove efficacious for controlling anthracnose of warm-season turfgrasses.

Selected References

Crouch, J. A. 2014. *Colletotrichum caudatum* s.l. is a species complex. IMA Fungus 5:17-30.

Crouch, J. A., Clarke, B. B., White, J. F., Jr., and Hillman, B. I. 2009. Systematic analysis of the falcate-spored graminicolous *Colletotrichum* and a description of six new species from warm-season grasses. Mycologia 101:717-732.

Crouch, J. A., and Tomaso-Peterson, M. 2012. Anthracnose disease of centipedegrass turf caused by *Colletotrichum eremochloae,* a new fungal species closely related to *Colletotrichum sublineola*. Mycologia 104:1085-1096.

Fuke, K., Hozumi, N., Enami, Y., Matsuura, K., and Tajimi, A. 2006. Anthracnose of centipede grass caused by *Colletotrichum caudatum*. J. Gen. Plant Pathol. 72:74-75.

Tomaso-Peterson, M., Crouch, J. A., and Balbalian, C. 2012. Anthracnose of centipedegrass turf. Appl. Turfgrass Sci. 9. doi:10.1094/ATS-2012-1126-01-MG

Zhang, W., Damm, U., Crous, P. W., Groenewald, J. Z., Niu, X., Lin, J., and Li, Y. 2020. Anthracnose disease of carpetgrass (*Axonopus compressus*) caused by *Colletotrichum hainanense* sp. nov. Plant Dis. 104:1744-1750.

(Prepared by M. Tomaso-Peterson)

Ascochyta Leaf Blight

Leaf spots and blights caused by *Ascochyta* and *Neoascochyta* spp. are common on many grasses. On turfgrasses, the principal symptom is a leaf blight that seldom causes widespread damage. These fungi are common colonizers of senescent leaf tissue and inflorescences.

Symptoms and Signs

Large areas of turf may assume a uniformly blighted appearance, or localized areas of heavy infection may cause a patchy appearance (Fig. 17). Individual leaves generally start dying back from the tips (Fig. 18), and uniform lesions affecting the entire leaf blade may develop. When leaf spots are present, they are dark brown and circular to elliptical and appear similar to those caused by *Pyrenophora* and *Bipolaris* spp. Less commonly, there may be minute chlorotic, purple, or brown areas where the pathogen has directly penetrated the leaf mid-blade. These spots enlarge into bleached lesions that girdle the leaf and cause it to become chlorotic and die. Round, brown pycnidia (fruiting bodies) of the pathogen generally form in the lesions or blighted leaf tissue (Fig. 19), where they can be seen with a hand lens.

Causal Agents

More than 15 species of *Ascochyta* and *Neoascochyta* are associated with turfgrasses as pathogens or saprophytes; refer to the appendix, Diseases of Turfgrasses, for a comprehensive list. On the basis of recent phylogenetic studies, certain species have been moved to the newly erected genus *Neoascochyta*. Several species associated with turfgrasses have not been studied and remain in *Ascochyta*. Morphological characteristics of pycnidia and conidia are highly variable and overlap across species; thus, they are no longer considered useful for identification to species.

Pycnidia formed by most *Ascochyta* and *Neoascochyta* species are 70–200 µm in diameter and yellow brown, rust brown, or brick red. Conidia (Fig. 20) are mostly hyaline to pale or straw yellow, fusoid, one to three septate, and 2.5–5.5 × 7–21 µm. Conidia of *A. rhodesii* are 8–10 × 24–32 µm, and this species forms pycnidia that are dark brown to black and up to 240 µm in diameter.

Disease Cycle

The pathogen survives as mycelium and pycnidia in host debris. The fungus generally enters plants soon after the grass is mowed and begins to grow from the freshly cut ends toward the leaf bases. Pycnidia form in leaf spot lesions or blighted tissues and produce conidia, which are released during wet periods and splashed or mechanically dispersed to uninfected leaves.

Epidemiology

Ascochyta leaf blight occurs throughout the year. Like Septoria and Stagonospora leaf spots, it can occur in late winter

Fig. 17. Ascochyta blight in a *Festuca arundinacea* landscape. (Courtesy J. P. Kerns—© APS)

Fig. 18. Leaf necrosis caused by an *Ascochyta* sp. on *Poa pratensis*. (Courtesy R. W. Smiley—© APS)

Fig. 19. Pycnidia of an *Ascochyta* sp. in a leaf spot on *Poa pratensis*. (Courtesy R. W. Smiley–© APS)

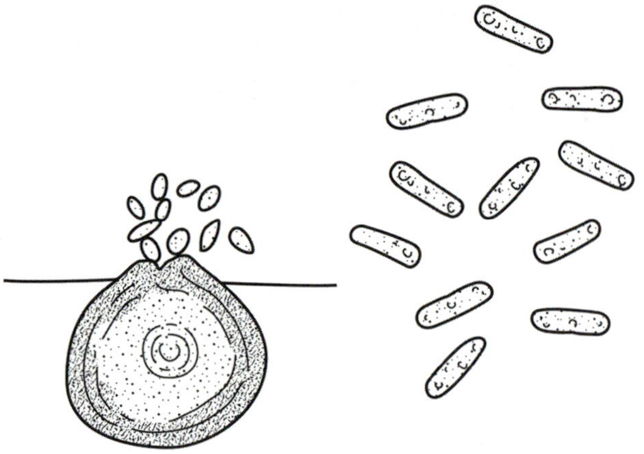

Fig. 20. Conidia emanating from a pycnidium (left) and conidia (right) of an *Ascochyta* sp. (Courtesy L. P. Tredway—© APS. Drawing by S. McKinley.)

and early spring. Periods of high humidity or frequent irrigation appear to favor the disease during midsummer. The infection process apparently requires the presence of a film of water or guttation fluids on the leaf surface. Frequent mowing favors the disease by creating more infection sites.

Management

Ascochyta leaf blight is generally not a severe disease, and fungicides are seldom recommended for control. Mowing appears to be an important factor favoring the disease. Grass should be mowed to the recommended height when the leaves are dry. Turf should be irrigated early in the morning when dew is present. Water should be applied thoroughly (deeply) and as infrequently as possible without causing moisture stress. Excessive applications of nitrogen fertilizer should be avoided, but enough fertilizer should be provided to maintain healthy, active growth of the turf. A curative application of a broad-spectrum fungicide may be warranted if the disease becomes severe. Information about fungicide selection and cultivar resistance is not available.

Selected References

Chen, Q., Jiang, J. R., Zhang, G. Z., Cai, L., and Crous, P. W. 2015. Resolving the *Phoma* enigma. Stud. Mycol. 82:137-217.

Punithalingam, E. 1979. Graminicolous *Ascochyta* species. Mycol. Pap. 142. Commonwealth Mycological Institute, Kew, England.

Smith, J. D., Gossen, B. D., and Evans, I. 2001. *Ascochyta phleina* causes a patch disease on turfgrass. Can. J. Plant Pathol. 23:181-183.

Sprague, R. 1950. Ascochyta leaf spots of cereals and grasses in the United States. Mycologia 42:523-553.

(Prepared by R. W. Smiley, P. H. Dernoeden, and B. B. Clarke; revised by L. P. Tredway)

Bacterial Diseases of Turfgrasses

While the vast majority of turfgrass diseases are caused by fungi, bacteria are sometimes implicated in significant turfgrass decline. Research has shown that turfgrasses are home to rich bacterial communities, but these organisms are often overlooked because of the numerous fungi that cause significant damage to turfgrasses worldwide. While turfgrass scientists have increased their research on bacteria (free-living, beneficial, and pathogenic) in turfgrass ecosystems, the etiology of bacterial turfgrass diseases is still poorly defined.

Research on bacterial diseases in turfgrasses can be linked to two significant periods beginning in the late 1900s. To begin, bacteria were first defined as pathogens of turfgrass in the early 1980s, when researchers at Michigan State University coined the term "bacterial wilt" to describe a disease on the *Agrostis stolonifera* cultivars Toronto, Cohansey, and Nimisilia. Later reports also showed bacterial wilt to occur on *Poa annua* putting greens and in an unidentified cultivar of a *Cynodon* sp. In 2012, researchers identified bacteria as the cause of etiolation, a poorly understood condition that has impacted multiple cool-season turfgrasses for decades, including those impacted by bacterial wilt. However, these reports also identified bacterial decline in several *A. stolonifera* cultivars, which involved a progression from etiolation resulting in significant loss of turfgrass density. Two bacterial diseases are described here: bacterial wilt and bacterial decline (Table 2). Symptoms associated with bacterial diseases of turfgrasses have been the subject of great debate because traditional inoculation protocols used for fungal diseases have proved unreliable in reproducing symptoms associated with bacterial infections.

Symptoms and Signs

Both bacterial wilt and bacterial decline of cool-season turfgrasses can initially cause etiolation, an abnormally rapid growth of turfgrass stem and/or leaf tissue (Fig. 21). Multiple names have been associated with etiolation, including etiolated tiller syndrome, mad tiller, and bacterial etiolation and decline. Etiolated turfgrass plants tower over nonetiolated neighboring plants (1.2–3.8 cm), with rapid growth occurring primarily during the overnight hours but sometimes throughout the day. Etiolated plants are often chlorotic, and etiolated stems may appear bleached. Stand symptoms appear as irregularly shaped patches of etiolated and chlorotic turf. Chlorosis of etiolated tissue is believed to result from the dilution of chlorophyll as leaf cells rapidly expand. In many cases, symptoms of etiolation are transient and dissipate with time.

Advanced cases of bacterial disease are defined as bacterial wilt or bacterial decline, and specific plant symptoms have been documented along with the presence of particular plant-pathogenic bacteria.

Bacterial wilt was first defined as a disease of *A. stolonifera* in 1981, when it was linked to irregular patterns of declining Toronto creeping bentgrass, a vegetatively propagated cultivar, primarily during wet periods in spring and autumn months (Figs. 22 and 23). Additional research showed that other vegetatively propagated cultivars (Cohansey and Nimisilla) were also susceptible, but it is important to note that use of vegetative *A. stolonifera* cultivars had greatly diminished by the end of the twentieth century. Close inspection of infected plants showed primary localization of bacteria in the xylem vessels of the roots with some extension into the crown and leaf tissue. Advanced colonization of xylem elements causes plants to wilt and turn blue green, followed by browning, shriveling, and death.

Bacterial wilt has also occurred on *P. annua* and *Lolium perenne*. Infections on *P. annua* can occur on golf course fairways, approaches, green collars, and greens, whereas reports of infection on *L. perenne* have been limited to golf course fairways. Infected *P. annua* plants initially appear wilted and yellow, lime green, or bluish green. Small, reddish brown spots then develop (Fig. 24) that fade to light brown or bleached tan as the turf begins to die. Stem bases often appear discolored and water soaked. In both *P. annua* and *L. perenne*, younger leaves can be etiolated.

Bacterial decline is a disease occurring mainly on intensely managed surfaces of *A. stolonifera* such as golf course greens.

TABLE 2. Comparison of Two Common Bacterial Diseases Associated with Turfgrasses[a]

	Bacterial Wilt	Bacterial Decline
Causal agent	*Xanthomonas translucens*	*Acidovorax avenae*
Hosts	*Agrostis, Dactylus, Festuca, Lolium,* and *Poa* spp.	*Agrostis, Festuca, Paspalum,* and *Poa* spp.
Symptoms	Etiolation—rapid growth	Etiolation—rapid growth
	General chlorosis (irregular stand patterns)	General chlorosis (irregular stand patterns)
	General decline (irregular stand patterns)	General decline (irregular stand patterns)
	Small, yellow leaf spots that may coalesce	Browning and twisting of leaf tissue
	Small, water-soaked spots (~1 mm in diameter)	Dieback to the leaf sheath
	Scattered, dark green, water-soaked leaf lesions	

[a] Prepared by J. A. Roberts—© APS.

Bacterial decline normally begins with symptoms of etiolation that progress to a general decline (Fig. 25). Etiolated plants develop browning and twisting of the leaves and dieback to the leaf sheath. Water-soaked lesions on leaf tissue have not been observed with plants suffering from bacterial decline. Close inspection of infected plants has also shown extensive colonization of xylem vessels that impede water movement through the plant (Fig. 26).

Bacterial decline, etiolation, and wilt are diagnosed by sectioning leaves, crowns, or roots followed by microscopic examination for rapid streaming of bacterial cells from vascular tissue (Fig. 27). In symptomatic *L. perenne* and *P. annua*, bacterial streaming is typically limited to the leaf, stem, and crown tissue (Fig. 28), whereas bacterial streaming associated with bacterial wilt of *A. stolonifera* can extend deeper to the roots. In rare cases, pathogenic bacteria can be found in symptomatic tissue in the absence of bacterial streaming. However, a slow ooze of bacterial cells from senescent turfgrass tissues is common and not a reliable diagnostic sign of a specific disease. Isolation and identification of bacteria associated with bacterial streaming is considered the best practice for proper diagnosis of bacterial diseases in turfgrasses.

Fig. 21. Etiolation of *Agrostis stolonifera* caused by infection with *Acidovorax avenae.* (Courtesy J. A. Roberts—© APS)

Fig. 22. Bacterial wilt of *Agrostis stolonifera* cv. Toronto on a golf green. (Courtesy D. J. Wehner—© APS)

Fig. 23. Bacterial wilt of *Agrostis stolonifera* cv. Toronto on a golf green. (Courtesy D. L. Roberts—© APS)

Causal Agents

The original bacterium associated with bacterial wilt of *A. stolonifera* was defined as *Xanthomonas campestris*. Similarities in morphology, biochemical properties, and host ranges of many *Xanthomonas* spp. contributed to the confusion surrounding early classifications, but newer DNA-based technologies have and will continue to improve our understanding of these plant-pathogenic bacteria. Prior to 1995, bacteria causing leaf streak diseases on members of Poaceae were differentiated as pathovars (pv.) under *X. campestris* before Vauterin

Fig. 24. Bacterial wilt of *Poa annua* on a golf green. (Courtesy P. H. Dernoeden—© APS)

Fig. 25. Thinning and decline of an *Agrostis stolonifera* putting green caused by infection with *Acidovorax avenae.* (Courtesy J. A. Roberts—© APS)

et al. used DNA reassociation rates and other analyses to divide *X. campestris* into multiple species. Through this research, all *X. campestris* isolates attacking grasses were reassigned to *X. translucens*. *X. translucens* (reported as *X. campestris*) was first discovered in Switzerland in 1975 and has since been reported elsewhere in northern Europe, North America, Japan, and New Zealand. *X. translucens* pv. *graminis* (formerly *X. campestris* pv. *graminis*) causes bacterial wilt in *Lolium*, *Festuca*, and *Dactylis* spp. in Europe. *X. translucens* pv. *poae* is the incitant of bacterial wilt of *P. trivialis* in Europe and *P. annua* in Japan and North America.

Acidovorax avenae (reported as *A. avenae* subsp. *avenae*) was identified as the incitant of bacterial decline of *A. stolonifera* in 2010. A later report showed that *A. canina* cv. SR720, *A. tenuis* cvs. Exeter and Alistar, and several *A. stolonifera* cultivars (Bengal, Declaration, Tyee, L-93, Penncross, Penn G-2, Penn A-4, Southshore, Crenshaw, Providence, MacKenzie, Cato, SR-1150, PennLinks, and PennEagle) were susceptible to *A. avenae* infection, although some (Declaration and Tyee) displayed lower disease severity. *Festuca arundinacea* cv. Pixie, *P. annua* var. *annua*, and *P. pratensis* cv. NuGlade were also susceptible in inoculation tests, but *A. avenae* has not been isolated from these turf species in the field. *A. avenae* can be traced back to *Pseudomonas avenae,* which was originally defined as the cause of blade blight of oats and has been reported to cause bacterial brown stripe of *Agropyron* and *Paspalum* spp. Improvements in DNA-based technologies resulted in reclassification of *P. avenae* to *Acidovorax,* which includes several plant-pathogenic bacteria and the type species, *A. facilis,* which was isolated from lawn soil (reported as *Hydrogenomonas facilis*) in 1952.

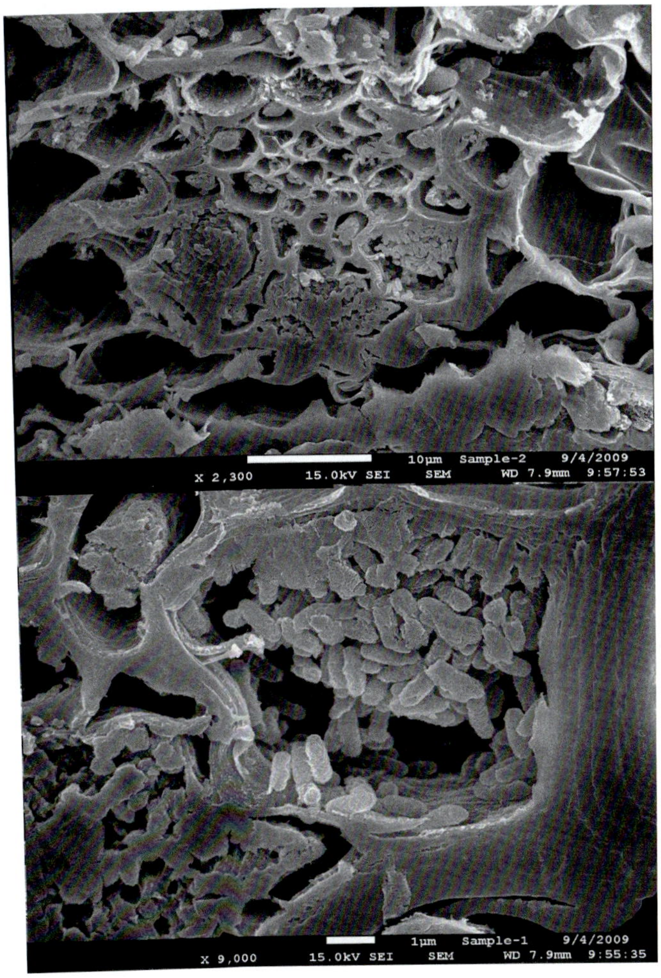

Fig. 26. Growth of *Acidovorax avenae* in the xylem vessels of *Agrostis stolonifera.* (Courtesy P. R. Giordano—© APS)

Additional bacteria have been reported as turfgrass pathogens, but reports of these are few compared with those of *A. avenae* or *X. translucens*. *Pantoea ananatis* has been shown to cause etiolation in *L. perenne* and chlorosis of *A. stolonifera* in North America. *Pseudomonas coronafaciens* var. *atropurpurea* causes halo blight of *Festuca, Lolium,* and *Poa* spp. in Japan. Bermuda stunting disease, caused by *Clavibacter xyli* subsp. *cynodontis,* is a rare condition observed in Asia that causes white leaves and witches'-broom symptoms. These bacteria are likely to occur on grasses in other humid regions of the world, but further research is necessary to improve our understanding of turf-associated bacteria.

All these bacteria consist of short rods (Fig. 29), measuring about 0.2–1 × 1–6 µm. High-resolution microscopy is required for accurate observation.

Disease Cycle

Bacteria are abundant on organic debris, in soil, and on foliage and root surfaces. Pathogenic bacteria gain entry into plant tissues as a result of mechanical damage to tissue or penetrating injuries initiated by other organisms (Fig. 30). Under very wet conditions, they can also enter plants through natural openings, such as stomata and hydathodes. Bacteria in guttation droplets can be drawn into leaf tissue through hydathodes as the foliage dries.

Plant-pathogenic bacteria produce hormones, toxins, or enzymes that injure plant tissue or affect plant metabolism. For instance, pure cultures of *A. avenae* have been shown to produce gibberellic acid (e.g., GA_1, GA_3, and GA_4), a phytohormone associated with the increased growth observed in etiolated

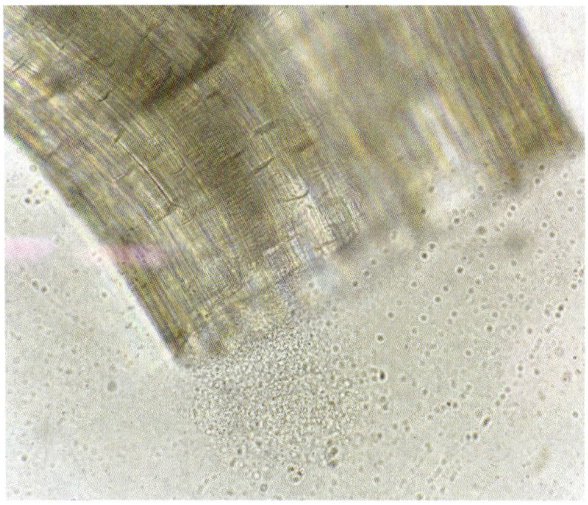

Fig. 27. Bacterial streaming from a cut leaf of *Agrostis stolonifera* infected with *Acidovorax avenae.* (Courtesy J. A. Roberts—© APS)

Fig. 28. *Xanthomonas translucens* pv. *poae* cells streaming from cut vascular bundles of a *Poa annua* leaf. (Courtesy N. A. Mitkowski—© APS)

turfgrass stems and/or leaves. Similar work has also shown higher levels of gibberellic acid in creeping bentgrass infected with *A. avenae*. While *X. translucens* has not been observed to produce gibberellic acid in culture, higher levels of GA_1, GA_3, and GA_{20} have been observed in etiolated annual bluegrass infected with *X. translucens* pv. *poae*. The bacterium may require in vivo conditions for production, or increased gibberellic acid could be a host response to bacterial infection. Regardless of the bacterium causing infection, it is likely that a high concentration of bacteria is necessary for phytohormone production and/or host responses.

X. translucens (and likely *A. avenae*) survive unfavorable periods in dead plant debris and infected plants. Outbreaks of bacterial wilt are likely to occur during periods of rainy weather, whereas bacterial decline is normally observed during hot and dry periods of the summer. Both bacteria apparently invade grasses through the cut ends of leaf blades or through wounds in roots caused by nematodes, fungi, or aeration equipment. Plant-pathogenic bacteria eventually become established in the xylem, where they may eventually interrupt the flow of water and nutrients to the foliage, leading to symptoms of wilt, dieback, and death.

X. translucens has been observed to track with mowers, while attempts to spread *A. avenae* with mowing equipment have been unsuccessful.

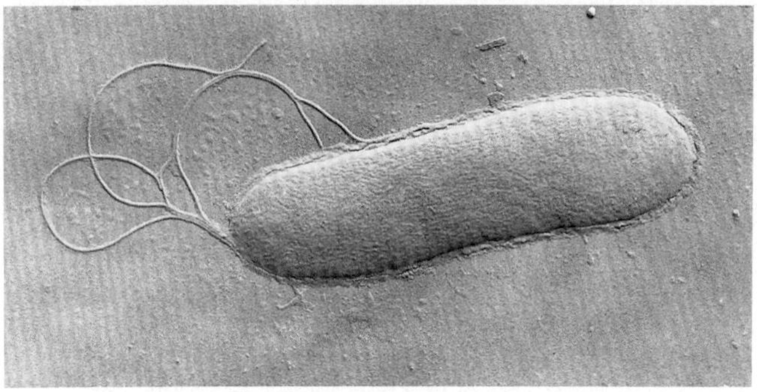

Fig. 29. Shadow-cast replica of a *Pseudomonas* sp., showing its typical rod-shaped morphology (1.1 x 5.4 µm) and polar tuft of flagella (13,300x). (Courtesy D. L. Charbonneau and W. C. Ghiorse—© APS)

Epidemiology

Moisture films on leaf surfaces are conducive to the transport of bacteria. Driving rain and streams of irrigation water are especially effective in moving bacteria across the turf and into stomatal cavities on leaves or into wounds. Bacterial etiolation, decline, and/or wilt are more severe in low-cut turf on putting greens or bowling greens than in higher-cut turf. Bacterial wilt generally occurs during warm to cool, wet periods in spring and autumn, but it may remain active in the summer. Bacterial decline is typically observed during stressful summer periods (i.e., high heat and/or drought).

Management

Given that turf loss associated with bacterial disease typically coincides with environmental stress periods, management to reduce turfgrass stress is critical. Bacteria can thrive and move in water films, so practices that reduce leaf wetness periods and/or waterlogged root zones promote turfgrass survival. Etiolation often involves elevated stem tissue, which may be more susceptible to damage from low mowing heights (i.e., less than 0.32 cm). Mowing height should be increased and mowing frequency decreased. Mowing practices performed later in the day, when the turf is dry, may limit spread of bacteria present in water films. Abrasive cultural practices like aeration and/or vertical cutting should be avoided if bacterial disease is suspected. Greens chronically affected by bacterial wilt or decline should be reestablished with resistant cultivars. Supporting turfgrass growth through application of recommended fertilization and irrigation can also limit turfgrass loss from bacterial infection.

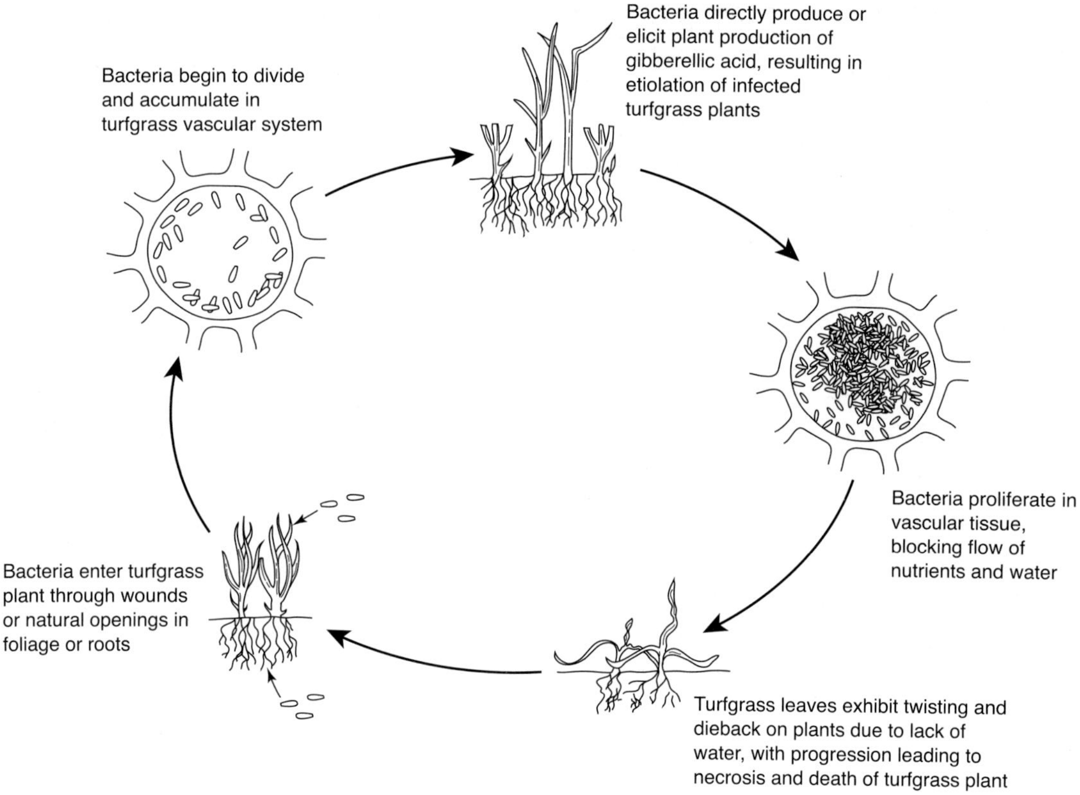

Fig. 30. Disease cycle of bacterial etiolation and decline caused by *Acidovorax avenae* in *Agrostis stolonifera*. (Courtesy J. A. Roberts—© APS. Drawing by S. McKinley.)

Plant growth regulators (PGRs) were implicated in etiolation outbreaks prior to research defining bacteria as causal agents, but more-recent research has shown adjustments in PGR programs to be effective at reducing symptoms. Particularly with infections caused by *A. avenae,* reducing trinexapac-ethyl (a late gibberellic acid inhibitor) application rates or eliminating it completely can reduce etiolation and therefore may reduce bacterial decline. Similarly, switching PGR active ingredients to early gibberellic acid inhibitors, such as paclobutrazol or flurprimidol, can also reduce etiolation. In contrast, etiolation caused by *X. translucens* has been reduced with trinexapac-ethyl applications. Differences in response to PGRs among bacterial pathogens are likely related to gibberellic acid production properties; *A. avenae* bypasses the regulation pathway of trinexapac-ethyl, whereas *X. translucens* has not been shown to produce gibberellic acid. Trinexapac-ethyl and prohexadione calcium reduce etiolation caused by *P. ananatis* on *L. perenne* fairway turf. Additional chemical control measures are limited because many fungicides do not have activity against bacterial pathogens, although some have proved efficacious when applied preventively. Certain demethylation inhibitor fungicides have gibberellic acid-regulating activity similar to that of paclobutrazol and flurprimidol and have been effective at reducing etiolation. Products containing acibenzolar-*S*-methyl can also reduce bacterial etiolation and wilt caused by *X. translucens* when applied preventively. Copper hydroxide has been shown to reduce bacterial plant pathogens residing on plant surfaces, but concerns about phytotoxicity have been reported with frequent use on closely mown *A. stolonifera.* While antibiotics are labeled for controlling plant-pathogenic bacteria in other crops, none are labeled for turfgrass use, have proved phytotoxic on turfgrass in some experiments, and are likely too expensive to be of practical value.

Selected References

Chaves, A., and Mitkowski, N. 2013. Virulence of *Xanthomonas translucens* pv. *poae* isolated from *Poa annua.* Plant Pathol. J. 29:93-98.

Egli, T., and Schmidt, D. 1982. Pathogenic variation among causal agents of bacterial wilt in forage grasses. Phytopathol. Z. 104:138-150.

Furuya, N., Ito, T., and Tsuchiya, K. 2009. Occurrence of bacterial brown stripe of creeping bentgrass on golf course green in Kyushu. J. Fac. Agric. Kyushu Univ. 54:13-17.

Giordano, P. R., Chaves, A. M., Mitkowski, N. A., and Vargas, J. M., Jr. 2012. Identification, characterization, and distribution of *Acidovorax avenae* subsp. *avenae* associated with creeping bentgrass etiolation and decline. Plant Dis. 96:1736-1742.

Giordano, P. R., Vargas, J. M., Jr., Detweiler, A. R., Dykema, N. M., and Yan, L. 2010. First report of a bacterial disease on creeping bentgrass (*Agrostis stolonifera*) caused by *Acidovorax* spp. in the United States. Plant Dis. 94:922.

Giordano, P. R., Zeng, Q., Dykema, N. M., Detweiler, A. R., and Vargas, J. M., Jr. 2015. First report of *Xanthomonas translucens* causing wilt disease on perennial ryegrass (*Lolium perenne*) in the United States. Plant Dis. 99:1270.

Imaizumi, S., and Fujimori, T. 1999. Persistence of *Xanthomonas campestris* pv. *poae* (JT-P482) in annual bluegrass. Ann. Phytopathol. Soc. Jpn. 65:52-56.

Liu, S., Vargas, J., and Merewitz, E. 2018. Temperature and hormones associated with bacterial etiolation symptoms in creeping bentgrass and annual bluegrass. J. Plant Growth Regul. 38:249-261.

Mitkowski, N. A., Browning, M., Basu, C., Jordan, K., and Jackson, N. 2005. Pathogenicity of *Xanthomonas translucens* from annual bluegrass on golf course putting greens. Plant Dis. 89:469-473.

Nishino, T., and Fujimori, T. 1998. Pathogenic diversity of *Xanthomonas campestris* pv. *poae* within the genus *Poa.* Ann. Phytopathol. Soc. Jpn. 64:1-6.

Roberts, D. L., Vargas, J. M., Jr., Detweiler, R., Baker, K. K., and Hooper, G. R. 1981. Association of a bacterium with a disease of Toronto creeping bentgrass. Plant Dis. 65:1014-1016.

Roberts, J. A., and Kerns, J. P. 2018. Etiolation and decline: An enigmatic condition in creeping bentgrass putting greens. Pages 89-100 in: Plant-Pathogenic *Acidovorax* Species. S. Burdman and R. R. Walcott, eds. American Phytopathological Society, St. Paul, MN.

Roberts, J. A., Kerns, J. P., and Ritchie, D. F. 2015. Bacterial etiolation of creeping bentgrass as influenced by biostimulants and trinexapac-ethyl. Crop Prot. 72:119-126.

Roberts, J. A., Ma, B., Tredway, L. P., Ritchie, D. F., and Kerns, J. P. 2018. Identification and pathogenicity of bacteria associated with etiolation and decline of creeping bentgrass golf course putting greens. Phytopathology 108:23-30.

Roberts, J. A., Ritchie, D. F., and Kerns, J. P. 2016. Plant growth regulator effects on bacterial etiolation of creeping bentgrass putting green turf caused by *Acidovorax avenae.* Plant Dis. 100:577-582.

Roberts, J. A., Soika, M. D., and Ritchie, D. F. 2014. Impact of antimicrobial compounds on etiolation caused by *Xanthomonas translucens* and on turf quality of creeping bentgrass putting-green turf. Appl. Turf Sci. 11. doi:10.2134/ATS-2014-0047-RS

Roberts, J. A., Tredway, L. P., and Ritchie, D. F. 2014. First report of *Xanthomonas translucens* causing etiolation on creeping bentgrass turf in Illinois, Kentucky, and North Carolina. Plant Dis. 98:839.

Schatz, A., and Bovell, C., Jr. 1952. Growth and hydrogenase activity of a new bacterium, *Hydrogenomonas facilis.* J. Bacteriol. 63:87-98.

Vauterin, L., Hoste, B., Kersters, K., and Swings, J. 1995. Reclassification of *Xanthomonas.* Int. J. Syst. Bacteriol. 45:472-489.

Willems, A., Goor, M., Thielemans, S., Gillis, M., Kersters, K., and Ley, J. D. 1992. Transfer of several phytopathogenic *Pseudomonas* species to *Acidovorax* as *Acidovorax avenae* subsp. *avenae* subsp. nov., comb. nov., *Acidovorax avenae* subsp. *citrulli, Acidovorax avenae* subsp. *cattleyae,* and *Acidovorax konjaci.* Int. J. Syst. Bacteriol. 42:107-119.

Zeng, Q., Wang, J., Bertels, F., Giordano, P. R., Chilvers, M. I., Huntley, R. B., Vargas, J. M., Sundin, G. W., Jacobs, J. L., and Yang, C.-H. 2017. Recombination of virulence genes in divergent *Acidovorax avenae* strains that infect a common host. Mol. Plant-Microbe Interact. 30:813-828.

(Prepared by J. A. Roberts)

Bipolaris and *Exserohilum* Diseases

The graminicolous species of *Helminthosporium* have been segregated into four genera: *Bipolaris, Curvularia, Pyrenophora* (syn. *Drechslera*), and *Exserohilum.* These fungi are darkly pigmented and share overlapping conidial characteristics, which previously resulted in the erroneous assignment of generic names. The generic assignments of these fungi have been resolved with the advancement of fungal molecular systematics. These fungi are economically important plant pathogens, inciting diseases in their asexual stages. The sexual stages of these fungi, *Cochliobolus* (*Bipolaris* and *Curvularia*) and *Setosphaeria* (*Exserohilum*), are no longer acknowledged by fungal taxonomists. For the exception see *Pyrenophora* (*Drechslera*) Diseases.

These genera share similar asexual life cycles. Conidia (asexual spores) are considered the primary inoculum sources, inciting polycyclic diseases of cool- and warm-season turfgrasses. Uniquely, these fungi can infect all plant parts (leaves, stems, crown, and roots) in a single epidemic. *Bipolaris* and *Exserohilum* spp. infect warm-season grasses, except for *B. sorokiniana* and *B. gigantea,* which are pathogenic to the majority of cool- and warm-season grasses. *Curvularia* spp. infect a broad range of cool- and warm-season grasses, whereas *Pyrenophora* (*Drechslera*) spp. are pathogens primarily of cool-season grasses.

Symptoms and Signs

Bipolaris and *Exserohilum* spp. are somewhat unique compared with most turf pathogens because they infect shoots, crowns, and roots and may be seedborne. They cause a preemergence blight and postemergence seedling disease. Symptoms may be limited to leaf spots on older tissue, while on younger tissue pinpoint leaf flecking can be observed (Fig. 31). Leaf spots are not observed on the youngest tissue. Under high disease pressure, lesions progress downward to the basal leaf sheaths, stems, crowns, and upper roots, leading to plant death.

Bipolaris cynodontis and *Exserohilum rostratum* incite similar symptoms of leaf, crown, and root rot diseases. *B. cynodontis* infects *Axonopus* and *Cynodon* spp., and *E. rostratum* infects *Agrostis, Cynodon, Bromus, Pennisetum, Stenotaphrum,* and *Zoysia* spp. *E. rostratum* is also a primary pathogen of forage grasses. The symptoms produced by these fungi first appear on older green leaves as random, reddish brown flecks or pinpoint flecks that enlarge to form irregularly shaped, brownish green to black spots that may advance to the leaf sheaths. Severely infected leaves appear light tan to straw colored and may die, leading to melting-out (loss of turf density) symptoms. Infected plants occur in irregular patches, ranging from 5 cm to 1 m in diameter, mimicking symptoms of drought stress. Extensive damage occurs when the pathogens attack crowns, stolons, or rhizomes.

B. gigantea causes leaf lesions on *Cynodon* spp. that begin as tiny, dark brown spots. As the spots enlarge, they become oblong or elongate, and the central areas bleach to a light tan. Lesions may completely girdle leaves. Rapid enlargement of lesions may sometimes give rise to roughly concentric bands of bleached and dark brown tissue, from which the name "zonate leaf spot" is derived (Fig. 32). Severely diseased turf becomes thinned as leaves are killed. The disease occurs in *Cynodon* spp. during the autumn, when the growth of this warm-season grass is slowed by cool weather. *B. gigantea* also causes leaf spot on *Agrostis* and *Poa* spp. during the summer, when their growth is slowed by warm to hot weather.

B. heveae infects the leaves of *Cynodon* and *Zoysia* spp. and causes brown stripe. Reddish brown, pinpoint flecks can be observed parallel to the leaf axis and progress with leaf maturation into brown to dark brown stripes, each surrounded by a chlorotic halo. Leaves may yellow and die under high disease pressure.

B. sorokiniana is the most destructive member of this group with the broadest host range among amenity turfgrasses. It causes diseases of the leaves (Fig. 33), crowns, and roots of cool- and warm-season grasses. Initial symptoms on leaves are similar to those caused by *B. cynodontis*; however, on cool-season grasses, chlorotic halos may be associated with foliar spots similar to those caused by *Pyrenophora* (*Drechslera*) spp. Leaf spots are prevalent under moderate temperatures, but as temperatures increase, leaf blight becomes more obvious. Leaf blight and plant death are observed when temperatures exceed 35°C (95°F). Infected crowns and roots appear black, resulting in irregular patches of dead grass.

Causal Agents

Bipolaris, Curvularia, and *Exserohilum* spp. are often observed in a complex when diseased grasses are examined. The *Curvularia* spp. are the most prolific sporulators; therefore, it is not uncommon to observe only *Curvularia* spp. associated with diseases caused by the complex. To determine the causal agent, a sample is incubated in a moist chamber to induce sporulation or a fungal isolation in pure culture may be necessary.

Conidia of *Bipolaris* and *Exserohilum* spp. are fusoid (gradually tapering toward both ends) and curved and often tend to be darkly pigmented. Conidia of *Exserohilum* spp. possess protuberant hila, and the septa of the terminal cells are conspicuous (Fig. 34). Conidia of *Bipolaris* and *Exserohilum* spp. are borne on brown conidiophores, which may form in abundance on necrotic tissue. The conidia produce germ tubes, which are formed only from the end cells in a direction roughly parallel

Fig. 31. Leaf spots caused by *Bipolaris cynodontis*. (Courtesy M. Tomaso-Peterson—© APS)

Fig. 32. Leaf spot of *Cynodon dactylon* caused by *Bipolaris gigantea.* (Courtesy J. L. Dale—© APS)

Fig. 33. Leaf spot of *Poa pratensis* caused by *Bipolaris sorokiniana.* (Courtesy R. W. Smiley—© APS)

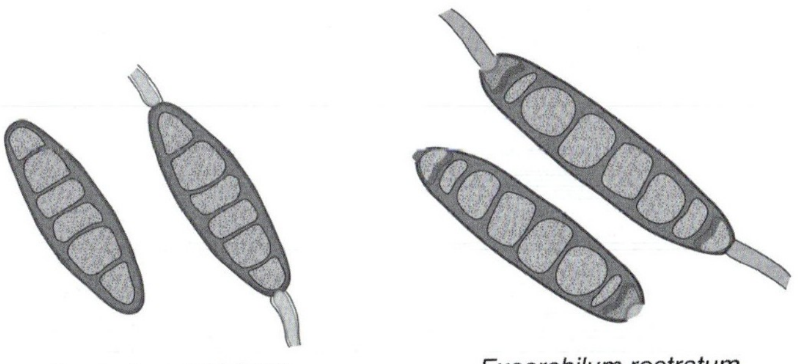

Fig. 34. Germ tubes develop from the terminal cells of *Bipolaris* spp. (left), while a protuberant hilum is present on conidia of *Exserohilum* spp. (right). (Courtesy L. P. Tredway—© APS. Drawing by S. McKinley.)

to the axis of the conidium. Conidial size and shape overlap among these fungi; therefore, molecular data is more reliable for identification of *Bipolaris, Curvularia, Pyrenophora* (*Drechslera*), and *Exserohilum* spp. than taxonomic keys. Analyses of the nuclear rDNA internal transcribed spacer region or glyceraldehyde-3-phosphate dehydrogenase is sufficient for species identification.

Disease Cycle, Epidemiology, and Management

The life cycles of *Bipolaris* and *Exserohilum* spp. closely resemble those of *Curvularia* and *Pyrenophora* (*Drechslera*) spp. *Bipolaris* and *Exserohilum* spp. infect the foliage of warm-season grasses during cool, wet periods from autumn through spring. Crown and root rots caused by *B. cynodontis* occur mostly during warm, dry periods in the summer.

B. sorokiniana causes leaf spot symptoms on warm-season grasses under moderate temperatures, but as temperatures exceed 30°C (86°F), leaf blight becomes more prominent. Foliar blight and plant death are observed when temperatures exceed 35°C (95°F). Leaf blights in cool-season grasses occur during warm, wet weather in midsummer. Disease severity increases as temperatures rise from 20 to 35°C (68 to 95°F). In contrast, leaf spots caused by *Pyrenophora* (*Drechslera*) spp. in cool-season grasses occur during cool, wet weather in spring and autumn.

Several fungicide chemistries are effective for controlling diseases caused by *Bipolaris* and *Exserohilum* spp. When crown and root rot are targeted, fungicides should be watered in (0.64 cm) immediately following the application. Cultural management practices such as vertical mowing may predispose ultradwarf Bermudagrasses to leaf spot. Under intense cultural management practices, a fungicide application is recommended. Host tolerance in some seeded Bermudagrass cultivars has been observed.

Selected References

Farr, D. F., and Rossman, A. Y. n.d. Fungal Databases, Systematic Mycology and Microbiology Laboratory, U.S. Department of Agriculture–Agricultural Research Service, Washington, DC. https://nt.ars-grin.gov/fungaldatabases/

Goh, T. K., Hyde, K. D., and Lee, D. K. L. 1998. Generic distinction in the *Helminthosporium*-complex based on restriction analysis of the nuclear ribosomal RNA gene. Fungal Diversity 1:85-107.

Lane, B., Stricker, K. B., Adhikari, A., Ascunce, M. S., Clay, K., Flory, S. L., Smith, M. E., Goss, E. M., and Harmon, P. F. 2020. Large-spored *Drechslera gigantea* is a *Bipolaris* species causing disease on the invasive grass *Microstegium vimineum*. Mycologia 112:921-931.

Manamgoda, D. S., Rossman, A. Y., Castlebury, L. A., Crous, P. W., Madrid, H., Chukeatirote, E., and Hyde, K. D. 2014. The genus *Bipolaris*. Stud. Mycol. 79:221-288.

Pratt, R. G. 2000. Diseases caused by dematiaceous fungal pathogens as potential limiting factors for production of bermudagrass on swine effluent application sites. Agron. J. 92:512-517.

Rossman, A. Y., Manamgoda, D. S., and Hyde, K. D. 2013. Proposal to conserve the name *Bipolaris* against *Cochliobolus* (*Ascomycota: Pleosporales: Pleosporaceae*). Taxon 62:1331-1332.

Sivanesan, A. 1987. Graminicolous species of *Bipolaris, Curvularia, Drechslera, Exserohilum* and their teleomorphs. Mycol. Pap. 158. CAB International, Wallingford, U.K.

Tomaso-Peterson, M., and Young, J. 2010. Cultivar response of seeded bermudagrass to leaf spot and the influence of nitrogen on disease severity. Appl. Turfgrass Sci. 7. doi:10.1094/ATS-2010-0326-01-RS

Tsukiboshi, T., Chung, W. H., and Yoshida, S. 2005. *Cochliobolus heveicola* sp. nov. (*Bipolaris heveae*) causes brown stripe of bermudagrass and zoysiagrass. Mycoscience 46:17-21.

Vines, P. L., Allen, T. W., DuBien, J., Stewart, B. R., and Tomaso-Peterson, M. 2017. Cultivar selection, cultural management practice, and fungicide application influence leaf spot and dollar spot on ultradwarf bermudagrass. Crop, Forage Turfgrass Manage. 3. doi:10.2134/cftm2017.02.0013

(Prepared by R. W. Smiley, P. H. Dernoeden, and B. B. Clarke; revised by M. Tomaso-Peterson)

Copper Spot (Zonate Leaf Spot)

Copper spot occurs mostly in *Agrostis* spp. It is particularly prevalent in *A. canina* and *A. capillaris* in the northeastern United States. The copper spot pathogen also causes zonate leaf spot in *Cynodon, Zoysia,* and other genera of Poaceae. Zonate leaf spot is an important foliar disease of sorghum in the southwestern and Gulf Coast regions of the United States.

Symptoms

Copper spot appears as scattered, roughly circular patches 3–8 cm in diameter that are orange to copper colored (Fig. 35). Small, red to brown, water-soaked lesions form on individual leaves and may coalesce to blight an entire leaf. During warm, wet weather, the color intensifies as leaves are covered with mycelium and tiny sporodochia containing a gelatinous mass of salmon-colored spores (Fig. 36). Sporodochia turn orange to red and crown shaped when dry. Dollar spot and copper spot often occur at the same time in susceptible turf (Fig. 37). These diseases can usually be distinguished by differences in lesion color (dollar spot lesions are a bleached color) and by differences in the outer margins of the patches; copper spot patches are typically more diffuse and the margins are not as well defined as those of dollar spot patches. During wet weather, copper spot infection centers may track in the direction of mowing or rolling, similar to Microdochium patch.

Fig. 35. Copper spot of *Agrostis canina*. (Courtesy N. Jackson—© APS)

Causal Agent

Microdochium sorghi (formerly called *Gloeocercospora sorghi*) produces sporodochia (fruiting bodies) that emerge through stomata in the leaf epidermis. Conidia, borne on the sporodochia, are hyaline, filiform (needle shaped), 1.4–3.2 × 20–195 µm (average 2.4 × 83 µm), and one to seven septate (Figs. 38 and 39). The conidia are produced in a gelatinous matrix and appear salmon colored en masse. Tiny, black, lenticular to spherical sclerotia, 0.1–0.2 mm in diameter, develop in necrotic leaf lesions but are often difficult to see without a hand lens. No sexual stage has been reported.

Disease Cycle

The pathogen survives adverse weather and overwinters as sclerotia or thick-walled mycelia embedded in thatch and infested leaves. When maximum soil temperatures at a depth of 2.5 cm exceed 17°C (62°F) for at least 1 week, sclerotia germinate to form sporodochia, which produce conidia during periods of warm, moist weather. Conidia can germinate within 5 hours to produce mycelium that infects leaves, reportedly by direct penetration or through stomata, producing visible lesions within 24 hours. Infection causes leaf necrosis, and the pathogen then produces more sporodochia with conidia. When conditions are conducive to infection, *M. sorghi* can produce large quantities of secondary inoculum in a short time (24–48 hours), resulting in very rapid disease development. Prolonged leaf wetness and the release of plant exudates from recently mowed leaf tips can also enhance disease severity. Conidia can be transferred to uninfected leaves by water, mowers, foot traffic, and equipment. Sclerotia are produced in large quantities as necrotic tissue ages.

Copper spot can cause severe thinning in susceptible swards, but the plants seldom die. Infected turf recovers slowly with the advent of cool, dry weather in autumn.

Epidemiology and Management

Copper spot occurs during warm, wet weather. Optimal air temperatures for infection are 20–30°C (68–86°F). In *A. canina,* the disease is especially damaging to turf that has been fertilized with excessive nitrogen or is grown in highly acidic soil.

Management practices recommended for the control of dollar spot also help to suppress copper spot. Dew should be removed in the morning by mowing, rolling, or dragging, and air circulation should be increased by pruning, removing trees, or installing fans to reduce the leaf wetness period. Applications of water-soluble nitrogen sources should be limited (no more than 0.25 lb per 1,000 ft^2 per application) when copper spot is active, and mowing of infested turf should be avoided when the foliage is wet. A soil pH of 6–7 should be maintained. The application of lime may help reduce disease severity in soils with pH less than 5.5.

Improved resistance to copper spot in new cultivars of *A. canina* has been reported. A turfgrass specialist from a local university extension service and reports published by the National Turfgrass Evaluation Program should be consulted for information on cultivars with increased tolerance to this disease.

Fig. 36. Sporodochia of *Microdochium sorghi* on leaves of *Agrostis stolonifera*. (Courtesy M. Tomaso-Peterson—© APS)

Fig. 37. Copper spot of *Agrostis capillaris*. Note smaller dollar spot infection centers interspersed among larger copper spots. (Courtesy M. L. Agnew—© APS)

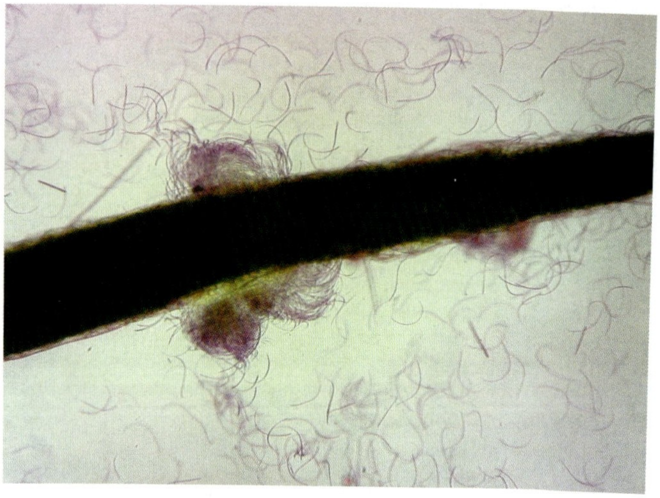

Fig. 38. Sporodochia and conidia of *Microdochium sorghi* on leaves of *Agrostis stolonifera* stained with acid fuchsin. (Courtesy M. Tomaso-Peterson—© APS)

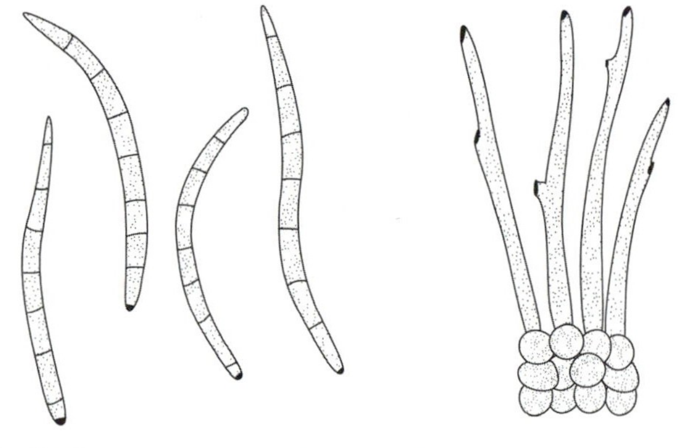

Fig. 39. Conidia and conidiophores of *Microdochium sorghi.* (Courtesy L. P. Tredway—© APS. Drawing by S. McKinley.)

Fungicides (e.g., chlorothalonil and demethylation inhibitors) can provide effective control of copper spot, whereas some succinate dehydrogenase-inhibiting fungicides have been reported to enhance the severity of copper spot.

Selected References

Braun, U. 1995. A Monograph of *Cercosporella, Ramularia* and Allied Genera (Phytopathogenic Hyphomycetes), vol. 1. IHW-Verlag, Munich.

Hernandez-Restrepo, M., Groenewald, J. Z., and Crous, P. W. 2016. Taxonomic and phylogenetic re-evaluation of *Microdochium, Monographella* and *Idriella*. Persoonia 36:57-82.

Mulder, J. L., and Holliday, P. 1971. *Gloeocercospora sorghi*. Descriptions of Pathogenic Fungi and Bacteria, no. 300. Commonwealth Mycological Institute, Kew, England.

Smith, J. D., Jackson, N., and Woolhouse, A. R. 1989. Fungal Diseases of Amenity Turf Grasses. E. & F. N. Spon, London.

(Prepared by R. W. Smiley, P. H. Dernoeden, and B. B. Clarke; revised by B. B. Clarke)

Cream Leaf Blight

Cream leaf blight, a disease similar to pink patch, has been observed in *Festuca arundinacea* and *Cynodon* spp. in subtropical and transition zone environments. The causal agent is described as a biotype of *Limonomyces roseipellis* that does not produce clamp connections and lacks the characteristic pink pigmentation normally associated with this pathogen.

Symptoms and Signs

In *F. arundinacea,* cream leaf blight causes grayish white or cream-colored patches up to 46 cm in diameter. Light yellow or white lesions form at the leaf tips and expand basipetally, eventually blighting entire tillers. The lesions are similar in appearance to those caused by *Melanotus phillipsii* but lack the abundant sporocarps (fruiting bodies) typically associated with white blight. Instead, infected leaf tissue is heavily colonized with cream-colored hyphae and threadlike hyphal aggregates. Leaves in patches may be matted together with cream-colored mycelium. The cream leaf blight pathogen does not produce the pink, gelatinous growth often associated with pink patch.

Cream leaf blight is also observed in hybrid Bermudagrasses (*Cynodon dactylon* × *transvaalensis*), especially ultradwarf putting greens, during winter dormancy or extended periods of slow growth. The disease is largely cosmetic, but in severe cases it can detract from the aesthetic quality or perceived uniformity of putting surfaces. Symptoms of the disease appear as cream-colored spots or patches 8–40 cm in diameter (Fig. 40). The surfaces of affected leaves are colonized by white hyphal growth, but no lesions or other symptoms are observed on the host plant. Upon incubation, the pathogen readily produces hyphal aggregates useful for diagnosis (Fig. 41).

Causal Agent

The cream leaf blight pathogen produces hyaline, binucleate hyphae characteristic of *L. roseipellis,* but clamp connections are absent. Basidiocarps have not been observed in association with cream leaf blight.

Disease Cycle and Epidemiology

In *F. arundinacea,* symptoms of cream leaf blight are typically observed in late spring or summer when high and low temperatures are approximately 30°C (86°F) and 15°C (59°F), respectively. Isolates of *L. roseipellis* from *F. arundinacea* grow in vitro at 10–32°C (50–90°F), with optimal growth at 25°C (77°F). Hyphae colonize the leaf surfaces, penetrate stomatal pores on the abaxial leaf surfaces, and colonize infected leaf tissue both intra- and extracellularly.

The epidemiology of cream leaf blight in Bermudagrass has not been investigated. Colonization of leaf surfaces may occur across a broad range of temperatures. Signs of the disease become visible when Bermudagrass growth is slow for extended periods of time. It is unknown whether the pathogen invades or infects Bermudagrass leaves because foliar symptoms have not been reported.

Management

Little is known about the management of cream leaf blight. In *C. dactylon* putting greens, application of succinate dehydrogenase inhibitor or demethylation inhibitor fungicides in late summer to fall, prior to winter dormancy, have effectively prevented cream leaf blight symptoms in the subsequent winter and spring. It is unclear whether these applications prevent a key step in the disease cycle occurring at that time or whether residual fungicide on or in the dormant foliage provides protection from colonization several months later.

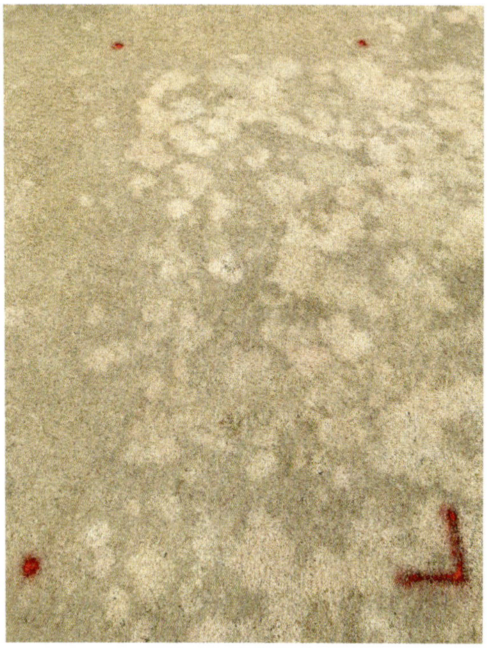

Fig. 40. Cream leaf blight symptoms in a dormant *Cynodon dactylon × transvaalensis* putting green. (Courtesy L. P. Tredway—© APS)

Fig. 41. Hyphal aggregates of *Limonomyces roseipellis* growing among Bermudagrass leaves following incubation in a moist chamber. (Courtesy E. L. Butler—© APS)

Selected References

Burpee, L. L., Mims, C. W., Tredway, L. P., Bae, J., and Jung, G. 2003. Pathogenicity of a novel biotype of *Limonomyces roseipellis* in tall fescue. Plant Dis. 87:1031-1036.

Kerns, J. P., and Butler, E. L. 2018. Cream leaf blight of ultradwarf bermudagrass. Golf Course Manage. 86:66-69.

(Prepared by L. P. Tredway)

Curvularia Diseases

Historically, *Curvularia* spp. were considered senectophytes, inciting disease on plants weakened by abiotic and biotic stresses. Weakly pathogenic *Curvularia* spp. may cause disease where turfgrasses are subjected to heat stress, fertility and moisture extremes, pesticide phytotoxicity, or reduced daily light periods. With the advent of fungal molecular systematics, a revision of the *Bipolaris–Curvularia* complex placed small-spored pathogenic *Bipolaris* spp. into the genus *Curvularia* as well as a sterile, nonsporulating species of *Curvularia*. Multiple species of *Curvularia* are frequently observed and isolated from the same plant, making diagnosis somewhat challenging at times.

Symptoms

Senectophytes. In stressed turfgrass, a general decline occurs and the grass becomes thin and ragged looking. Thinning of the grass may occur in irregularly shaped patches or streaks, which may coalesce into larger areas.

The predominant initial symptom on individual leaves of *Poa* and *Festuca* spp., hybrid Bermudagrass (*Cynodon dactylon* × *C. transvaalensis*), and *Zoysia* spp. is slight leaf tip necrosis and an indistinct, green and yellow dappled pattern extending downward from the leaf tip. The affected tissue then turns brown and ultimately gray as it shrivels and dies. A reddish brown margin separating the diseased tissue from the healthy tissue may or may not be present. In addition to tip dieback, tan leaf lesions with red or brown margins may develop.

Symptoms in *Agrostis* spp. are similar, except that blighted leaves are tan rather than brown or gray and leaf lesions have not been observed.

In the cool-season grasses, crown and leaf sheath infections result in dark brown dry rots without distinct boundaries. Prolific sporulation of the fungus may cause lesions to appear blackened.

Curvularia blight (dog footprint). *C. lunata, C. geniculata,* and *C. verruculosa* can independently or in a complex cause a blighting of *Cynodon* and *Zoysia* spp., resulting in tan to light brown patches. Foliar symptoms begin as small, dark brown spots that progress to foliar dieback, resulting in blighted plants. Under severe disease conditions, plants do not recover. Dark brown to black sporulation may be observed on infected plant material during periods of high humidity and wet conditions.

Ink spot. Ink spot is a foliar disease of *Cynodon* and *Zoysia* spp. maintained under golf course conditions. Ink spot is caused by *C. malina,* which is sterile since no sexual or asexual reproductive structures have been observed. Symptoms first appear as distinct chocolate brown to black spots (2–5 cm in diameter) in the fall and spring months. The dark spots may coalesce to form large, irregular areas of blighted turf (Fig. 42). Foliar symptoms appear as small, purplish black spots that develop into eyespot lesions on the leaves of *Zoysia* spp. and develop round to elliptical lesions on *Cynodon* spp. The leaf lesions have necrotic centers with dark brown to black margins. As the disease progresses, blighted leaves at the base of the plant turn chocolate brown to black, giving rise to the symptoms observed in the turfgrass sward.

Leaf spot and stem and crown rots. *C. australiensis* causes a leaf spot of *Cynodon* spp. that appears as irregular, water-soaked, brown lesions on leaves and sheaths. *C. spicifera* causes stem and crown rots of *Cynodon* and *Zoysia* spp. Infected plants appear unthrifty, and their leaves become chlorotic and then turn brown. Stem lesions are purple to black. An extensive rot of stems, crowns, and roots may occur. A general thinning of the grasses may at first be the only overall symptom.

Causal Agents

The majority of *Curvularia* species are weakly pathogenic to turfgrasses, except for *C. australiensis, C. malina,* and *C. spicifera.* The pathogenic *Curvularia* spp. were formerly classified as *Bipolaris* spp., with the exception of *C. malina.*

These fungi produce dark brown mycelium, conidiophores, and conidia, with the exception of the sterile *C. malina.* Black, netlike mycelium and darkly pigmented appressoria of *C. malina* may be observed colonizing leaves and stems of affected grasses. The mycelium of *Curvularia* spp. grows inside leaf tissue or along the leaf surface. The conidia are usually abruptly curved or straight (*C. australiensis, C. buchloës, C. hawaiiensis,* and *C. spicifera*), 20–40 µm long, and three or four septate. The curved conidia have enlarged and darkened central cells (Fig. 43). The characteristics and sizes of conidia differ slightly for each species. Infection pads produced by *C. lunata* may be observed on foliage of warm-season turfgrasses when environmental conditions are favorable (Fig. 44). Molecular data are more reliable for identification of *Curvularia* spp. Analyses of the nuclear rDNA internal transcribed spacer region or glyceraldehyde-3-phosphate dehydrogenase are sufficient for species identification.

Disease Cycle, Epidemiology, and Management

Curvularia spp. closely resemble *Bipolaris* spp. in host spectrum, morphology, and epidemiology. *Curvularia* spp. attack all common turfgrass species and cause the most damage during periods of high temperature (27–35°C [80–95°F]) or drought stress. Disease caused by senectopathic species generally develops only after high temperatures accelerate leaf aging (senescence) and thereby predispose the host to infection. All *Curvularia* spp. are capable of growing as saprophytes on plant debris and survive unfavorable weather as mycelia and/or co-

Fig. 42. Ink spot symptoms in zoysiagrass. (Courtesy M. Tomaso-Peterson—© APS)

nidia on or in organic debris. During warm, wet weather, infection proceeds rapidly. Profuse sporulation occurs on dead plant tissues, and conidia are often associated with necrotic plant tissues during the summer. *C. malina* is most active in the spring and autumn during warm periods (21–29°C [70 84°F]) associated with prolonged or high-precipitation events. The pathogen overwinters as mycelium on or in dead plant debris. Ink spot may persist throughout the growing season if environmental conditions remain favorable. Leaf spot (*C. australiensis*) and stem and crown rots (*C. spicifera*) generally occur during early and late summer.

These diseases are managed through the improvement of growing conditions and the use of sound cultural practices. Fungicides may be recommended in some situations. The effects of moisture, mowing, thatch, shade, and pesticides on diseases caused by *Curvularia* spp. are similar to their effects on diseases caused by *Drechslera, Bipolaris,* and *Exserohilum* spp. (see *Bipolaris* and *Exserohilum* Diseases and *Pyrenophora* (*Drechslera*) Diseases).

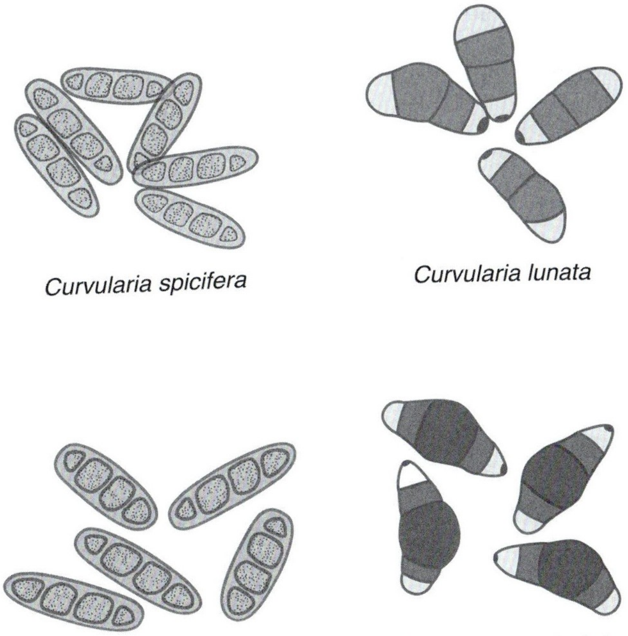

Fig. 43. Conidia of select *Curvularia* spp. associated with turfgrasses. (Courtesy L. P. Tredway—© APS. Drawing by S. McKinley.)

Fig. 44. Infection pads produced by *Curvularia lunata*. (Courtesy M. Tomaso-Peterson—© APS)

Selected References

Brecht, M. O., Stiles, C. M., and Datnoff, L. E. 2007. Evaluation of pathogenicity of *Bipolaris* and *Curvularia* spp. on dwarf and ultradwarf bermudagrasses in Florida. Plant Health Prog. 8. doi:10.1094/PHP-2007-0119-02-RS

Couch, H. B. 1995. Diseases of Turfgrasses. 3rd ed. Krieger Publishing, Malabar, FL.

Fang, K. F., Huang, J. B., and Hsiang, T. 2007. First report of brown leaf spot caused by *Bipolaris australiensis* on *Cynodon* spp. in China. Plant Pathol. 56:349.

Hodges, C. F., and Campbell, D. A. 1995. Growth of *Agrostis palustris* in response to adventitious root infection by *Curvularia lunata* J. Phytopathol. 143:639-642.

Kim, J.-C., Choi, G. J., Kim, H. T., Kim, H.-J., and Cho, K. Y. 2000. Pathogenicity and pyrenocine production of *Curvularia inaequalis* isolated from zoysia grass. Plant Dis. 84:684-688.

Manamgoda, D. S., Cai, L., McKenzie, E. H. C., Crous, P. W., Madrid, H., Chukeatirote, E., Shivas, R. G., Tan, Y. P., and Hyde, K. D. 2012. A phylogenetic and taxonomic re-evaluation of the *Bipolaris-Cochliobolus-Curvularia* complex. Fungal Diversity 56:131-144.

Manamgoda, D. S., Rossman, A. Y., Castlebury, L. A., Crous, P. W., Madrid, H., Chukeatirote, E., and Hyde, K. D. 2014. The genus *Bipolaris*. Stud. Mycol. 79:221-288.

Muchovej, J. J., and Couch, H. B. 1987. Colonization of bentgrass turf by *Curvularia lunata* after leaf clipping and heat stress. Plant Dis. 71:873-875.

Roberts, J. A., and Tredway, L. P. 2008. First report of Curvularia blight of zoysiagrass caused by *Curvularia lunata* in the United States. Plant Dis. 92:173.

Smith, J. D., Jackson, N., and Woolhouse, A. R. 1989. Fungal Diseases of Amenity Turf Grasses. E. & F. N. Spon, London.

Tani, T., and Beard, J. B. 1997. Curvularia leaf blight (dog foot-print). Pages 68-72 in: Color Atlas of Turfgrass Diseases. John Wiley & Sons, Hoboken, NJ.

Tomaso-Peterson, M., Jo, Y. K., Vines, P. L., and Hoffman, F. G. 2016. *Curvularia malina* sp. nov. incites a new disease of warm-season turfgrasses in the southeastern United States. Mycologia 108:915-924.

(Prepared by R. W. Smiley, P. H. Dernoeden, and B. B. Clarke; revised by M. Tomaso-Peterson)

Damping-Off (Seedling Diseases)

When seed germination and seedling growth are slowed by suboptimal environmental conditions, seedlings become susceptible to infection by many common soil fungi. Diseases of young plants are often referred to as "seed decay" or "damping-off." These diseases occasionally cause extensive losses of plants in newly seeded turfs.

Symptoms

When environmental conditions favor the growth of pathogens more than the germination of seeds, the pathogens may cause seed decay and death. If conditions are less heavily weighted in favor of the pathogens, seeds may germinate to produce weak seedlings, which then may be attacked and are easily killed, either before or after they emerge from the soil. When seedlings do not emerge, it is typically referred to as "preemergence damping-off." When seedlings are killed after emergence, it is referred to as "postemergence damping-off." Damping-off is a broad term used to describe the loss of juvenile plants during or after emergence from the seed.

The incidence of seed and seedling death is generally not uniform across a seedbed. Some areas may have fairly dense and vigorous seedling growth, whereas other areas have few or no healthy seedlings. The overall patchy appearance of a new stand is sometimes associated with areas where seeds were buried too deep or too shallow.

Individual affected plants or seeds must be inspected microscopically to determine the cause of thinning in a stand. Seedlings killed after they emerge from the soil usually have water-soaked lesions at the soil surface. The lesions rapidly girdle the plant, causing it to collapse, shrivel, and turn yellow to brown as it dies.

Causal Agents

Many soil-inhabiting fungi cause seed rot and seedling diseases. The most prevalent are *Pythium* spp., *Fusarium* spp., *Rhizoctonia* spp., and *Microdochium nivale*. Numerous other fungi also cause seed decay and seedling diseases, including *Microdochium bolleyi* and species of *Alternaria, Bipolaris, Botrytis, Cladosporium, Colletotrichum, Curvularia, Drechslera, Nigrospora, Pyrucularia, Septoria, Stagonospora,* and *Trichoderma,* among others.

Disease Cycle

Pathogens capable of causing seedling diseases are likely to be present in all seedbeds. It should be assumed that they are present whenever seed is planted, even in fumigated soil. When pathogen growth is active and seedling growth is slow, pathogens attack and easily kill entire plants. Seedlings in seed lots contaminated with spores of these pathogens are particularly prone to disease, because the pathogens are in an ideal location for infection at the very beginning of seedling development.

Epidemiology

Seeds of different grass species require different lengths of time for germination. Under favorable conditions, *Poa pratensis* generally emerges in about 2–3 weeks; *Lolium perenne* in less than 1 week; *Festuca rubra, F. arundinacea,* and *Agrostis stolonifera* in about 1 week; *Eremochloa* spp. in about 2 weeks; *Cynodon dactylon* in about 3 weeks; and *Buchloë dactyloides* in 3–4 weeks. Adequate but not excessive moisture in the seedbed during the entire germination period is critical to ensure seedling vigor. Sunlight and adequate air circulation are also required for the establishment of healthy turfgrass. Cyclic temperatures of about 15–30°C (59–86°F) favor rapid germination of cool-season turfgrass species, and temperatures of 20–35°C (68–95°F) favor warm-season species.

If temperature, moisture, air, or light impose limiting conditions on the growth of seedlings, the risk of seed decay and damping-off increases. Each potential pathogen also has requirements for optimal growth. In general, infection by any of these pathogens requires wet soils or plant surfaces, but the temperature and moisture conditions favoring pathogenic activity vary among species. *Pythium myriotylum* and *P. aphanidermatum,* for instance, have a high optimal temperature (35°C [95°F]), whereas *P. vanterpoolii* and *P. ultimum* have a lower optimal temperature (22°C [72°F]). Several other *Pythium* spp. can cause disease at both temperatures. *M. nivale* and *Fusarium culmorum* are capable of pathogenesis at low temperatures, and the latter species can also cause seedling death at high temperatures. *Rhizoctonia solani, Pyricularia grisea,* and *Bipolaris sorokiniana* are warm-weather pathogens, whereas *M. nivale* functions best when it is cool. These varying capabilities of potential pathogens make it necessary for the turfgrass manager to plant at a time when seedling vigor is strongly favored during the establishment phase.

Other conditions that favor disease include poor contact of seeds with soil. Seeds that lie on the soil surface are fully exposed to variations in the atmospheric environment, and those placed too deep in the soil use excessive amounts of their food reserves before they reach sunlight. Both extremes weaken seedlings. Old seeds germinate more slowly and in lower percentages than recently collected seeds and are particularly vulnerable. Lastly, seeds sown in existing stands of turfgrasses are fully exposed to (1) high inoculum densities of several pathogens; (2) competition with larger plants for nutrients; (3) shading by taller plants; and (4) generally poor growing conditions for seedlings in thatch. The seed and seedling environment in an existing turf, or one killed by a nonselective herbicide, is much poorer than that in a properly prepared seedbed. Establishment of overseeded grasses into existing stands generally is less than certain, particularly if attempted at times other than late summer to early autumn.

Management

Good surface and subsurface drainage and aeration should be provided in the seedbed. Planting should be avoided in areas where air movement is greatly reduced by trees, shrubs, or other barriers. Recently harvested and certified seeds should be planted at the recommended depth when the soil is cool and dry. Rolling after seeding ensures good contact between the seed and the soil so that germination will be rapid. A new seedbed should not be overwatered.

Several fungicides are available for use as seed treatments or soil drenches to control seed decay or as foliar sprays or soil drenches to control seedling blights. It is difficult to determine or predict what fungi will infect and colonize seeds or seedlings. Therefore, mixing active ingredients that have activity on the pathogens listed above is suggested to maximize control of damping-off.

Selected References

Couch, H. B. 1995. Diseases of Turfgrasses. 3rd ed. Krieger Publishing, Malabar, FL.

Freeman, T. E. 1980. Seedling diseases of turfgrasses incited by *Pythium.* Pages 41-44 in: Advances in Turfgrass Pathology. P. O. Larsen and B. G. Joyner, eds. Harcourt Brace Jovanovich, Duluth, MN.

Smith, J. D., Jackson, N., and Woolhouse, A. R. 1989. Fungal Diseases of Amenity Turf Grasses. E. & F. N. Spon, London.

(Prepared by R. W. Smiley, P. H. Dernoeden, and B. B. Clarke; revised by J. P. Kerns)

Dollar Spot

Dollar spot is a common and economically important disease of most warm- and cool-season turfgrass species throughout the world. The disease can cause extensive damage on susceptible cultivars of *Agrostis, Cynodon, Festuca, Lolium, Paspalum, Poa, Stenotaphrum,* and *Zoysia* spp. Fungicide resistance has been a persistent problem in the attempt to control this disease, particularly on golf courses where products with a moderate to high risk of inciting resistance are frequently applied. Despite the aesthetic and economic impact of dollar spot on turfgrasses, the taxonomy of the causal agent was in a state of flux for almost eight decades. Recent phylogenetic analysis has shown that there are not one, but six distinct species of fungi that cause dollar spot and that they belong to the newly described genus *Clarireedia.* This resolved taxonomy has put to rest a major controversy among plant pathologists and highlights the tremendous biological diversity present within this new genus.

Symptoms and Signs

In closely mowed grasses (for example, on golf greens and fairways), dollar spot appears as small, circular, sunken spots of blighted turf (Figs. 45 and 46) that rarely exceed 5 cm in diameter. These spots may coalesce into larger, irregular patches when conditions are conducive to disease development (Fig. 47). Residential lawns and other taller turfs may exhibit irregularly shaped, bleached patches that are 2–15 cm or more across (Fig. 48). Patches may coalesce to form large areas of blighted turf (Fig. 49).

Leaves affected by the pathogen develop lesions that are at first chlorotic, then water soaked, and finally bleached white or straw colored. An individual leaf blade may have a single lesion, many small lesions, or be entirely blighted (Fig. 50). Dollar spot lesions are characteristically delimited by tan to reddish brown margins (Figs. 51 and 52), and they generally enlarge to extend across the entire leaf, except on grasses with coarse leaves such as *Eremochloa* spp. Lesions often appear hourglass shaped, but on some warm- and cool-season grass species, lesions may be oblong or oval and separated from healthy tissue by brown borders. Dieback from leaf tips is also common. Leaf symptoms may be confused with symptoms of red thread, copper spot, Nigrospora blight, brown patch, and Pythium blight.

When dew is present on grass blades and the pathogen is active, a white, cottony, or cobwebby growth of fungal mycelium may be seen on diseased turf (Figs. 53 and 54). The mycelium disappears as the leaves dry. This aerial mycelium can be confused with that produced by species of *Pythium, Nigrospora,* and *Rhizoctonia*.

Causal Agent

Dollar spot was first reported in 1927; however, the causal agent was not formally classified until 1937 when F. T. Bennett

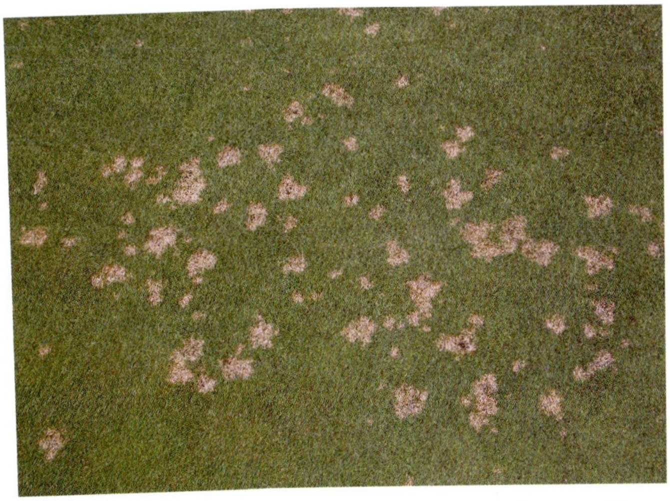

Fig. 45. Dollar spot of *Agrostis stolonifera* on a golf green. (Courtesy L. P. Tredway—© APS)

Fig. 46. Dollar spot on a *Cynodon dactylon* fairway. (Courtesy M. Tomaso-Peterson—© APS)

Fig. 47. Dollar spot on *Agrostis capillaris* (top left) and three *Agrostis stolonifera* cultivars (top right, bottom left, and bottom right) maintained as fairway turf. Note the difference in susceptibility between the two *Agrostis* spp. (Courtesy J. Hempfling—© APS)

Fig. 48. Dollar spot of a *Zoysia* sp. turf. (Courtesy T. E. Freeman)

Fig. 49. Dollar spot on an *Eremochloa ophiuroides* lawn. (Courtesy L. P. Tredway—© APS)

Fig. 50. Dollar spot of *Lolium perenne*. (Courtesy B. B. Clarke—© APS)

assigned the name *Sclerotinia homoeocarpa*. For the next 80 years, the taxonomy of the dollar spot organism remained in a state of flux. In 1945, the genus *Sclerotinia* was restricted to include only fungi that produce apothecia from tuberoid sclerotia. Because the dollar spot pathogen produces stromata rather than sclerotia, it was no longer accommodated by the genus *Sclerotinia*. However, the pathogen was not reclassified at that time.

During the following years, the dollar spot pathogen was the subject of numerous taxonomic studies. Morphological analysis suggested placement within the genus *Poculum, Lanzia,* or *Rutstroemia,* but attempts to rename it were limited by the lack of sexual structures. In nature, the dollar spot fungus exists almost exclusively as mycelium, with fertile ascocarps reported only from the United Kingdom. North American strains of the fungus have yielded only sterile apothecia. Morphological examination of this tissue and biochemical and immunological studies of stromatal proteins data suggested that the fungus was not a true *Sclerotinia* species, but the pathogen was not renamed.

With the advancements of DNA-based technologies in the 1990s, it became possible to study the genetics of the dollar spot fungus. Sequencing of the internal transcribed spacer region confirmed that the pathogen was not a true *Sclerotinia* species and may belong to *Rutstroemia* or *Poculum*. Unfortunately, small sample sizes and limited reference isolates prevented the formal renaming of this fungus.

More recently, a collaborative effort to determine the true taxonomic placement of the dollar spot fungus was completed. Using a large collection of isolates from North America, Europe, and Asia and representative members of the Rutstroemiaceae and Sclerotiniaceae families, researchers sequenced and compared three genetic regions from all isolates. Phylogenetic analysis of these genetic markers revealed that the dollar spot pathogen does not belong to any known genus within Rutstroemiaceae or Sclerotiniaceae. As such, a new genus, *Clarireedia,* was erected to accommodate the dollar spot pathogen.

Within the genus *Clarireedia,* six species of the dollar spot fungus have been identified. *C. homoeocarpa* has been found only on *Festuca rubra* in the United Kingdom; *C. bennettii* has been isolated only from cool-season grass hosts in the Netherlands, United Kingdom, and United States; and *C. hainanense* and *C. paspali* have been reported on *Paspalum vaginatum* in China. The two remaining species, *C. jacksonii* and *C. monteithiana,* are responsible for most modern-day epidemics of dollar spot around the world. *C. jacksonii* appears to be restricted to cool-season grass species, while *C. monteithiana* is limited to warm-season grass species.

Fig. 51. Dollar spot of *Poa pratensis*. (Courtesy L. P. Tredway—© APS)

Fig. 52. Dollar spot of *Cynodon dactylon*. (Courtesy M. Tomaso-Peterson—© APS)

Fig. 53. Foliar mycelium of the dollar spot fungus on *Poa pratensis*. (Courtesy A. Windham—© APS)

Fig. 54. Cobwebby mycelium of the dollar spot fungus and guttation droplets on leaves of *Agrostis stolonifera*. (Courtesy L. P. Tredway—© APS)

In culture, *Clarireedia* spp. produce fluffy, white, aerial mycelium that collapses and becomes brown, tan, olive, gray, or slightly pink. *C. jacksonii* and *C. monteithiana* grow quickly at room temperature. Hyphae are hyaline and septate. Apothecia arise from thin, dark stromata and are typically brown to light orange. Apothecia are typically sterile and devoid of asci or ascospores. Apothecial production has not been observed in all isolates, and photoperiod and hypovirulence are known to influence apothecial production in vitro.

In nature, stromata are occasionally produced on infected leaves late in the growing season under very moist conditions, especially on *Festuca* spp.

Disease Cycle

During periods unfavorable for their development, the dollar spot fungi survive as mycelium in infected plants and plant debris, in thatch, and as stromata on leaf surfaces. They are disseminated in infected leaf debris transported by equipment, people, animals, water, or wind. Pathogenic isolates have also been recovered from commercial seed lots. When environmental conditions favor the activity of the fungi (warm, humid days, cool nights, and heavy dew), mycelium within previously infected tissue or from stromata colonizes the foliage. The prolific growth of the mycelium may extend outward into the humid air. When it contacts a moist leaf surface, aerial mycelium can penetrate the leaf and cause infection. The pathogen also enters leaves through stomata, cut leaf tips, and other mechanical wounds.

A toxin produced by the fungus has been associated with necrosis of the apical meristems and roots of *Agrostis stolonifera* grown in sand culture. The importance of this toxin in the infection process remains to be determined.

Although spores have not been observed in North America, reports of genetic variation within populations of the pathogen provide evidence that genetic recombination does occur. Apothecia are rarely observed in nature.

Epidemiology

Prolonged periods of high humidity in the turfgrass canopy are required for growth of the dollar spot fungus. The fungus may commence activity from late spring through late autumn or early winter. Symptoms that develop late in the growing season may persist into the spring as depressed pockets of necrotic tissue (Fig. 45). Disease development is favored by warm, humid weather and cool nights that produce heavy dews. Temperatures of 15–30°C (59–86°F) are conducive to disease development; however, the optimal temperature for mycelial infection varies considerably among biotypes collected from different geographical locations and at different times of the year. Dollar spot is more severe in dry soils, even though high moisture levels in the aerial canopy are required for infection.

Several epidemiological models have been developed to assist with the timing of fungicide applications for dollar spot control. The earliest models used simple temperature and moisture thresholds to predict dollar spot development in bentgrass turf. The Mills–Rothwell model recommended a preventive fungicide application be made when air temperature exceeded 25°C (77°F) and relative humidity exceeded 90% for 3 of 7 days. The Hall model predicted dollar spot activity when average air temperature exceeded 22°C (72°F) combined with 3 consecutive days of measurable rainfall or 25°C (77°F) with 2 consecutive days of rainfall. Later research by Burpee and Goulty found the Mills–Rothwell model to overpredict and the Hall model to underpredict dollar spot development.

C. P. Ryan and colleagues developed a growing degree day (GDD) model to predict the initial onset of dollar spot development. Using a base air temperature of 15°C (59°F) and a start date of April 1, they identified disease thresholds of 60–70 GDD for highly susceptible cultivars such as Crenshaw and 105–115 GDD for moderately susceptible cultivars like L-93 or 007.

More recently, D. L. Smith, J. P. Kerns, and colleagues developed a logistic regression model that uses 5-day average air temperature and relative humidity to estimate the probability (risk index) of new dollar spot infections in creeping bentgrass turf. Field research in several locations showed model-based fungicide applications, with a 20% risk threshold, provided control similar to that of a calendar-based approach, often with fewer applications. For creeping bentgrass cultivars with higher tolerance to dollar spot, such as 007 and Declaration, model accuracy was improved at risk thresholds greater than 20% when the risk index over a 5-day period was predicted to increase. The Smith–Kerns model is available for use by turfgrass managers through several online platforms.

Turfgrasses grown with low nitrogen fertility are typically more susceptible to dollar spot. Soil pH and phosphorus levels apparently have no effect on disease severity. Responses to potassium have been variable, with some studies indicating that potassium applications increase dollar spot in *Poa annua* and *A. stolonifera* turf but only when soil levels are very low. ELISA disease-detection kits have been used in the field to identify the pathogen in infected plant tissue, and real-time PCR can be used to diagnose samples in the laboratory.

Management

At present, cultural and chemical control methods are not known to differ among species of *Clarireedia*. Turf should be watered thoroughly (deeply) and as infrequently as possible without allowing moisture stress to occur. Turf should not be irrigated in the late afternoon or evening because doing so prolongs leaf wetness and may intensify the disease on cool nights when dew is likely to form. Dew and guttation fluids should be removed from leaf surfaces by mowing, rolling, or poling greens early in the morning. Good air circulation over the turf can be promoted by pruning or removing trees, shrubs, and other barriers. Turfgrass should be mowed at the frequency and cutting height recommended for the species. Turf can be aerated to reduce compaction and control thatch. Lightweight rolling several times per week has been shown to reduce dollar spot pressure. Sand topdressing has also been reported to reduce disease severity. Adequate to high nitrogen fertility should be maintained when dollar spot is active. Light, frequent applications of nitrogen help reduce disease severity and enhance recovery once infection has been arrested. Ferrous sulfate has been shown to suppress dollar spot. Applications may be beneficial when used in conjunction with a fungicide program.

Some organic fertilizers and composted products have been reported to reduce the severity of dollar spot in *Agrostis* spp. and other cool-season turfgrasses. The suppression of dollar spot by these products may result in part from their effects on populations of soil microbes, but disease control has been found to be associated more with nitrogen availability than with enhanced microbial activity. Hypovirulent isolates of *C. jacksonii* and strains of the biological control agents *Enterobacter cloacae, Fusarium heterosporum,* and *Bacillus subtilis* have been shown to suppress dollar spot in the field. Microbial products (biofungicides) containing *Trichoderma harzianum* (strain 1295-22), *T. virens, Bacillus lichenoformens,* or *Pseudomonas aureofaciens* (strain Tx-1) can also reduce the incidence and severity of dollar spot but often not to commercially acceptable levels. New cultivars of *A. stolonifera* have been developed with significantly higher levels of disease resistance than older cultivars such as Penncross, a standard in the golf industry. Some *Festuca* spp. are naturally infected with the *Epichloë festucae* endophyte, a beneficial fungus that enhances host plant resistance to dollar spot by producing an antifungal protein.

Many fungicides are available for the control of dollar spot. Local variations in pathogen sensitivity to different active ingredients, however, precludes a guarantee of success with any one product. Fungicide resistance in the dollar spot pathogen has been a persistent problem for more than 50 years, especially in areas that receive frequent fungicide applications, such as golf courses. Resistance to several fungicide chemistries, including the benzimidazoles, dicarboximides, succinate dehydrogenase inhibitors, and demethylation inhibitors, has been reported. To reduce the potential for development of fungicide-resistant biotypes, label directions must be followed, repeated use of chemically similar fungicides should be avoided, products with different biochemical modes of action can be tank mixed or alternated, and sound cultural practices that reduce stress and enhance plant vigor should be implemented. Disease predictive models, which utilize average daily temperature and relative humidity values to determine the likelihood of dollar spot outbreaks, can help time fungicide applications more precisely. Turfgrass extension specialists or county agricultural agents can be consulted about turfgrass cultivars that are tolerant to dollar spot and fungicides that are recommended for control of the disease in specific geographic regions.

Selected References

Aynardi, B. A., Jiménez-Gasco, M. M., and Uddin, W. 2019. Effects of isolates of *Clarireedia jacksonii* and *Clarireedia monteithiana* on severity of dollar spot in turfgrasses by host type. Eur. J. Plant Pathol. 155:817-829.

Baldwin, N. A., and Newell, A. J. 1992. Field production of fertile apothecia by *Sclerotinia homoeocarpa* in *Festuca* turf. J. Sports Turf Res. Inst. 68:73-76.

Bennett, F. T. 1937. Dollar spot diseases of turf and its causal organism *Sclerotinia homoeocarpa* n. sp. Ann. Appl. Biol. 24:236-257.

Burpee, L. L., and Goulty, L. G. 1986. Evaluation of two dollar spot forecasting systems for creeping bentgrass. Can. J. Plant Sci. 66:315-351.

Davis, J. G., and Dernoeden, P. H. 2002. Dollar spot severity, tissue nitrogen, and soil microbial activity in bentgrass as influenced by nitrogen source. Crop Sci. 42:480-488.

Giordano, P. R., Nikolai, T. A., Hammerschmidt, R., and Vargas, J. M. 2012. Timing and frequency effects of light weight rolling on dollar spot disease in creeping bentgrass putting greens. Crop Sci. 52:1371-1378.

Hempfling, J. W., Ward, D. L., Lalancette, N., Clarke, B. B., and Murphy, J. A. 2021. Novel action thresholds of a logistic regression model to forecast dollar spot on bentgrasses. Crop Sci. 61:3124-3133.

Jackson, N. 1973. Apothecial production in *Sclerotinia homoeocarpa*. J. Sports Turf Res. 49:58-63.

Kohn, L. M., and Grenville, D. J. 1989. Anatomy and histochemistry of stromatal anamorphs in the Sclerotiniaceae. Can. J. Bot. 67:371-393.

McCall, D. S., Ervin, E., Shelton, C. C., Reams, N., and Askew, S. D. 2016. Influence of ferrous sulfate and its elemental components on dollar spot suppression. Crop Sci. 57:581-586.

Miller, G. L., Stevenson, K. L., and Burpee, L. L. 2002. Sensitivity of *Sclerotinia homoeocarpa* isolates to propiconazole and impact on control of dollar spot. Plant Dis. 86:1240-1246.

Monteith, J. 1927. Can you identify brown patch? Natl. Greenskeeper 6:7-11.

Orshinksy, A. M., and Boland, G. J. 2011. *Ophiostoma mitovirus 3a,* ascorbic acid, glutathione, and photoperiod affect the development of stromata and apothecia in *Sclerotinia homoeocarpa*. Can. J. Microbiol. 57:398-407.

Popko, J. T., Jr., Sang, H., Lee, J., Yamada, T., Hoshino, Y., and Jung, G. 2018. Resistance of *Sclerotinia homoeocarpa* field isolates to succinate dehydrogenase inhibitor fungicides. Plant Dis. 102:2625-2631.

Rioux, R. A., Shultz, J., Garcia, M., Willis, D. K., Casler, M., Bonos, S., Smith, D., and Kerns, J. 2014. *Sclerotinia homoeocarpa* overwinters in turfgrass and is present in commercial seed. PLoS One 9(10):e110897.

Ryan, C. P., Dernoeden, P. H., and Grybauskas, A. P. 2012. Seasonal development of dollar spot epidemics in six creeping bentgrass cultivars in Maryland. HortScience 47:422-426.

Salgado-Salazar, C., Beirn, L. A., Ismaiel, A., Boehm, M. J., Carbone, I., Putman, A. I., Tredway, L. P., Clarke, B. B., and Crouch, J. A. 2018. *Clarireedia:* A new fungal genus comprising four pathogenic species responsible for dollar spot disease of turfgrass. Fungal Biol. 122:761-773.

Shelton, C. D., Askew, S. D., Ervin, E. H., and McCall, D. S. 2021. Impact of ferrous sulfate concentration on *Clarireedia* isolate growth and dollar spot development. Crop Sci. 61:2148-2155.

Smith, D. L., Kerns, J. P., Walker, N. F., Payne, A. F., Horvath, B., Inguagiato, J. C., Kaminski, J. E., Tomaso-Peterson, M., and Koch, P. L. 2018. Development and validation of a weather-based warning system to advise fungicide applications to control dollar spot on turfgrass. PLoS One 13(3):e0194216.

Tian, Z., Wang, R., Ambrose, K. A., Clarke, B. B., and Belanger, F. C. 2017. The *Epichloë festucae* antifungal protein has activity against the plant pathogen *Sclerotinia homoeocarpa,* the causal agent of dollar spot disease. Sci. Rep. 7:5643.

Walsh, B., Ikeda, S. S., and Boland, G. J. 1999. Biology and management of dollar spot (*Sclerotinia homoeocarpa*): An important disease of turfgrass. HortScience 34:13-21.

Williams, D. W., Powell, A. J., Vincelli, P., and Dougherty, C. T. 1996. Dollar spot on bentgrass influenced by displacement of leaf surface moisture, nitrogen, and clipping removal. Crop Sci. 36:1304-1309.

Zhang, H., Dong, Y., Zhou, Y., Hu, J., Lamour, K., and Yang, Z. 2022. *Clarireedia hainanense:* A new species is associated with dollar spot of turfgrass in Hainan, China. Plant Dis. 106:996-1002.

(Prepared by R. W. Smiley, P. H. Dernoeden, and B. B. Clarke; revised by L. A. Beirn and B. B. Clarke)

Gray Leaf Spot

Gray leaf spot has long been associated with *Stenotaphrum secundatum.* It has also become a very destructive disease of *Lolium perenne* and *L. multiflorum,* and it can cause severe damage to turfgrasses in the genera *Festuca, Eremochloa,* and *Pennisetum.* The disease has been reported to occur on *Cynodon* and *Paspalum* spp. and was artificially induced in *Agrostis* spp., but turfgrasses in these genera are not considered primary hosts in the field. The disease of turfgrasses, known as gray leaf spot in North America and many other regions of the world, is also referred to as "blast" disease of forage grasses. Blast disease of rice (*Oryza sativa*) and other cereal grains such as wheat (*Triticum aestivum*) and finger millet (*Eleusine coracana*) is caused by a closely related pathogen.

Symptoms and Signs

In warm-season turfgrasses, dark, water-soaked lesions on leaves and stems enlarge rapidly into round to diamond-shaped spots. Spots occur frequently on the edges of a leaf blade or on the midrib and may extend across almost the entire leaf. Spots that coalesce wither, distort, and eventually kill heavily damaged leaves. Leaf spots have tan to gray centers and distinct, thin, reddish brown borders (Fig. 55). During periods of warm, humid weather, the spots may be covered with grayish, felty masses of conidia (Fig. 56). When the disease is severe, the entire planting thins and may appear scorched, as if it were suffering from severe drought.

In *Lolium* spp. and *Festuca arundinacea,* symptoms first appear as small, water-soaked leaf lesions, which quickly become necrotic. The leaf spots can vary considerably in color, size, and shape, but they are often oblong and gray to light brown (Fig. 57) and may be surrounded by purple to dark brown borders (Fig. 58). As the disease progresses, lesions may coalesce, forming irregular shapes and causing partial to complete blighting of the leaf blades. On fine-textured turfgrasses, blighted leaf tips can have a conspicuous twisted or fishhook shape (Fig. 59), and thick masses of grayish conidia and conidiophores may give leaves a feltlike appearance.

From a distance, symptoms may initially look like those of drought or heat stress. These are followed by the development of small patches of blighted turf (Fig. 60) similar to those caused by dollar spot or Pythium blight, except for a lack of aerial mycelium. Depending on environmental conditions, the infected patches may remain small or may rapidly expand into large, irregularly shaped, necrotic areas (Figs. 61 and 62), which can easily be confused with symptoms of brown patch. In severe cases, the entire planting dies, leaving behind resistant grass species or weeds. Once conditions are no longer conducive to the development of the disease, *F. arundinacea* generally recovers more rapidly than *Lolium* spp.

Causal Agent

Pyricularia oryzae is a devastating fungal pathogen of over 50 monocot hosts with a worldwide distribution. This pathogen causes rice blast and wheat blast, two of the most economically important diseases of cereal crops. On turfgrass hosts, populations of *P. oryzae* that cause disease on *Lolium* and *Festuca*

Fig. 55. Gray leaf spot of *Stenotaphrum secundatum.* (Courtesy M. Tomaso-Peterson—© APS)

Fig. 56. Gray leaf spots on *Stenotaphrum secundatum* covered with grayish masses of conidia. (Courtesy P. F. Harmon—© APS)

Fig. 57. Gray leaf spot lesions on *Lolium perenne.* (Courtesy P. H. Dernoeden—© APS)

Fig. 58. Gray leaf spot lesions on *Festuca arundinacea.* (Courtesy L. P. Tredway—© APS)

Fig. 59. Fishhook symptom of gray leaf spot on *Lolium perenne.* (Courtesy P. F. Harmon—© APS)

spp. overlap and are genetically distinct from lineages isolated from *S. secundatum*. In 2002, *P. grisea,* the gray leaf spot pathogen that infects species of the weed genus *Digitaria,* was shown to be genetically distinct from *P. oryzae,* which is associated with rice; however, the taxonomic placement of turfgrass isolates was unclear. Although *P. grisea* and *P. oryzae* both appear in the voluminous literature on gray leaf spot diseases of amenity turfgrass, comparisons of whole-genome sequences have shown that isolates from turfgrass are *P. oryzae.*

Fig. 60. Early symptoms of gray leaf spot of *Lolium perenne,* which appear similar to symptoms of dollar spot. (Courtesy P. H. Dernoeden—© APS)

Fig. 61. Severe symptoms of gray leaf spot in a *Lolium perenne* rough. (Courtesy P. F. Harmon—© APS)

Fig. 62. Severe damage to *Festuca arundinacea* caused by gray leaf spot. (Courtesy L. P. Tredway—© APS)

During periods of warm, humid weather, *P. oryzae* produces masses of conidiophores and conidia, which appear grayish in mass, on infected leaves. The conidiophores are gray to olive and produce hyaline, pyriform conidia, which measure 6–9 × 17–28 µm and are mostly two septate when mature (Fig. 63).

Only the asexual stage of the pathogen has been found in nature, but mating studies with compatible isolates of *P. oryzae* have yielded mature perithecia of the sexual stage, *Magnaporthe oryzae,* a heterothallic ascomycete. The perithecia are dark brown to black, with spherical to subspherical bases (80–260 µm in diameter) and long, cylindrical necks (500–1,200 µm long and 55–160 µm in diameter). The asci are unitunicate, cylindrical to clavate (55–110 × 8–15 µm), and hyaline. Each ascus contains eight ascospores. At maturity, the ascospores are hyaline, fusiform (16–25 × 4–8 µm), curved, three septate, and sometimes slightly constricted at the septa.

Disease Cycle

The pathogen survives adverse periods primarily as dormant mycelium in infested leaves and plant debris. When environmental conditions become conducive to pathogen growth, the fungus produces conidiophores on necrotic tissue. The conidiophores sporulate, and conidia are disseminated by wind, water, machinery, and animals. Conidia are produced in abundance under warm, humid conditions. Germination and infection of host tissue occur when the relative humidity is high and the leaf surface is moist, but excessive leaf wetness may inhibit spore production. Alternating wet and dry cycles over a period of several days provide ideal conditions for spore production and subsequent infection of new host tissue.

Epidemiology

Gray leaf spot is most severe in warm, humid weather. The optimal temperatures for disease development are 28–32°C (82–90°F). In *Lolium* spp., the disease typically occurs in the summer from the transition zone through the mid-Atlantic region and epidemics may continue into October in the eastern United States. The disease generally subsides following a heavy frost. In the southern United States where *L. perenne* is primarily used to overseed Bermudagrass and other C4 turfgrass species, the disease is active in late spring and early summer during transition of the warm-season turf. On *S. secundatum,* gray leaf spot is common in early summer through December, or dormancy, with peak disease pressure occurring during regular daily rainfall of July and August. Gray leaf spot is more severe in newly established plantings than in mature stands, especially when nitrogen fertility is moderate to high. Disease severity typically increases with the nitrogen rate. Water-soluble sources of nitrogen that promote rapid growth can enhance the development of gray leaf spot compared with slow-release nitrogen sources, which often do not stimulate

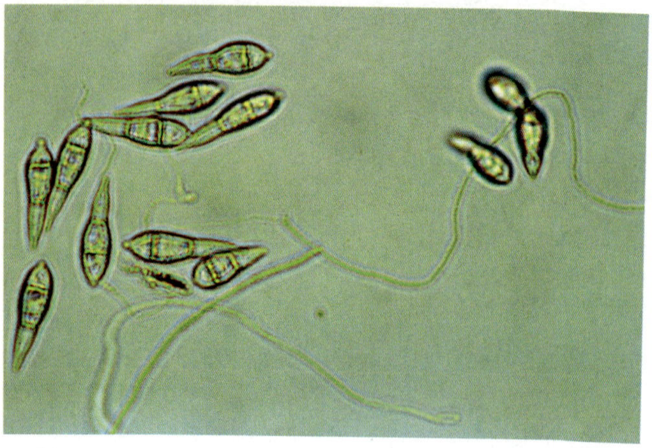

Fig. 63. Conidia of *Pyricularia oryzae.* (Courtesy B. B. Clarke—© APS)

Fig. 64. Differences among *Lolium perenne* cultivars in susceptibility to gray leaf spot caused by *Pyricularia oryzae*. (Courtesy B. B. Clarke—© APS)

disease development. There is also a positive relationship between mowing height and disease severity in the eastern United States, where golf course roughs regularly sustain more disease than fairways. Under environmental conditions favorable for disease development, however, severe outbreaks can occur even on fairways of *L. perenne*. Herbicides, plant growth regulators, soil compaction, drought, excessive fertility, and other stress factors may predispose turfgrasses to gray leaf spot. Air temperature and leaf wetness also are critical factors for the development of gray leaf spot epidemics. At suboptimal temperatures (20–23°C [68–73°F]), a period of 21–36 hours of leaf wetness is typically required for infection; however, at optimal temperatures (28–32°C [82–90°F]), peak infection can occur with as little as 9 hours of continuous leaf moisture. A disease predictive model, based on temperature and leaf wetness duration, is available.

Management

Management practices that limit drought stress, extended periods of leaf wetness, excessive application of nitrogen, soil compaction, and plant growth regulator- or herbicide-induced stress can help to reduce disease severity. In hot, humid weather, the use of nitrogen should be reduced, particularly quick-release, water-soluble types. Turf should be watered thoroughly (deeply) and as infrequently as possible to avoid drought stress. If possible, overseeding in the autumn should be delayed until the onset of cooler weather, when disease outbreaks are likely to be less severe and of shorter duration. Removal of infested clippings generally has little effect in reducing gray leaf spot, unless disease severity is low.

Several cultivars of *S. secundatum* and *F. arundinacea* have fairly high levels of resistance to gray leaf spot, but some are also more susceptible to attack by *Blissus* spp. (chinch bugs) and *Rhizoctonia solani*, respectively. Cultivars of *L. perenne* with gray leaf spot resistance have been introduced into the market and are being evaluated with promising early results (Fig. 64). A turfgrass specialist at a university extension service can provide advice on cultivar selection for a specific need.

On *S. secundatum*, fungicides can effectively limit disease damage, particularly on juvenile turfgrass during sod grow-in or in problematic, shady, moist areas of some lawns. For managers of *L. perenne* on golf courses, preventive applications of fungicides are recommended where the disease has been a chronic problem. Isolates of *P. oryzae* from *Lolium* spp. that are resistant to strobilurin (quinone outside inhibitor) fungicides have been reported, and therefore appropriate fungicide resistance management strategies should be followed.

Selected References

Bonos, S. A., Kubik, C., Clarke, B. B., and Meyer, W. A., 2004. Breeding perennial ryegrass for resistance to gray leaf spot. Crop Sci. 44:575-580.

Couch, B. C., and Kohn, L. M. 2002. A multilocus gene genealogy concordant with host preference indicates segregation of a new species, *Magnaporthe oryzae*, from *M. grisea*. Mycologia 94:683-693.

Farman, M. L. 2002. *Pyricularia grisea* isolates causing gray leaf spot on perennial ryegrass (*Lolium perenne*) in the United States: Relationship to *P. grisea* isolates from other host plants. Phytopathology 92:245-254.

Freeman, T. E. 1967. Diseases of southern turfgrasses. Fla. Agric. Exp. Stn. Tech. Bull. 713.

Gladieux, P., Condon, B., Ravel, S., Soanes, D., Maciel, J. L. N., Nhani, A., Jr., Chen, L., Terauchi, R., Lebrun, M. H., Tharreau, D., Mitchell, T., Pedley, K. F., Valent, B., Talbot, N. J., Farman, M., and Fournier, E. 2018. Gene flow between divergent cereal- and grass-specific lineages of the rice blast fungus *Magnaporthe oryzae*. mBio 9. doi:10.1128/mBio.01219-17

Harmon, P. F., and Latin, R. 2005. Winter survival of the perennial ryegrass pathogen *Magnaporthe oryzae* in north central Indiana. Plant Dis. 89:412-418.

Landschoot, P. J., and Hoyland, B. F. 1992. Gray leaf spot of perennial ryegrass turf in Pennsylvania. Plant Dis. 76:1280-1282.

Uddin, W., Serlemitsos, K., and Viji, G. 2003. A temperature and leaf wetness duration-based model for prediction of gray leaf spot of perennial ryegrass turf. Phytopathology 93:336-343.

Uddin, W., Viji, G., and Vincelli, P. 2003. Gray leaf spot (blast) of perennial ryegrass turf: An emerging problem for the turfgrass industry. Plant Dis. 87:880-889.

Viji, G., Wu, B., Kang, S., Uddin, W., and Huff, D. R. 2001. *Pyricularia grisea* causing gray leaf spot of perennial ryegrass turf: Population structure and host specificity. Plant Dis. 85:817-826.

Williams, D. W., Burrus, P. B., and Vincelli, P. 2001. Severity of gray leaf spot in perennial ryegrass as influenced by mowing height and nitrogen level. Crop Sci. 41:1207-1211.

Yaegashi, H., and Udagawa, S. 1978. The taxonomical identity of the perfect state of *Pyricularia grisea* and its allies. Can. J. Bot. 56:180-183.

(Prepared by R. W. Smiley, P. H. Dernoeden, and B. B. Clarke; revised by P. F. Harmon)

Leaf Smuts

Smut fungi are a diverse group of highly specialized pathogens that cause diseases of leaves, sheaths, stems, or inflorescences. The three most common leaf smuts are stripe smut, the most widespread and destructive of these diseases, caused by *Ustilago salweyi*; flag smut, caused by *Urocystis agropyri*; and blister smuts, caused by species in several genera including *Entyloma, Ustilentyloma,* and *Jamesdicksonia*. Less prevalent leaf smuts caused by *Tilletia* spp. have been reported in Europe and New Zealand.

Stripe smut is most prevalent in *Poa pratensis,* occasionally observed in *Agrostis* spp., rarely seen in *Lolium* or *Festuca* spp., and unreported in *Zoysia* and *Cynodon* spp. Flag smut has the same host range, except for *Lolium* spp. Blister smut is less common than stripe smut and flag smut in turfgrasses in North America.

Symptoms and Signs

Symptoms of leaf smuts vary widely, depending on the causal agent, host species and cultivar, and climatic conditions. Symptoms of stripe smut and flag smut are nearly identical in managed turfgrasses and are discussed together. Blister smuts are described separately.

Stripe smut and flag smut. Individual infected plants may be distributed generally throughout the turf, or large patches may form in which most plants are infected. In cool weather in spring and autumn, infected plants may appear pale green or yellow. Affected plants are stunted, root growth is reduced, and leaves remain stiff and erect rather than relaxed and spreading (Fig. 65). These symptoms are often difficult to observe.

Inspection of individual infected plants reveals the presence of short to long, narrow, yellow green streaks parallel to the vein of the leaf and leaf sheaths. The streaks quickly turn from yellow green to gray. These silver gray, parallel streaks are sori (fruiting bodies) of the pathogen, which form below the epidermis and contain large numbers of spores (teliospores). Production of spores continues inside the leaf until the pressure causes the leaf epidermis to rupture along the streaks, releasing masses of black, dusty spores. The water balance of the leaf is disrupted when its epidermis is ruptured. Leaves soon turn brown, split or shred between the veins (Fig. 66), curl from the tips downward, and die.

In newly infected turfgrasses, only widely scattered plants show symptoms, and they are nearly impossible to detect. The disease becomes more apparent several years after the first infections occur. Turfgrasses are usually 4 years old or older before heavily infected areas are noticed. Smut infection weakens plants and renders them susceptible to injury and death from high temperature, drought stress, or infection by other fungi. In properly maintained stands, infected plants often appear healthy during the summer. However, during long hot and dry periods, individual tillers or even large groups of infected plants may die, causing a distinct thinning of the turf. During a moist summer, large areas of badly smutted stands may die when plants are infected simultaneously by *Puccinia* spp. (rust fungi) or other pathogens.

Most *Ustilago* and *Urocystis* spp. cause symptoms similar to those described above. One exception is *Ustilago buchloës,* which causes sausage-shaped pustules up to 10 mm long.

Blister smuts. The symptoms of blister smuts vary considerably. Infected turf may have an overall chlorotic appearance when viewed from a distance (Fig. 67).

Blister smuts differ from stripe and flag smuts mainly in that the epidermis is not ruptured. Instead, blister smut fungi cause the epidermis to stretch, forming sori that resemble blisters and range in color from black to yellow or white. *Jamesdicksonia dactylidis,* a smut fungus with worldwide distribution, produces sori that form oval blisters on the lower leaf surface only. The blisters are seldom more than 2 mm long and are black to lead gray (Fig. 68). On lightly infected leaves, the blisters are usually surrounded by chlorotic halos. As the disease becomes more severe, leaves become entirely chlorotic or bleached to white, with readily apparent black spots. Another blister smut fungus, *Ustilentyloma brefeldii,* produces long, yellow to blue gray, striped sori along leaf veins.

Causal Agents

Similar to rust fungi, the smut fungi have a complex life cycle involving multiple spore and hyphal stages with different ploidy levels. From a diagnostic standpoint, the most prominent signs of smut fungi are dikaryotic resting spores (teliospores) produced in sori embedded in leaf tissue.

Teliospores of *Ustilago* and *Jamesdicksonia* spp. are borne singly, although they can be seen adhering to each other when examined microscopically (Fig. 69). In contrast, teliospores of *Urocystis* spp. are borne in "spore balls" (Fig. 69). The spores of the most prevalent species are described below.

Teliospores of *J. dactylidis* are variously polyhedral, oblong, ovoid, or irregular in shape as a result of their growth in a confined sorus in an infected leaf. They tend to adhere in irregular groups. The spores are olivaceous to brown and up to 7–11 × 8–17 µm, and the spore walls appear laminated, thick, and smooth.

Fig. 65. Stripe smut of *Poa pratensis.* (Courtesy R. S. Byther)

Fig. 67. Entyloma blister smut of *Poa pratensis.* (Courtesy R. S. Byther)

Fig. 66. Shredding of a leaf between vascular bundles caused by the stripe smut fungus, *Ustilago salweyi.* (Courtesy R. W. Smiley—© APS)

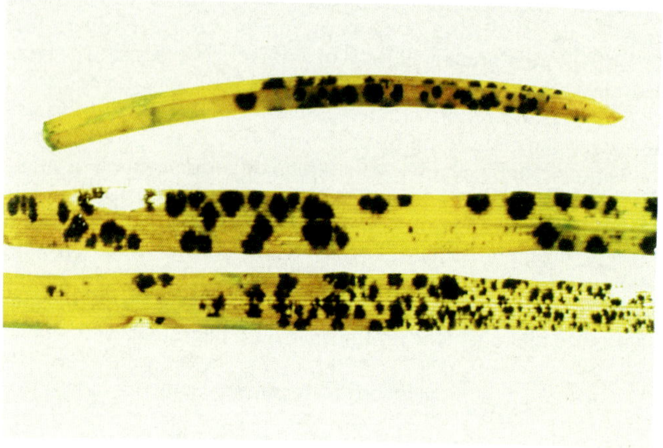

Fig. 68. Sori of *Jamesdicksonia dactylidis* on *Poa pratensis.* (Courtesy R. S. Byther)

In contrast, the spores of *Ustilago salweyi* are mostly globose, dark olive brown, usually prominently echinulate (spiny), and 9–11 µm in diameter. The spores of other *Ustilago* species vary in size, and the spore walls may be verrucose (having flat or rounded warts) or may have little ornamentation.

Urocystis agropyri spores are globose, reddish brown, smooth walled, and 12–18 µm in diameter. They are formed in inseparable spore balls consisting of one to six (usually one to three) teliospores surrounded by smaller, hyaline to brown, sterile cells. The spore balls are globose to elongate and measure 18–35 × 35–40 µm.

Strains of *Ustilago salweyi* and *Urocystis agropyri* have high levels of host specificity. Pathogenic isolates from one host species do not necessarily infect another host species. For example, both stripe smut and flag smut occur in *Agrostis tenuis,* but only stripe smut occurs in *A. stolonifera* (especially in older vegetative cultivars). Isolates of *Ustilago salweyi* from *Agrostis* spp. do not infect *Poa annua* or *P. pratensis,* although these *Poa* spp. are susceptible to infection by other strains of *Ustilago salweyi.* Similarly, flag smut occurs in *P. pratensis* but not in *P. annua.*

Disease Cycle

Leaf smut fungi survive the winter as teliospores in soil, thatch, or infected foliage. In addition, they may become systemically distributed throughout infected plants and survive as mycelium in the crowns, nodes, and foliage. Teliospores may lie dormant for up to 3 years and can be spread on seed, by water or wind, or by any process that moves infected plant tissue or infested soil or thatch.

During spring or autumn, teliospores germinate to produce a mycelium-like basidium, which bears haploid basidiospores. The basidiospores germinate to form germ tubes of a single mating type. Conjugation of two germ tubes of opposite mating types produces an infection hypha capable of penetrating the coleoptiles of seedling plants and buds on rhizomes, stolons, and crowns of older plants. Infected plants remain infected for the rest of their lives. The pathogen grows systemically in newly forming leaves, tillers, and rhizomes and eventually produces spores in these infected tissues. Teliospores are released from the plant when the leaf or leaf sheath ruptures, shreds, and dies. They may also be liberated underground from sori produced on rhizomes.

Jamesdicksonia spp. apparently do not grow systemically throughout an infected plant, and most of the infection is thought to result from sporidia spreading onto leaves. Basidiospores may be transmitted in raindrop splash, by physical contact of leaves, on equipment or animals, or by other such physical means.

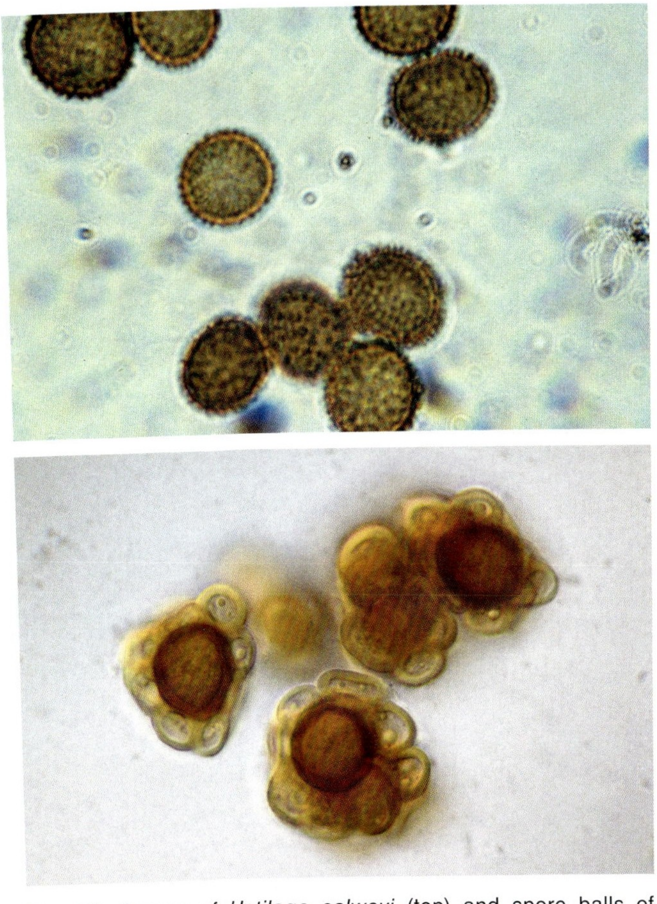

Fig. 69. Spores of *Ustilago salweyi* (top) and spore balls of *Urocystis agropyri* (bottom). (Top, courtesy R. W. Smiley—© APS; and bottom, courtesy L. M. Carris—© APS)

Epidemiology

Leaf smuts are most apparent during spring and autumn when the host and pathogen grow most rapidly. Plants survive infections best in cool to mild weather with adequate moisture and nutrition. Tillers and occasionally entire plants die when exposed to drought or high temperatures or when the turf is over- or underfertilized. Excess thatch and acidic soils promote disease development. Infection by *Ustilago* and *Urocystis* spp. is thought to be favored by low soil moisture and temperatures of 10–20°C (50–68°F).

Management

The use of seed mixtures and blends or sod that includes cultivars resistant to these pathogens is the best means of control. However, because there are numerous races of smut fungi, a cultivar that exhibits resistance at one location may not be resistant at other locations.

Excessive or deficient nitrogen fertilization can enhance smut severity. Smut diseases may also be favored by high nitrogen levels alone or high nitrogen plus low phosphorus and potassium fertility. Watering thoroughly at the onset of drought stress helps to reduce smut injury. Badly smutted stands should be renovated by introducing resistant cultivars. Fall applications of demethylation inhibitor or benzimidazole fungicides can provide effective prevention of leaf smuts.

Selected References

Dernoeden, P. H. 1989. Residual stripe smut control in Kentucky bluegrass with reduced fungicide levels. HortScience 24:796-798.

Fischer, G. W., and Holton, C. S. 1957. Biology and Control of the Smut Fungi. Ronald Press, New York.

Hodges, C. F. 1976. Comparative primary infection characteristics of *Ustilago striiformis* and *Urocystis agropyri* on cultivars of *Poa pratensis.* Phytopathology 66:1111-1115.

Hull, R. J., Jackson, N., and Skogley, C. R. 1979. Influence of nutrition on stripe smut in Kentucky bluegrass turf. Agron. J. 71:553-555.

Smith, J. D., Jackson, N., and Woolhouse, A. R. 1989. Fungal Diseases of Amenity Turf Grasses. E. & F. N. Spon, London.

Vánky, K. 2002. Illustrated Genera of Smut Fungi. 2nd ed. American Phytopathological Society, St. Paul, MN.

(Prepared by R. W. Smiley, P. H. Dernoeden, and B. B. Clarke; revised by L. P. Tredway)

Pink Patch

Pink patch was once considered a form of red thread disease. The pink patch fungus and the red thread fungus are similar in appearance, and they often develop concurrently in stands of turfgrass. Hosts of the pink patch pathogen include both warm- and cool-season grasses of the genera *Agrostis, Cynodon, Festuca, Lolium, Paspalum,* and *Poa.*

Symptoms and Signs

Symptoms of pink patch, like those of red thread, appear as roughly circular, pinkish red to tan patches, 5–10 cm in diameter (Fig. 70). Pink patch, however, spreads much more slowly than red thread and is much less severe, because the pathogen damages only a comparatively small number of leaves. In addition, pink patch does not significantly reduce the growth rate of the grass and affected tissue may remain green. The damage generally is superficial.

Individual leaves affected by pink patch are covered with a pink, gelatinous growth of fungal mycelium. The mycelium initially develops along the leaf margins but then expands to cover the entire width of the blade (Fig. 71). The pathogen enters the plant through stomata, by direct penetration of the epidermis, or through wounds. Infection results in the production of chlorotic lesions. The lesions expand and become water soaked, and the leaf dies from the tip toward the base (basipetally). As leaves dry, affected sheaths and dead leaves become matted together and turn straw brown or tan. The pink patch fungus does not produce threadlike sclerotia (red threads) or pink, cottony flocks of arthroconidia, two diagnostic signs of red thread.

Fig. 70. Pink patch of *Agrostis stolonifera*. (Courtesy P. H. Dernoeden—© APS)

Fig. 71. Mycelium of *Limonomyces roseipellis* on Bermudagrass. (Courtesy M. Tomaso-Peterson—© APS)

Causal Agent

Limonomyces roseipellis causes pink patch. The pathogen was previously known by other names, including *Corticium fuciforme* and *Athelia fuciformis*. Its hyphae (Fig. 72) are hyaline, binucleate, and 2–6 µm wide and have clamp connections at nearly all septa (Fig. 73). In contrast, hyphae of the red thread fungus, *Laetisaria fuciformis,* are multinucleate, slightly wider, and lack clamp connections. *L. roseipellis* does not produce arthroconidia or sclerotia. Like the red thread fungus, it may produce extensive, effuse basidiocarps on hymenia on leaf surfaces, but the role of basidiospores in the infection process is unclear.

Disease Cycle and Epidemiology

The disease cycle and epidemiology of pink patch are apparently similar to those of red thread, except that the pink patch pathogen does not produce arthroconidia or sclerotia. Pink patch is often observed concurrently with red thread in the same stand of turf or even in the same plant. Both diseases are more severe in turf maintained with low nitrogen fertility. In culture, *L. roseipellis* grows at temperatures of 4–32°C (39–90°F), with optimal growth at 20–28°C (68–82°F). Isolates from warm-season grasses tend to have higher temperature optima than those from cool-season grasses. In nature, pink patch is most common during cool, moist weather in the spring and autumn when temperatures are 18–24°C (64–75°F), but it may occur at any time of year in the presence of adequate surface moisture. The pathogen has a slow rate of growth and therefore generally causes greater damage to unmowed or slow-growing turf. It can survive for long periods as mycelium in infested plant debris.

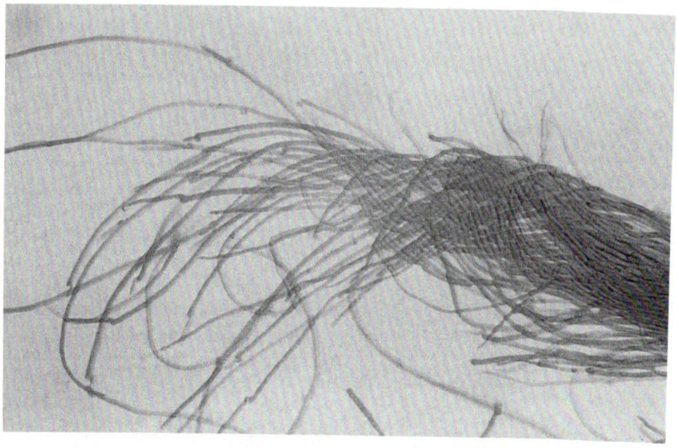

Fig. 72. Aggregates of hyphae with clamp connections produced by *Limonomyces roseipellis*. (Courtesy N. R. O'Neill)

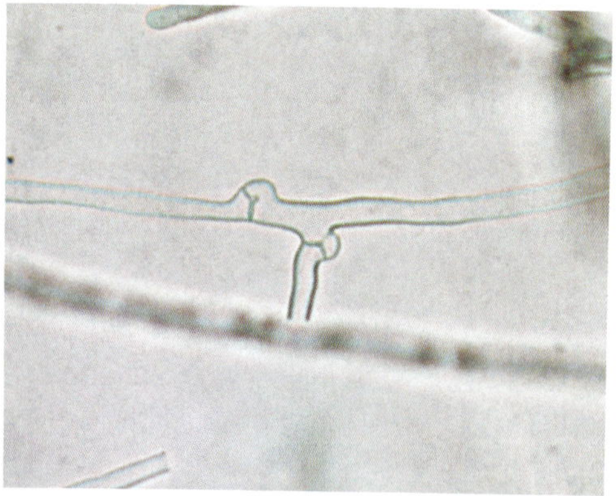

Fig. 73. Clamp connections at hyphal septa, characteristic of many basidiomycete fungi. (Courtesy G. L. Miller—© APS)

Management

Adequate and balanced fertility, as determined by soil or tissue tests, should be maintained. Applications of water-soluble, quick-release nitrogen often reduce pink patch severity and promote turf recovery. Soil pH should be in the range recommended for the turfgrass species grown (generally pH 6.0–6.5). Water should be applied as needed to prevent drought stress. Irrigation should be applied thoroughly (deeply), as infrequently as possible, and early in the day. Light, frequent irrigation or application in late afternoon and early evening, which can allow leaves to remain wet for a prolonged period, should be avoided.

Trees or shrubs can be selectively pruned or the landscape design can be arranged to increase penetration by light and the movement of air over the turf. The use of sharp mower blades minimizes wounding of leaf tips, which may reduce infection. Collecting turfgrass clippings during periods when pink patch is active may remove some inoculum and help to slow the spread of the disease.

Selected References

Kaplan, J. D., and Jackson, N. 1983. Red thread and pink patch diseases of turfgrasses. Plant Dis. 67:159-162.

Smith, J. D. 1954. Fungi and turfgrasses. IV. *Corticium* diseases. J. Sports Turf Res. Inst. 8:365-377.

Stalpers, J. A., and Loerakker, W. M. 1982. *Laetisaria* and *Limonomyces* species (Corticiaceae) causing pink diseases in turfgrasses. Can. J. Bot. 60:529-537.

Zhang, W., Hu, M., Liu, G., Gao, Z., Li, M., and Nan, Z. 2015. Investigation and characterization of red thread and pink patch on warm-season turfgrasses in Hainan Province, tropical China. Eur. J. Plant Pathol. 141:311-325.

(Prepared by R. W. Smiley, P. H. Dernoeden, and B. B. Clarke; revised by L. P. Tredway)

Powdery Mildew

Powdery mildew is an important disease of susceptible cultivars of *Poa* and *Festuca* spp. that are subjected to extended periods of low-intensity light, especially in shady locations or in areas with poor air circulation.

Symptoms and Signs

Powdery mildew is first observed as isolated colonies of fine, whitish mycelium on leaves and leaf sheaths (Fig. 74). The colonies enlarge rapidly and coalesce to cover much of the leaf surface with superficial, gray to white wefts of mycelium that appear powdery or dusty. Large, infected areas look as if they have been dusted with flour or lime. Older leaves generally are infected more severely than younger leaves. Heavily infected leaves turn yellow and then tan or brown as they die. Mycelial colonies of powdery mildew darken as they age, and dark cleistothecia (fruiting structures) of the pathogen occasionally form in the weft of mycelium (Fig. 75). Infected plants are weakened and sometimes killed by other diseases and stresses such as drought and low temperatures.

Fig. 74. Colonies of powdery mildew on *Poa pratensis,* coalescing to cover entire leaves with mycelium and conidia. (Courtesy J. J. Benelli—© APS)

Causal Agent

Blumeria graminis (formerly *Erysiphe graminis*) is a widely distributed obligate parasite that produces chains of ellipsoidal conidia (14–17 × 25–33 µm) (Fig. 76) in such profusion that infected leaves appear powdery. Cleistothecia are dark brown to black, immersed in the mycelial mat, and 135–280 µm (average 200 µm) in diameter. A cleistothecium produces 15–20 asci, which bear ellipsoidal ascospores (10–13 × 20–23 µm).

Fig. 75. Fluffy, white mycelium and black cleistothecia of *Blumeria graminis.* (Courtesy R. W. Smiley—© APS)

Fig. 76. Chains of *Blumeria graminis* conidia. (Courtesy W. W. Bockus—© APS. Reproduced, by permission, from Bockus, W. W., et al., eds. 2010. Compendium of Wheat Diseases and Pests, 3rd ed. American Phytopathological Society, St. Paul, MN.)

B. graminis comprises highly specialized physiologic races, many of which infect only certain cultivars of certain turfgrass species. Races that infect *Poa pratensis* are unlikely to infect *Festuca* spp. In addition, powdery mildew fungi that infect grasses are different from those that cause diseases of ornamentals, vegetables, and cereal grains.

Disease Cycle

The pathogen overwinters primarily as mycelium within living infected plants. In areas with harsh winter climates, the fungus survives as ascospores in cleistothecia embedded in infected plants or debris. Conidia are produced on surviving mycelia in the spring and become airborne, as do ascospores that were not discharged the previous autumn. Conidia are the dominant spore form, serving as primary inoculum in turfgrasses. When conidia land on a susceptible plant, they germinate quickly and can infect the host. New conidia may be produced within 1 week on newly infected leaves, and therefore the disease can spread rapidly. Conidia live only a few days, but they can germinate over a wide range of temperatures (1–30°C [34–86°F]), and hyphal growth does not require a film of free moisture on the leaf surface. Cleistothecia develop as infected leaves mature or die prematurely, and ascospores generally are discharged following rain in the autumn.

Epidemiology

Powdery mildew is generally observed in the spring and autumn, because the production of conidia and infection of leaves are favored by cool, humid, cloudy periods with temperatures of 5–22°C (42–72°F). The disease is most severe in shaded areas and in areas where air circulation is poor. Under these conditions, *P. pratensis* is highly susceptible to powdery mildew, especially when heavily fertilized with nitrogen.

Management

The best strategy for controlling powdery mildew is to include shade-tolerant cultivars in the seed mixture when the turf is established. Selective pruning and placement of ornamental shrubs and trees to increase sunlight penetration to the turfgrass canopy can help to control the disease. This measure also improves air circulation and thus helps to reduce humidity levels in the turf canopy. Where the disease is severe, fungicide treatment or replacement of turf with shade-adapted ground covers may be required. The grass should be mowed relatively high in mildew-prone areas, watered as needed to avoid drought, and fertilized at appropriate times of the year to promote vigorous, but not lush, growth.

Selected References

Bélanger, R. R., Bushnell, W. R., Dik, A. J., and Carver, T. L. W., eds. 2002. The Powdery Mildews: A Comprehensive Treatise. American Phytopathological Society, St. Paul, MN.

Butt, D. J. 1978. Epidemiology of powdery mildews. Pages 51-81 in: The Powdery Mildews. D. M. Spencer, ed. Academic Press, London.

Smith, J. D., Jackson, N., and Woolhouse, A. R. 1989. Fungal Diseases of Amenity Turf Grasses. E. & F. N. Spon, London.

(Prepared by R. W. Smiley, P. H. Dernoeden, and B. B. Clarke; revised by L. P. Tredway)

Pyrenophora (*Drechslera*) Diseases

In the dual fungal nomenclature system, the generic name *Drechslera* refers to the asexual stage of *Pyrenophora*. The International Mycological Association no longer supports dual nomenclature, and the Nomenclature Committee for Fungi has given priority to the sexual stage, *Pyrenophora*. Phylogenetic analysis also places three species of *Marielliottia* in the genus *Pyrenophora*. As of this writing, *D. andersenii* remains unresolved and is the only member this group that retains the generic name *Drechslera*.

Pyrenophora spp. cause root, crown, and rhizome rots, leaf spots, and melting-out of cool-season grasses: (1) leaf spot, melting-out, and root, crown, and rhizome rots of *Lolium* and *Poa* spp., caused by *P. poae*; (2) leaf spot and net blotch of *Festuca, Lolium, Poa,* and *Zoysia* spp., caused by *P. dictyoides*; (3) leaf spot and brown blight of *Festuca, Lolium,* and *Poa* spp., caused by *P. lolii*; (4) leaf streak of *Agrostis* and *Lolium* spp., caused by *P. grahamii*; (5) leaf blight and crown rot of *Agrostis, Festuca,* and *Lolium* spp., caused by *P. catenaria*; and (6) red leaf spot of *Agrostis* spp., caused by *P. erythrospila*.

Important diseases caused by *Pyrenophora* spp. are somewhat specific to the hosts listed above. However, several of these fungi also cause minor leaf spotting (flecking) in other grass species.

Symptoms and Signs

***Poa* spp.** Leaf and sheath lesions caused by *P. poae* (Figs. 77 and 78) in *Poa* spp. usually resemble those caused by *Bipolaris sorokiniana* (Fig. 33). They first appear as small, water-soaked areas that soon become uniformly dark, ranging from reddish brown to purplish black. The lesions are often surrounded by yellow zones that fade into the adjacent healthy green tissue. As the lesions enlarge, they become elongated parallel to the leaf axis, and their centers become necrotic and turn brown and then white to straw colored. Spots of various sizes, at different stages of development, may be present on a single leaf. Lesions may coalesce to form large necrotic areas or blighting of entire leaves and sheaths. The lesions on leaves and leaf sheaths

Fig. 77. Pyrenophora leaf spot of *Poa pratensis*. (Courtesy R. S. Byther)

Fig. 78. Symptoms of melting-out caused by *Pyrenophora poae* on a tiller of *Poa pratensis*. (Courtesy B. B. Clarke—© APS)

are usually distinct and symmetric. Atypical lesions develop in certain cultivars of *P. pratensis*. Older leaves are much more susceptible to attack than younger leaves. When lesions extend all the way across the leaf blade or leaf sheath, the vascular system becomes girdled, and the entire leaf or tiller dies and drops from the plant. In severe cases, nearly all leaves and tillers are killed (Fig. 79); this severe thinning of turf is known as "melting-out."

P. poae also invades roots, rhizomes (Fig. 80), and crowns of *P. pratensis*, causing a reddish brown dry rot. Symptoms on the foliage of plants with crown and root rot include wilting and chlorosis and occasionally some leaf spotting, followed by death and browning of the plants.

Leaf spots followed by melting-out occur mainly during cool, humid, overcast periods, usually in spring and autumn. Melting-out and root, rhizome, and crown rots also occur during warm, dry periods or during a wet period immediately following a dry period.

***Festuca* spp.** Leaf lesions caused by *P. dictyoides* in fine-leaf fescues appear as small, reddish brown, irregularly shaped spots. The lesions quickly girdle the small leaves of these grasses, causing yellowing and dieback from the tips. Small brown patches, 3–10 cm in diameter, or pockets of dead foliage commonly occur over large areas when the disease is severe in turfs containing a high proportion of fine-leaf fescues. The pathogen also causes root and crown rots, resulting in melting-out; the symptoms are similar to those described for *Poa* spp. The diseases occur primarily during cool, wet, overcast periods in spring and autumn.

In *F. arundinacea* and *Lolium perenne*, *P. dictyoides* causes a fine network of very small, brown lesions that often appear as short, brown streaks in netlike patterns, parallel to and perpendicular to the leaf axis (Fig. 81). The netlike patterns coalesce to form dark brown spots. The name "net blotch" is derived from this symptom on these wide-leaved grasses. Heavily infected turf turns yellow brown, leaves progressively die back from the tips, and large areas of turf turn tan or brown and thin out.

***Lolium* spp.** Leaf lesions caused by *P. lolii* in *Lolium* spp. vary considerably in shape and size. Numerous small, oval, chocolate brown spots are common. As the lesions enlarge, their centers may turn light brown to white. Large, dark brown, elongated streaks along the axes also appear on leaves, usually at the same time as the smaller, oval lesions. Faint net symptoms are occasionally present, but the absence of streaks perpendicular to the leaf axes distinguishes this disease from that caused by *P. dictyoides*. Girdled leaves become yellowed and die back from the tips, and the turf can become severely thinned (Fig. 82). Root and crown rots also occur; the symptoms are similar to those described for *Poa* spp.

***Agrostis* spp.** Leaf lesions caused by *P. erythrospila* (Fig. 83) in *Agrostis* spp. are small, brown to reddish brown, circular to oval spots, which may or may not have tan to straw-colored centers. The lesions may become numerous and coalesce to

Fig. 79. Dieback and thinning of *Poa pratensis*, symptomatic of melting-out caused by *Pyrenophora poae*. (Courtesy P. H. Dernoeden—© APS)

Fig. 80. Rhizome root rot of *Poa pratensis* infected by *Pyrenophora poae*. (Courtesy R. W. Smiley—© APS)

Fig. 81. Net blotch of *Festuca arundinacea*, caused by *Pyrenophora dictyoides*. (Courtesy N. R. O'Neill)

Fig. 82. Pyrenophora leaf spot of *Lolium perenne*, caused by *Pyrenophora lolii*. (Courtesy R. W. Smiley—© APS)

give an overall reddish cast to infected turf. The disease, red leaf spot, often develops in randomly distributed, irregular patches, in which the turf appears reddish brown (Fig. 84). Infected leaves are easily girdled and then wither and die, giving the turf a smoky blue cast; the overall appearance is like that of drought-stressed turf, even when soil moisture is plentiful. Because the leaves of *Agrostis* spp. grown on golf and bowling greens are small, leaf spot symptoms may be indistinct on individual leaves. *P. erythrospila* is more common on *A. capillaris* and is only occasionally found on improved cultivars of *A. stolonifera*. Red leaf spot is most likely to appear during warm, humid weather. *Agrostis* spp. on older greens commonly segregate into clones or patches of particularly competitive plants originally derived from one seed. Some clones may develop a reddish cast in cold weather, and these reddish patches may be confused with symptoms of red leaf spot.

P. catenaria causes leaf blight and crown rot of *Agrostis* spp. on golf greens. Red leaf lesions, tip dieback, and shoot blight occur in small patches 2–3 cm in diameter. The infection centers are sunken and reddish brown, and they may coalesce to involve large areas. Affected greens appear reddish brown, similar to those infected by *P. erythrospila*. The causal agents of these diseases are easily distinguished by their conidial morphology.

Causal Agents

Although the sexual stages of these fungi, *Pyrenophora* spp., take priority for this genus, they are not commonly observed in turfgrasses. When pathogenic to turfgrass, these fungi readily undergo asexual reproduction, producing dark brown mycelia, conidiophores, and conidia. The mycelium grows inside leaf tissue or along the leaf surface and gives rise to conidiophores, 50–80 µm long, which emerge through or arise from the leaf surface. The conidiophores bear cylindrical, three- to 10-septate conidia (Fig. 85). The sizes and shapes of conidia are characteristic for each species. They range from 8 to 23 µm in width (with an average width of 12–20 µm) and from 25 to 130 µm in length (with an average length of 60–70 µm). Germ tubes are produced from any or all conidial cells in directions perpendicular to the conidium axis. Taxonomic keys are useful for species identification; however, molecular data is more reliable. Similar to *Bipolaris, Curvularia,* and *Exserohilum* spp., analysis of the nuclear rDNA internal transcribed spacer region or glyceraldehyde-3-phosphate dehydrogenase is sufficient for species identification.

Pyrenophora spp. are somewhat host specific in their ability to cause important diseases in different genera of cool-season grasses.

Fig. 83. Red leaf spot of *Agrostis stolonifera,* caused by *Pyrenophora erythrospila*. (Courtesy L. P. Tredway—© APS)

Fig. 84. Red leaf spot of *Agrostis stolonifera,* caused by *Pyrenophora erythrospila*. (Courtesy J. P. Kerns—© APS)

Disease Cycle

Pyrenophora spp. survive unfavorable conditions as conidia and as dormant mycelium in infected plant tissue and in debris. These fungi can grow saprophytically on dead tissue and sporulate profusely when dry debris is rewetted. Leaf clippings are the primary source of *P. poae* conidia in *Poa* spp.

In cool, moist weather, conidia can be spread to healthy leaves or leaf sheaths by wind, rain, irrigation water, equipment, people, or animals. When leaf surfaces are moist and temperatures are favorable (see Epidemiology), conidia germinate quickly and produce germ tubes that can infect leaves. Lesions are produced, and additional sporulation may occur on dead tissue in the larger lesions. Leaf spots occur mainly during the spring and autumn, except those caused by *P. erythrospila,* which occur in warmer weather.

P. dictyoides, P. poae, and *P. lolii* also infect lower stems, roots, and rhizomes at any time from spring through autumn and, in some areas, during mild periods in winter. Infection of stems and roots results in dry rots. These pathogens can also infect seeds and cause seedling blights in newly established turfgrass.

Epidemiology

Pyrenophora spp. are active throughout the year, except when the soil is frozen or excessively hot. Roots, rhizomes, and lower stems may be infected directly by mycelium growing from nearby infected tissues or plant debris. Conidia are produced at temperatures of 3–27°C (37–80°F) (optimal 15–18°C

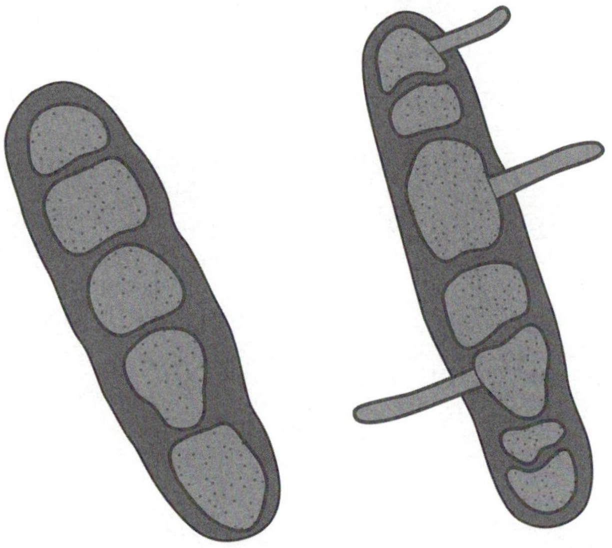

Fig. 85. Conidium of *Pyrenophora poae* (left) and germinating conidium of *P. poae* with germ tubes emanating from terminal and central cells (right). (Courtesy L. P. Tredway—© APS. Drawing by S. McKinley.)

[59–64°F]). A film of moisture on the leaf surface is necessary for infection. Leaf spots, therefore, occur primarily when the microenvironment around leaves is cool and wet. The exception is red leaf spot of *Agrostis* spp., caused by *P. erythrospila,* which occurs mostly during warm, wet periods.

Diseases caused by *Pyrenophora* spp. are favored by prolonged leaf wetness on overcast or foggy days or when irrigation is applied late in the day or in the evening. In areas where air movement is impeded by trees, shrubs, or structures, the resulting high humidity is favorable for leaf infections. Physiological changes in plants heighten disease susceptibility when the intensity of the light is low (because of shading by trees or clouds, for example), when the mowing height is lower than that recommended for the turfgrass species being grown, when the turf is fertilized with excessive amounts of nitrogen, and when growth-regulating fungicides or herbicides have been applied. In addition, large amounts of thatch and leaf clippings provide a substrate on which the pathogens can grow and sporulate. *Pyrenophora* spp. are most active at 15–18°C (59–64°F) and are curtailed at temperatures above 27°C (80°F).

Management

Fertilizers should be applied as needed to promote moderate plant growth but not at rates that cause excessive growth. It is critical to avoid overstimulating the growth of grass in early spring and midsummer. Water should be applied thoroughly in the morning to wet the soil deeply but as infrequently as possible without causing stress. Frequent short irrigations should be avoided, especially in the evening. Grass should be mowed as high as is practical. For example, for lawns and commercial cool-season turfgrasses, the minimum mowing height is 5–6 cm. Overuse of demethylation inhibitor fungicides and synthetic auxin herbicides such as dicamba, MCPP, and 2,4-D should be avoided when these diseases are present (see Nontarget Effects of Pesticides in Part III). Thatch should be reduced if it accumulates to depths of more than 1 cm. In cool-season grasses, thatch should be removed in early spring or autumn; in most warm-season grasses, it should be removed during the summer. Landscape plantings should be pruned or designed so that they do not cause dense shade or impede the movement of air.

Most cultivars of *P. pratensis* and many of *F. rubra* and *L. perenne* have been selected or bred for resistance to these diseases.

Several classes of fungicides are available for controlling leaf spot and melting-out diseases caused by *Pyrenophora, Bipolaris,* and *Exserohilum* spp. and for reducing spore production on organic debris.

Selected References

Alcorn, J. L. 1983. Generic concepts in *Drechslera, Bipolaris* and *Exserohilum.* Mycotaxon 17:1-86.

Ellis, M. B. 1971. Dematiaceous Hyphomycetes. Commonwealth Mycological Institute, Kew, England.

Hagan, A. K., and Larsen, P. O. 1985. Source and dispersal of conidia of *Drechslera poae* in Kentucky bluegrass turf. Plant Dis. 69:21-24.

Rossman, A. Y., Crous, P. D., Hyde, K. D., Hawksworth, D. L., Aptroot, A., Bezerra, J. L., Bhat, J. D., Boehm, E., Braun, U., Boonmee, S., Camporesi, E., Chomnunti, P., Dai, D. Q., D'souza, M. J., Dissanayake, A., Gareth Jones, E. B., Broenewald, J. Z., Hernández-Restrepo, M., Hongsanan, S., Jaklitsch, W. M., Jayawardena, R., Jing, L. W., Kirk, P. M., Lawrey, J. D., Mapook, A., McKenzie, E. H. C., Monkai, J., Phillips, A. J. L., Phookamsak, R., Raja, H. A., Seifert, K. A., Senanayake, I., Slippers, B., Suetron, S., Tanaka, K., Taylor, J. E., Thanbugala, K. M., Tian, Q., Tibpromma, S., Wanasinghe, D. N., Wijayawardene, N. N., Wikee, S., Woudenberg, J. H. C., Wu, H. X., Yan, J., Yang, T., and Zhang, Y. 2015. Recommended names for pleomorphic genera in *Dothideomycetes.* IMA Fungus 6:507-523.

Shoemaker, R. A. 1962. *Drechslera* Ito. Can. J. Bot. 40:809-846.

Sivanesan, A. 1987. Graminicolous species of *Bipolaris, Curvularia, Drechslera, Exserohilum* and their teleomorphs. Mycol. Pap. 158. CAB International, Wallingford, U.K.

Spilker, D. A., and Larsen, P. O. 1985. Characterization and host range of *Drechslera catenaria,* the pathogen of leaf blight and crown rot of creeping bentgrass. Plant Dis. 69:331-333.

Wingfield, M. J., De Beer, Z. W., Slippers, B., Wingfield, B. D., Groenewald, J. Z., Lombard, L., and Crous, P. W. 2012. One fungus, one name promotes progressive plant pathology. Mol. Plant Pathol. 13:604-613.

Zhang, G., and Berbee, M. L. 2001. *Pyrenophora* phylogenetics inferred from ITS and glyceraldehyde-3-phosphate dehydrogenase gene sequences. Mycologia 93:1048-1063.

(Prepared by R. W. Smiley, P. H. Dernoeden, and B. B. Clarke; revised by M. Tomaso-Peterson)

Pythium Blight

Pythium spp. cause numerous diseases of turfgrasses, which have been referred to as Pythium blight, grease spot, spot blight, crown and root rot, root dysfunction, cottony blight, and snow blight. Each of these names refers to only one aspect of the wide spectrum of damage and symptoms caused by *Pythium* spp. All turfgrasses are susceptible to attack by *Pythium* spp.

Pythium blight is commonly associated with golf course turfs and athletic fields, but it also occurs in landscape turfgrass settings when conditions are favorable for disease development. Cool-season grasses are prone to attack during hot, humid weather, especially when soil moisture is excessive. Susceptible cool-season grasses can sustain severe and widespread damage very quickly under ideal conditions.

In warm-season grasses, Pythium blight occurs during cold or cool, wet weather, and symptoms may resemble those of other diseases such as Bipolaris leaf spot. Putting greens cultured with ultradwarf *Cynodon* spp. are particularly susceptible to Pythium blight during periods of low light intensity and high soil moisture.

Symptoms

Cool-season grasses. Circular spots 2–15 cm in diameter usually appear suddenly during hot, humid weather. In closely mowed grasses, the patches initially appear as small, discolored spots (Fig. 86), but they can enlarge at an alarming rate. In higher-cut grasses, the patches tend to be larger in diameter and less regular in shape (Fig. 87). Blighting may also appear in streaks that follow the water drainage pattern or the movement of mowers or other equipment (Fig. 88). In affected patches,

Fig. 86. Pythium blight of *Agrostis stolonifera* on a golf green. (Courtesy P. H. Dernoeden—© APS)

leaves appear water soaked and dark in the early morning. When rubbed between the fingers, these water-soaked patches feel oily. The term "grease spot" refers to this symptom.

In *Agrostis* spp. or *Poa annua* on golf greens, tees, and fairways, the disease initially appears as orange or bronze-colored spots. Gray "smoke rings" may be present, especially early in the morning (Fig. 86). The spots are typically 3–8 cm in diameter. In higher-cut *Lolium, Festuca,* and other grass species, the spots may be copper colored or may have a gray, water-soaked appearance. Infected leaves turn light tan to brown and become shriveled and matted when dry (Fig. 89). When humidity remains high, the collapsed leaves become matted and covered with a fluffy, white or gray mass of mycelium (Figs. 90 and 91). The blighting may occur in discrete patches 15–30 cm in diameter or as "frog-eyes," in which the turf is affected in a ring with living turf in the center of the ring. When numerous infection centers occur, patches may coalesce to form large, irregularly shaped areas of dead turf (Fig. 92). These larger areas are often elongated or concentrated in wet areas. Blighting often develops in a streaking pattern associated with the flow

Fig. 87. Pythium blight of *Festuca arundinacea*. (Courtesy L. P. Tredway—© APS)

Fig. 88. Pythium blight of *Agrostis stolonifera* spreading in a pattern that follows water drainage. (Courtesy S. Bauer—© APS)

Fig. 89. Shriveled, matted leaves of *Lolium perenne* caused by Pythium blight. (Courtesy B. B. Clarke—© APS)

Fig. 90. Pythium blight of *Festuca arundinacea* with mycelium developing at the perimeters of the affected areas. (Courtesy B. B. Clarke—© APS)

Fig. 91. Pythium blight of *Lolium perenne* seedlings. (Courtesy B. B. Clarke—© APS)

Fig. 92. Pythium blight of *Agrostis stolonifera* on a golf green. (Courtesy P. H. Dernoeden—© APS)

of water carrying mycelium or spores or with the movement of equipment spreading mycelium.

Warm-season grasses. On warm-season turf such as *Cynodon* spp. grown for putting greens, Pythium blight first appears as small, black or purple spots that can expand into larger irregular areas, especially during prolonged periods of humid, rainy, and/or cloudy weather any time of the year. These symptoms can easily be mistaken for those of leaf spot, so microscopic evaluation of the affected leaf tissue is necessary to diagnose Pythium blight. Leaves within affected areas are matted, black or dark gray, and greasy in appearance (Figs. 93 and 94). Abundant foliar mycelium is rarely seen on warm-season grasses afflicted with Pythium blight but may occur following incubation in a humid chamber.

Causal Agents

Pythium spp. are no longer considered true fungi and instead are classified in the kingdom Chromista or kingdom Stramenopila because of their unique cellular structure and modes of reproduction. *Pythium* spp. are uniquely adapted to aqueous environments and require free water to infect their hosts and disseminate to new host tissue.

Numerous pathogenic species of *Pythium* and *Globosporangium* are associated with Pythium blight, including *P. aphanidermatum, P. arrhenomanes, P. graminicola, P. myriotylum, P. torulosum, P. vanterpoolii, G. irregulare,* and *G. ultimum.*

Most *Pythium* spp. that infect grasses are homothallic and generally produce oospores and/or sporangia on artificial media. All species produce coenocytic mycelia. They are distinguished by characteristics of their antheridia, oogonia, and sporangia (Figs. 95–97). Oospores observed in tissue may be 15–30 µm in diameter, depending on the species.

Fig. 93. Pythium blight on a *Cynodon* sp. golf course putting green. (Courtesy J. P. Kerns—© APS)

Fig. 94. Symptoms and signs of Pythium blight on a cup cutter plug collected from a *Cynodon* sp. putting green. (Courtesy J. P. Kerns—© APS)

Disease Cycle

The occurrence of Pythium blight (Fig. 98) depends primarily on environmental conditions and the host species. Running water can transport mycelium, sporangia, zoospores, and oospores over considerable distances. Zoospores are very small (10 µm in diameter) and move rapidly in free water, and in this form, the pathogen can spread in water drainage patterns. These pathogens also spread from leaf to leaf by rapid growth of mycelium, which directly penetrates leaves. *Pythium* spp. can be transported over large areas when mycelium, infected plant tissues, or infested soil is moved in water or by maintenance equipment.

Epidemiology

Pythium blight causes extensive damage only under wet conditions. On cool-season grasses the disease is most severe during hot (30–35°C [86–95°F]), humid, or rainy weather with warm nights (above 20°C [68°F]). A weather-based forecasting model identified the most favorable conditions for Pythium blight as relative humidity exceeding 90% for at least 14 hours and a minimum temperature of 20°C (68°F). The researchers demonstrated that Pythium blight usually developed on cool-season grasses when these conditions occurred.

Warm-season turfgrasses are susceptible to Pythium blight during extended periods of humid, rainy, and/or cloudy weather almost any time of the year when temperatures are above 10°C (50°F). Lush, dense grass growing under high nitrogen fertilization is especially susceptible to attack, and the disease may become more severe in alkaline soils than in acid soils. Some species are most active at temperatures of 11–21°C (52–70°F), and other species are dominant at 23–34°C (73–93°F).

Management

Pythium blight is most severe on wet soils, which promote continual leaf wetness; therefore, water management is of utmost importance in reducing disease incidence. When new turf is established, good surface and subsurface drainage is necessary. Fungicide-treated seed should be used when planting during warm, moist periods. Overwatering of new plantings should be avoided, and water should not be applied late in the day. On established turf, water should be applied early in the morning, so that leaves are dry by late morning or midday. Turf should be watered as infrequently and thoroughly as possible to provide adequate soil water throughout the root zone.

Mechanical removal of thatch is recommended if the thatch depth exceeds 1.25 cm. Overfertilization with nitrogen should be avoided. Slow-release forms of nitrogen or light, frequent spoon-feeding applications do not enhance Pythium blight severity, even during the summer months. If possible, turf should not be mowed during moist, hot weather when foliar mycelium is evident. In the development of new landscape designs, the penetration of light into the turf canopy and the circulation of air should be considered. Trees and shrubs can be selectively pruned and fans can be used around golf greens to eliminate local pockets of excessively humid and still air.

Fungicides are often required for the control of Pythium blight on golf and bowling greens, stadium athletic fields, and other intensively managed turfs. Among lawn and athletic field species, *Lolium perenne* and any cool-season turfgrass being established in a late spring or summer seeding are most likely to benefit from fungicide treatment. Fungicides targeting *Pythium* spp. should be rotated or applied in recommended tank-mix combinations; repeated treatment with fungicides of the same mode of action favors the development of resistance within populations of *Pythium* spp.

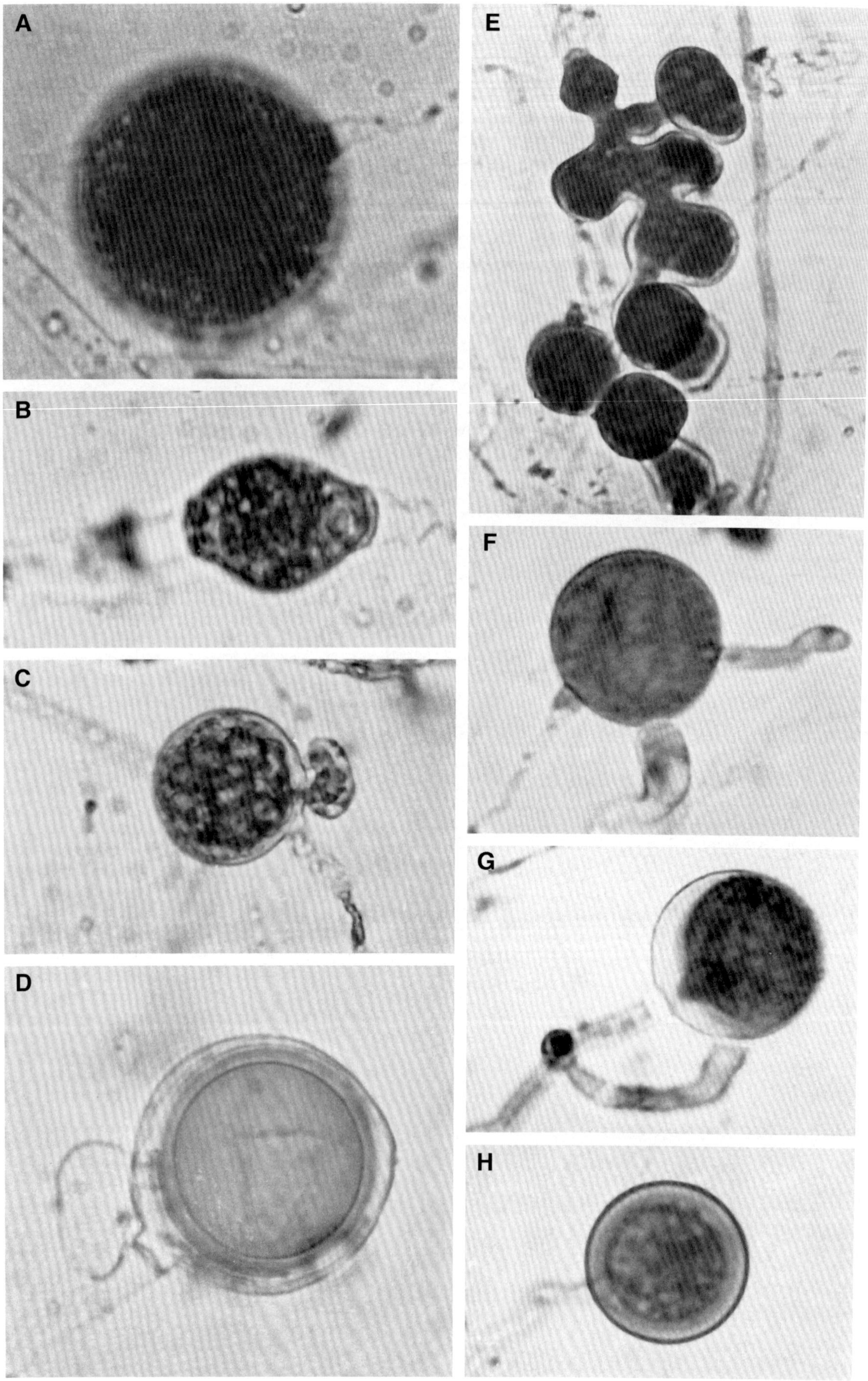

Fig. 95. Characteristics of *Pythium* spp. **A–D,** *P. ultimum:* **A,** spherical terminal sporangium; **B,** intercalary lemon-shaped sporangium; **C,** young oogonium with sessile, monoclinous antheridium; and **D,** aplerotic oospore. **E–H,** *P. vanterpoolii:* **E,** lobulate sporangial complex; **F and G,** oogonia with monoclinous antheridia arising some distance from the oogonium; and **H,** plerotic oospore. (Reproduced, by permission, from Robertson, G. I. 1980. The genus *Pythium* in New Zealand. N.Z. J. Bot. 18:73-102.)

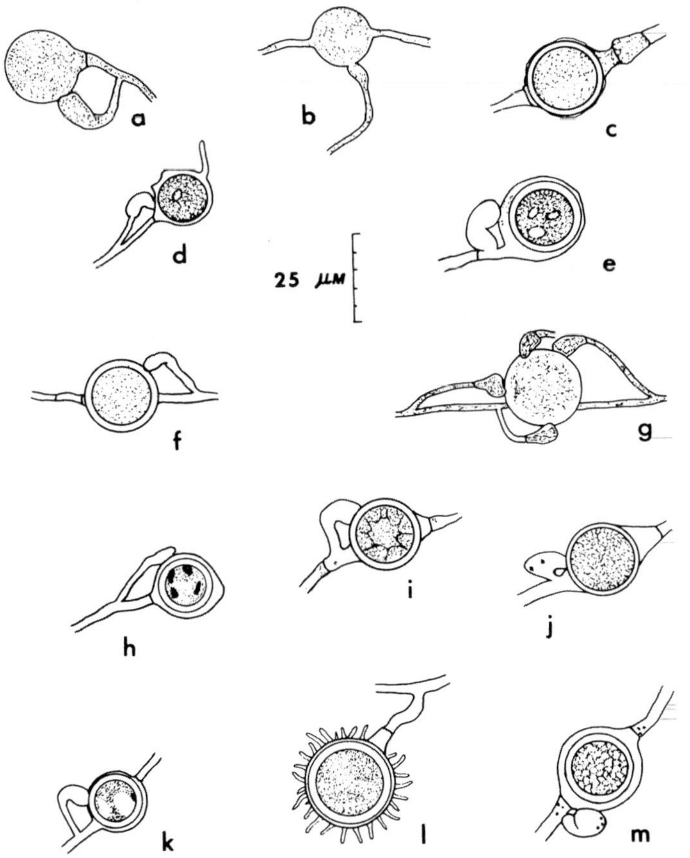

Fig. 96. *Pythium* spp. sex organs. Antheridia may be monoclinous (a), diclinous (b), or hypogynous (c). They may be stalked (d) or sessile (e). The number of antheridia per oogonium ranges from one (f) to several (g). Oogonia may be terminal (h) or intercalary (i), with walls that are smooth (k) or ornamented (l). Oospores may be plerotic (j), completely filling the oogonium, or aplerotic (m), not filling the oogonium but with a distinct space between the oospore and the oogonium wall. (Reproduced, by permission, from G. I. Robertson, 1980, N.Z. J. Bot. 18:73-102.)

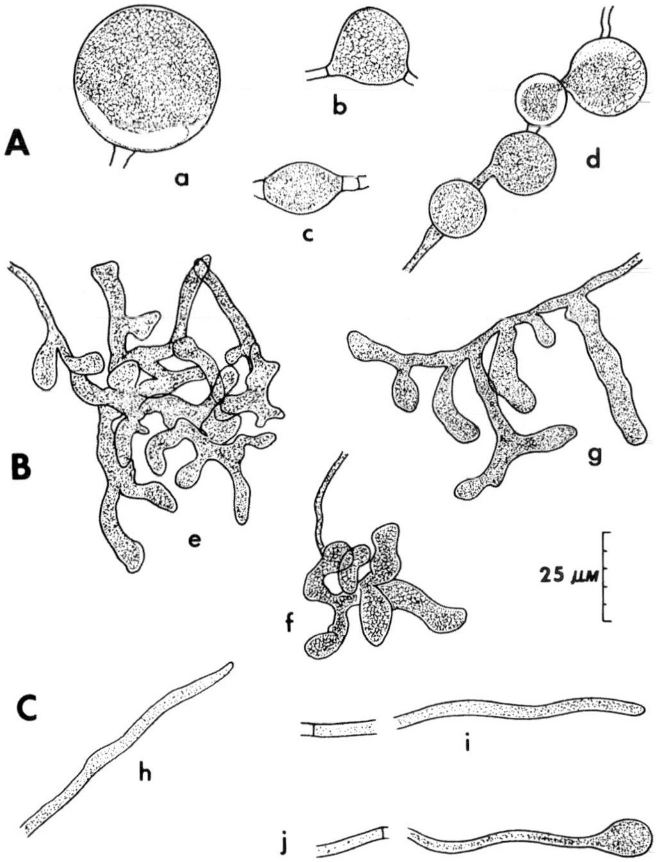

Fig. 97. Sporangial types in *Pythium* spp. **A,** Spherical sporangia may be terminal (a) or intercalary (b). Intercalary sporangia are usually subspherical (b) or lemon shaped (c) and may be grouped to form a catenulate sporangium (d). **B,** Lobulate sporangia (e–g) are usually indistinguishable from vegetative hyphae **(C,** h and i**)** but, when produced in water culture, usually form readily visible vesicles (j). (Reproduced, by permission, from G. I. Robertson, 1980, N.Z. J. Bot. 18:73-102.)

Selected References

Chang, T., and Lee, Y. S. 2013. Occurrence of Pythium blight caused by *Pythium aphanidermatum* on chewing fescue. Weed Turfgrass Sci. 2:306-311.

Ichitani, T., Tani, T., and Umakoshi, T. 1986. Identification of *Pythium* spp. pathogenic on manila grass (*Zoysia matrella* Merr.). Trans. Mycol. Soc. Jpn. 27:41-50.

Nutter, F. W., Cole, H., Jr., and Schein, R. D. 1983. Disease forecasting system for warm weather Pythium blight of turf grass. Plant Dis. 67:1126-1128.

Parker, S. R. 2004. Identification and pathogenicity of *Pythium* species from bermudagrass and effects of spring-applied fungicides on bermudagrass putting greens. M.S. thesis. Clemson University, Clemson, SC.

Plaats-Niterink, A. J. van der. 1981. Monograph of the genus *Pythium*. Stud. Mycol. 21:1-242.

Sanders, P. L., Houser, W. J., Parish, P. J., and Cole, H., Jr. 1985. Reduced-rate fungicide mixtures to delay fungicide resistance and to control selected turfgrass diseases. Plant Dis. 69:939-943.

Schmitthenner, A. F. 1980. *Pythium* species: Isolation, biology and identification. Pages 33-36 in: Advances in Turfgrass Pathology. P. O. Larsen and B. G. Joyner, eds. Harcourt Brace Jovanovich, Duluth, MN.

Schroeder, K. L, Martin, F. N., de Cock, A. W. A. M., Lévesque, C. A., Spies, C. F. J., Okubara, P. A., and Paulitz, T. C. 2013. Molecular detection and quantification of *Pythium* species: Evolving taxonomy, new tools, and challenges. Plant Dis. 97:4-20.

Stiles, C. M., Datnoff, L. E., and Rayside, P. A. 2007. *Pythium* spp. isolated from bermudagrass during overseed transitions in Florida and pathogenicity of *Pythium irregulare* on *Poa trivialis*. Plant Dis. 91:1237-1244.

(Prepared by R. W. Smiley, P. H. Dernoeden, and B. B. Clarke; revised by J. P. Kerns)

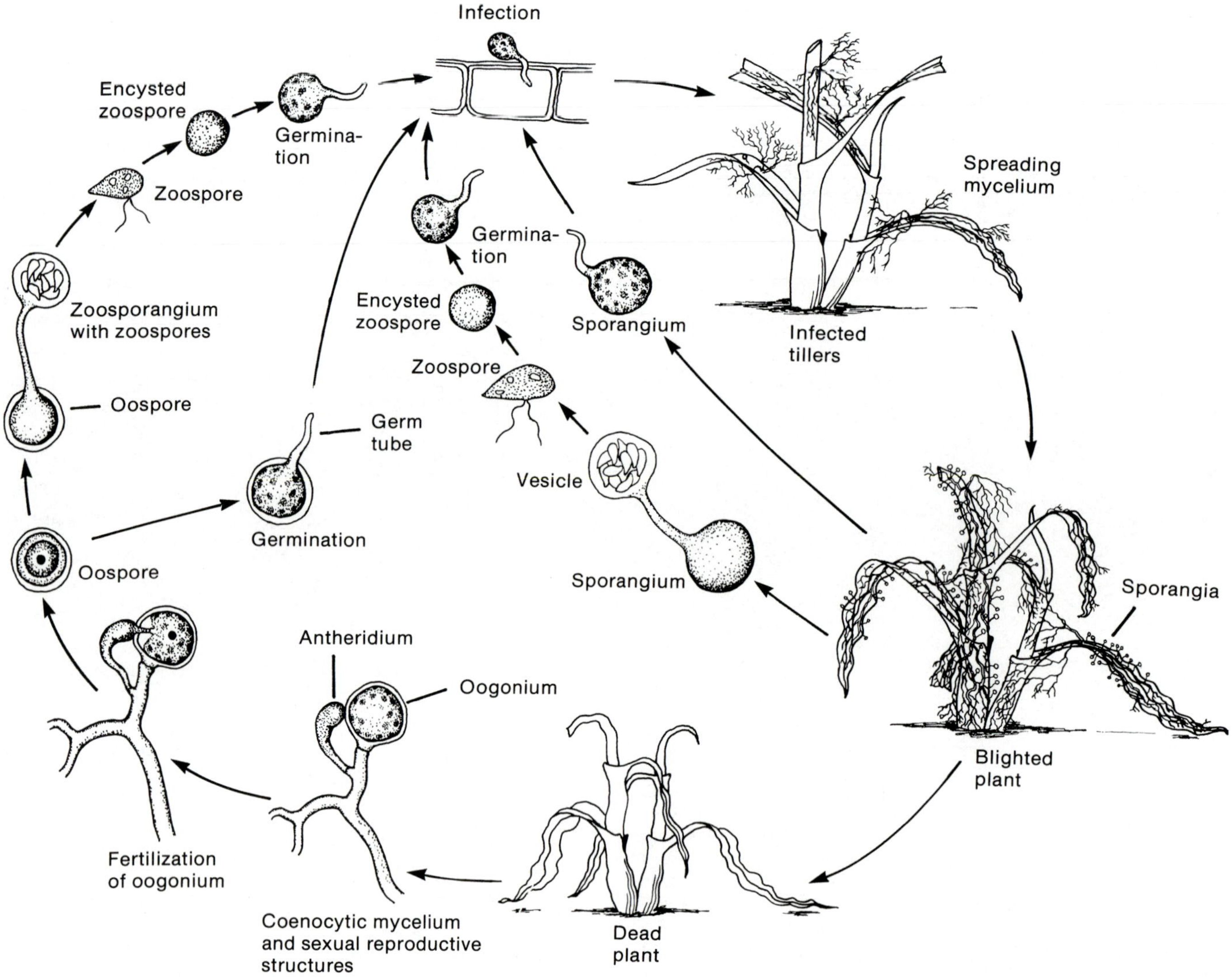

Fig. 98. Disease cycle of Pythium blight. (Courtesy C. R. Whittaker—© APS. Adapted from G. N. Agrios, 1978, Plant Pathology, 2nd ed., Academic Press, New York.)

Rapid Blight

Rapid blight is a disease of cool-season grasses that occurs where drought and accumulation of sodium salts coincide. Rapid blight was first named as a disease of turfgrasses in 2002, although symptoms of the disease were reported on *Poa annua* golf course greens as early as 1995 in California. Rapid blight has caused damage to golf greens, tees, fairways, and roughs cultured with *Poa annua, P. trivialis, Lolium perenne,* and sometimes *Agrostis stolonifera*. It has also been identified in *L. perenne* overseedings on athletic fields. In the United States, rapid blight has been reported in Arizona, California, Colorado, Florida, Georgia, Louisiana, Nevada, North Carolina, South Carolina, Texas, and Utah. Worldwide, rapid blight has been reported in Argentina, Ireland, Spain, and the United Kingdom.

Rapid blight is caused by *Labyrinthula terrestris,* also referred to as a net or marine slime mold. In the taxonomic system used in this compendium, *Labyrinthula* is in the phylum Labyrinthulomycota in the kingdom Chromista. *Labyrinthula* spp. are single-celled organisms (Fig. 99) that occur in netlike aggregations in culture, with spindle-shaped cells that glide in a matrix excreted from vegetative cells (Fig. 100).

The ability of the pathogen to survive and cause disease in saline environments is one of its major characteristics. Rapid blight is similar in many respects to "wasting disease" of eelgrass (*Zostera marina* L.), caused by *L. zosterae,* the only other known pathogenic species of *Labyrinthula*.

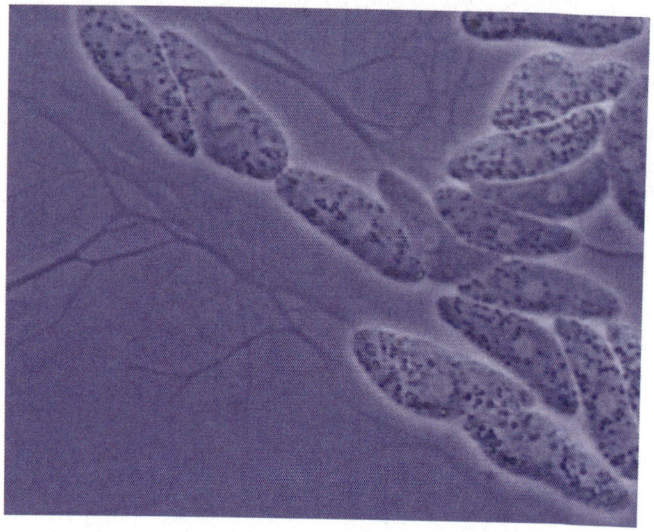

Fig. 99. Spindle-shaped cells of *Labyrinthula terrestris,* the cause of rapid blight of turfgrasses. (Courtesy S. B. Martin—© APS)

Symptoms and Signs

Symptoms appear as chlorotic or darkened turf in irregularly shaped patches 2–3 m in diameter (Figs. 101–103). Sometimes patches are distinctly circular. In overseeded grasses on low-cut greens, patches may have dark borders of water-soaked turf. Upon close examination, chlorotic leaves appear mottled and water soaked. Affected leaves can exhibit tip dieback and eventually become shriveled and necrotic. Rapid blight can affect turf in the seedling stage at first mowing, causing extensive damage that can eliminate a stand within a week. Mature stands of turf, such as putting greens of *P. annua,* can be severely affected. Microscopic examination reveals hyaline, spindle-shaped vegetative cells in leaf epidermal and mesophyll cells. Chloroplasts in infected leaf tissues degrade as the disease progresses.

Causal Agent

Rapid blight is caused by *L. terrestris,* a eukaryote. Cells are hyaline, fusiform or spindle shaped, contain vacuoles and numerous lipid droplets, and vary in size (13.0–17.5 × 4.5–7.0 µm). Current knowledge indicates that cells of *L. terrestris* divide mitotically; no identifiable sexual reproductive cells or structures have been found. Cells move in tubular ectoplasmic networks, excreted by specialized organelles in cells known as the "bothrosome." In these networks, cells move or "glide" at about 3.5–9 mm per hour on agar plate cultures. These networks branch and anastomose, which gives actively growing colonies their characteristic labyrinth, forking, or digitate shape. Cells multiply by dividing longitudinally or obliquely. In older colonies, cells aggregate into clumps that appear hyaline to pale yellow. Cells within the clumps are motionless and ovoid, but, if not too old, aggregated cells can reinitiate movement via slimeways when placed on a suitable medium. The function of the aggregates is unknown, although it is presumed that they serve as long-term survival structures.

Disease Cycle

Spindle-shaped vegetative cells divide mitotically and appear to be the major propagules disseminated by mowing, traffic, and other means. The disease cycle begins with infection through natural openings or wounds. In studies conducted on infection, wounding accelerates the appearance of infected leaves, although cell penetration through leaf stomata has been observed.

The organism does not produce mycelium, but slime pathways produced may serve to move cells within a turf canopy. Some *Labyrinthula* spp. form thick-walled cysts, which may aid in their survival, but cyst formation has not been confirmed in isolates from turfgrasses. Cell aggregates formed in aging cultures of *L. terrestris* may serve as survival structures. The pathogen has been isolated from symptomless hosts such as Bermudagrass and seashore paspalum, which are relatively tolerant of salinity stress compared with most cool-season grasses. In situations where these warm-season hosts occur, they may serve as sources of inoculum to susceptible cool-season grasses used for overseeding.

Epidemiology

Rapid blight has been associated with variable environmental conditions. Epidemics may occur during any season when dry periods occur and soil and water salinity are 1–3.5 dS/m (640–2,240 ppm or 1,000–3,500 mmhos/cm). Typically, the disease occurs during mild to warm conditions when cool-season turfgrasses are more susceptible to drought and salinity

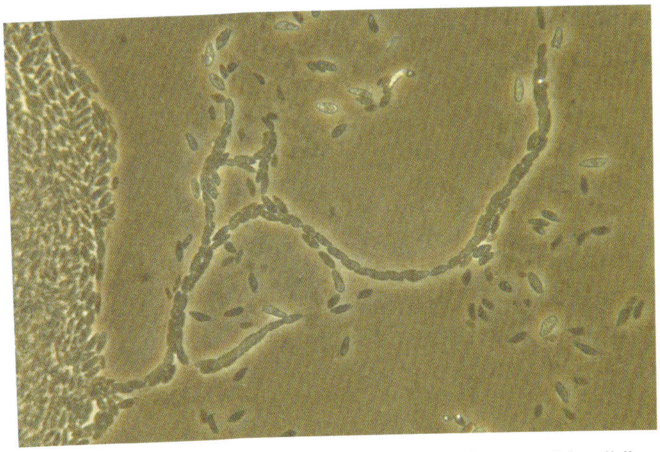

Fig. 100. Spindle-shaped cells of *Labyrinthula terrestris* gliding on ectoplastic networks. (Courtesy S. B. Martin)

Fig. 101. Rapid blight of *Poa trivialis* on a golf green, caused by *Labyrinthula terrestris.* (Courtesy S. B. Martin)

Fig. 102. Rapid blight on a *Poa annua* putting green. (Courtesy S. B. Martin—© APS)

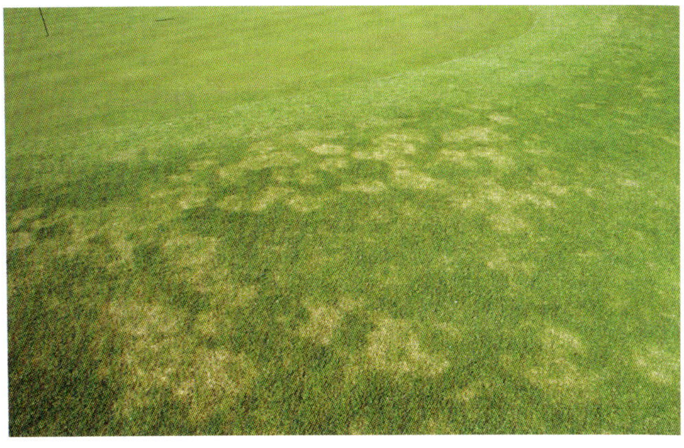

Fig. 103. Rapid blight on *Lolium perenne* overseeded into *Cynodon* spp. (Courtesy S. B. Martin—© APS)

stress. Increased precipitation that leads to leaching of sodium cations from the soil decreases rapid blight severity. However, if dry conditions recur for as little as 2 weeks and irrigation water with a salinity level of 1.5 dS/m or higher is applied, the disease can develop in highly susceptible hosts such as *P. trivialis* or *P. annua*.

After initial infections occur through wounds or natural openings in susceptible hosts, local movement of the pathogen can occur through slime pathways. Movement on a larger scale likely occurs from spread of infected tissues by mowing, transfer of the pathogen on equipment, or foot traffic.

Management

Strategies for disease management, as with other diseases, involve manipulation of host, pathogen, and environmental conditions. Strategies associated with host manipulation include the choice of more-tolerant cool-season grasses or even eliminating overseeding where it has been previously practiced and control has been ineffective. Use of paints and pigments to color dormant warm-season turfgrasses has lessened reliance on overseeding. If cool-season grasses are used, species and cultivars are available that are more tolerant to rapid blight under conditions typically encountered. One of these, slender creeping red fescue (*Festuca rubra* subsp. *litoralis*), may be mixed with more-susceptible but quickly germinating species such as perennial ryegrasses. In greenhouse and field experiments, salt-tolerant cultivars of perennial ryegrass have been shown to be more tolerant to rapid blight. Alkali grasses (*Puccinellia* spp.) are very tolerant to rapid blight and may be mixed with more-susceptible *P. trivialis* cultivars or used as a single species for overseeding, although its appearance and slower germination rate make it less desirable as a single-component overseeding grass.

Manipulation of environmental conditions includes strategies to reduce salinity in irrigation water or soil. The best way to control rapid blight is to reduce salinity by leaching with less-saline water. This may be useful if alternative sources of water are available or if saline soils can be leached with an excess leaching fraction. Application of chemicals such as gypsum to reduce sodium in the soil may reduce disease but does not stop it altogether. Some evidence exists that increased potassium fertilization may reduce rapid blight by competition with sodium. Leaching and the use of gypsum to displace sodium rely on the presence of internal drainage that is adequate to move displaced sodium from the root zone. Besides promoting rapid blight, sodium has detrimental effects on soil structure, and sodium-affected soils typically exhibit poor drainage resulting from deflocculation of organic matter and clay colloids. On golf greens affected by rapid blight, a typical observation is that turf around aerification holes becomes infected and symptomatic more slowly than areas in which surface drainage is impeded. Sodium can also be removed from irrigation water by use of reverse osmosis systems, where they are affordable and can be permitted. This process is expensive but effectively eliminates excess salinity from irrigation water.

Options to manipulate the pathogen directly are limited. Only a few fungicides are effective in controlling rapid blight, including pyraclostrobin, trifloxystrobin, and mancozeb. Fluazinam has some activity against the disease. Mancozeb and trifloxystrobin alone or in a mixture can control moderate levels of disease for 2–3 weeks. The best control has been provided by pyraclostrobin or a mixture of pyraclostrobin plus mancozeb. It should be noted that under very high disease pressure, rapid blight overcomes the effectiveness of fungicides alone.

Selected References

Bigelow, D. M., Olsen, M. W., and Gilbertson, R. L. 2005. *Labyrinthula terrestris* sp. nov., a new pathogen of turf grass. Mycologia 97:185-190.

Camberato, J. J., Peterson, P. D., and Martin, S. B. 2006. Salinity and salinity tolerance alter rapid blight in Kentucky bluegrass, perennial ryegrass, and slender creeping red fescue. Appl. Turfgrass Sci. 3. doi:10.1094/ATS-2006-0213-01-RS

Chitrampalam, P., Goldberg, N., and Olsen, M. W. 2015. *Labyrinthula* species associated with turfgrasses in Arizona and New Mexico. Eur. J. Plant Pathol. 143:485-493.

Entwistle, C. A., Olsen, M. W., and Bigelow, D. M. 2006. First report of a *Labyrinthula* spp. causing rapid blight of *Agrostis capillaris* and *Poa annua* on amenity turfgrass in the UK. Plant Pathol. 55:306.

Martin, S. B., Olsen, M. W., Peterson, P. D., and Camberato, J. J. 2004. Evaluation of fungicides for control of rapid blight on cool season grasses. (Abstr.) Phytopathology 94(suppl.):S66.

Muehlstein, L. K., Porter, D., and Short, F. T. 1991. *Labyrinthula zosterae* sp. nov., the causative agent of wasting disease of eelgrass, *Zostera marina*. Mycologia 83:180-191.

Olsen, M. W. 2007. *Labyrinthula terrestris:* A new pathogen of cool-season turfgrasses. Mol. Plant Pathol. 8:817-820.

Olsen, M. W., Bigelow, D. M., Gilbertson, R. L., Stowell, L. J., and Gelernter, W. D. 2003. First report of a *Labyrinthula* sp. causing rapid blight disease of rough bluegrass and perennial ryegrass. Plant Dis. 87:1267.

Peterson, P. D., Martin, S. B., and Camberato, J. J. 2005. Tolerance of cool-season turfgrasses to rapid blight disease. Appl. Turfgrass Sci. 2. doi:10.1094/ATS-2005-0328-01-RS

Stowell, L. J., Martin, S. B., Olsen, M., Bigelow, D., Kohout, M., Peterson, P. D., Camberato, J., and Gelernter, W. D. 2005. Rapid blight: A new plant disease. doi:10.1094/APSnetFeature/2005-0705

Yadagiri, K. K., Kerrigan, J. L., and Martin, S. B. 2012. Improved methods for axenic culture of *Labyrinthula terrestris,* causal agent of rapid blight of turfgrasses. Can. J. Microbiol. 58:1230-1235.

(Prepared by S. B. Martin)

Red Thread

Red thread is especially prevalent during the spring and autumn in slow-growing, nitrogen-deficient turf. It may cause severe damage to *Agrostis, Festuca, Lolium,* and *Poa* spp. in temperate climates during periods of cool and humid weather; *L. perenne* and *F. rubra* are particularly susceptible to damage. Red thread has also been documented in warm-season grasses including *Cynodon, Paspalum,* and *Zoysia* spp.

Symptoms and Signs

Grass becomes water soaked and rapidly dies in circular or irregularly shaped patches 5–60 cm in diameter (Figs. 104 and 105). The straw brown to tan color of dead leaves may be the first symptom observed. Necrotic leaves may be interspersed with symptomless leaves, causing an overall diffuse, scorched,

Fig. 104. Red thread of *Lolium perenne*. (Courtesy P. H. Dernoeden—© APS)

or ragged appearance to the patch. The patches may be widely scattered or close together and may coalesce to form large areas of infected turf.

Inspection of individual plants reveals only foliar symptoms. Necrosis usually proceeds from the leaf tip toward the base. When ample moisture is present, the pathogen produces colorful, threadlike sclerotia (red threads), which protrude from affected leaves (Fig. 106). The sclerotia are branched and pink to pale red or orange and may extend 10 mm beyond the end of the leaf tip (Fig. 107). In dry weather, sclerotia often turn bright red and become brittle. In this condition, they easily are detached and may be spread mechanically or fall into the thatch. Mycelial mats and pink cottony flocks of mycelium (Fig. 108) may also be produced during humid weather. When sclerotia or mycelial flocks are present, the patches take on a reddish or pink cast. The disease can be difficult to distinguish from dollar spot or pink patch when sclerotia or cottony flocks are not present. In this case, microscopic examination of the fungal hyphae can aid in diagnosis.

Causal Agent

Red thread is caused by *Laetisaria fuciformis*. Hyphae are multinucleate and 3.5–10 µm wide and do not have clamp connections. The fungus forms a weblike, pale reddish mycelium that surrounds and connects leaf blades. Antlerlike sclerotia (Fig. 107) develop in infected leaves, protruding beyond the tip of the leaf blade. Sclerotia may be pink, orange, or red. Cottony flocks of mycelium (Fig. 108) are pink and up to 10 mm in diameter. Upon drying, mycelial flocks become brittle and fragment into masses of arthroconidia. The arthroconidia are hyaline and ellipsoidal to cylindrical, measuring 5–17 × 10–47 µm. Basidia and basidiospores may also be produced on a hyaline to pink hymenium on the surface of infected leaf tissue.

Fig. 105. Red thread of *Festuca arundinacea* and *Poa pratensis*. (Courtesy L. P. Tredway—©APS)

Fig. 106. Sclerotia of *Laetisaria fuciformis* on blighted leaves of *Festuca arundinacea* and *Poa pratensis*. (Courtesy L. P. Tredway—© APS)

Disease Cycle

L. fuciformis survives unfavorable periods as sclerotia on infected leaves or in the thatch. Sclerotia survive temperatures as high as 32°C (90°F) and as low as –20°C (–4°F) and may remain viable for up to 2 years when dry. The pathogen is spread locally as arthroconidia or sclerotia moved by running water, equipment, people, or animals. Arthroconidia and infected plant debris may also become windborne and thus may be disseminated over long distances. The importance of basidiospores in the disease cycle is uncertain, but the pathogen can produce them in abundance. *L. fuciformis* has been reported to be seedborne in *Lolium perenne*.

A film of moisture over the surfaces of leaves or leaf sheaths is necessary for the germination of sclerotia and arthroconidia (and possibly basidiospores). Infection occurs when the fungus enters the plant through open stomata or wounds or by direct penetration of the epidermis. The fungus can kill leaves within 2 days of initial infection.

Epidemiology

Red thread can occur year round, because the pathogen is capable of growth at temperatures of 3–30°C (37–86°F). The optimal temperature for growth varies among strains of *L. fuciformis*, ranging from 15 to 25°C (59 to 77°F). Prolonged

Fig. 107. Sclerotia of *Laetisaria fuciformis* protruding beyond the tips of affected leaves. (Courtesy J. E. Kaminski—© APS)

Fig. 108. Pink, cottony flocks of *Laetisaria fuciformis* mycelium on turfgrass. (Courtesy J. E. Kaminski—© APS)

periods of moisture-saturated air favor rapid growth of and infection by *L. fuciformis*. Heavy dew, light rain, and fog are especially favorable for disease development when they occur at the optimal temperature of the fungal strain that is present. Red thread is most severe during periods when the grass is growing slowly because of low temperatures, lack of sunlight, drought, inadequate fertility, or excessive use of plant growth regulators. Disease severity is enhanced when potassium, phosphorus, calcium, and especially nitrogen are deficient.

Management

Adequate and complete fertility, as determined by soil or tissue tests, should be maintained. Applications of nitrogen help to reduce disease severity, but excessive rates should be avoided. Water-soluble (quick-release) sources of nitrogen, which reduce the severity of red thread more effectively than slow-release sources, should be applied. Soil pH should be in the range recommended for the turfgrass species grown (generally pH 6.0–6.5). Water should be applied as needed to prevent drought stress and should be applied thoroughly (deeply), as infrequently as possible, and early in the day. Light, frequent irrigation or applications in late afternoon and early evening, which can allow leaves to remain wet for a prolonged period, should be avoided.

Trees or shrubs can be selectively pruned or the landscape design can be arranged to increase sunlight penetration and air movement. Since the pathogen can enter leaves through wounds, sharp blades should be used for mowing to minimize wounding of leaf tips. Collecting turfgrass clippings during periods when red thread is active may reduce the number of sclerotia that are incorporated back into the turf.

Cultivars of the major turfgrass species vary in susceptibility to red thread. Several fungicides provide effective control of the disease. Local extension personnel may be consulted for a list of turfgrass cultivars and fungicides recommended in a specific geographic region.

Selected References

Cahill, J. V., Murray, J. J., O'Neill, N. R., and Dernoeden, P. H. 1983. Interrelationships between fertility and red thread fungal disease of turfgrass. Plant Dis. 67:1080-1083.

Kaplan, J. D., and Jackson, N. 1983. Red thread and pink patch diseases of turfgrasses. Plant Dis. 67:159-162.

Smith, J. D. 1954. Fungi and turfgrasses. IV. Corticium diseases. J. Sports Turf Res. Inst. 8:365-377.

Stalpers, J. A., and Loerakker, W. M. 1982. *Laetisaria* and *Limonomyces* species (Corticiaceae) causing pink diseases in turfgrasses. Can. J. Bot. 60:529-537.

Tredway, L. P., Soika, M. D., and Clarke, B. B. 2001. Red thread development in perennial ryegrass in response to nitrogen, phosphorus, and potassium fertilizer applications. Int. Turfgrass Soc. Res. J. 9:715-722.

Zhang, W., Hu, M., Liu, G., Gao, Z., Li, M., and Nan, Z. 2014. Investigation and characterization of red thread and pink patch on warm-season turfgrasses in Hainan Province, tropical China. Eur. J. Plant Pathol. 141:311-325.

(Prepared by R. W. Smiley, P. H. Dernoeden, and B. B. Clarke; revised by L. P. Tredway)

Rusts

Rust diseases are caused by multiple fungal species with unique optimal infection conditions and identifying characteristics. Rusts occur in all turfgrass species and are some of the most common turfgrass diseases observed in home lawns, sod production, and sports turf. Although they rarely result in turfgrass death, they can significantly stunt its growth, and they form an orange covering on the leaves that is aesthetically unpleasing and can stain shoes and clothes. Grasses growing under suboptimal conditions such as those caused by heat, drought, nutrient deficiencies, and compacted soils are most likely to be parasitized by rust fungi.

Symptoms and Signs

Early symptoms of rust are light yellow flecks on leaves or culms (Fig. 109). The yellowed areas of infected spots enlarge and eventually expand to cover much of the leaf surface (Fig. 110). Inside maturing lesions, the invading fungus pro-

Fig. 109. Uredinia of *Puccinia graminis* on leaves of *Festuca arundinacea*. (Courtesy L. P. Tredway—© APS)

Fig. 110. Uredinia of *Puccinia graminis* on leaves of *Poa pratensis*. (Courtesy L. P. Tredway—© APS)

Fig. 111. Stem rust of *Poa pratensis*, with uredinia of *Puccinia graminis*. (Courtesy L. P. Tredway—© APS)

Fig. 112. Cultivars of *Poa pratensis* with a wide range of susceptibility to stem rust. (Courtesy B. B. Clarke—© APS)

Fig. 113. Uredinia with teliospores of *Puccinia coronata* on *Lolium perenne.* (Courtesy R. W. Smiley—© APS)

duces cushions of spores, which enlarge to cause the epidermis to rupture. The resulting pustules are known as the uredinia (fruiting bodies) of the pathogen (Fig. 111). Spores produced within the uredinia, called "urediniospores," may be yellow, orange, brownish yellow, chestnut brown, or brick red, and they appear powdery in large numbers. These spores are easily distributed by wind and can move many miles before landing on susceptible plants. Mowing through a stand of heavily diseased turf can produce large plumes of urediniospores.

When rust is severe, areas of infected turf may appear thin and weak and irregular patches of red, brown, or yellow turf may form (Fig. 112). Heavily infected plants may wither and die from excessive loss of moisture and blockage of sunlight from rust-covered leaves. Juvenile turfgrass seedlings are more susceptible to rust than mature turfgrass plants. Teliospores are the overwintering spores produced by rust fungi (Fig. 113). They are normally black and are excellent identifying characteristics for differentiating among the various rust fungi when placed under a compound microscope.

Causal Agents

Rust fungi that occur on turfgrasses include many species of *Puccinia* and *Uromyces* and a single species of *Physopella.* The genera can be differentiated by their teliospore morphology with the use of a compound microscope. Teliospores of *Uromyces* and *Puccinia* spp. (Fig. 114) are one and two celled, respectively, and have pedicels (stalks). Teliospores of *Physopella* spp. have no pedicels and are catenulate (occurring in chains), with two or three spores in a chain.

Descriptions of the various rust fungi on turfgrasses are beyond the scope of this compendium. Descriptions and keys to identification of species of rust fungi are available in publications by G. B. Cummins and G. B. Cummins and Y. Hiratsuka. An abbreviated summary is presented in Table 3.

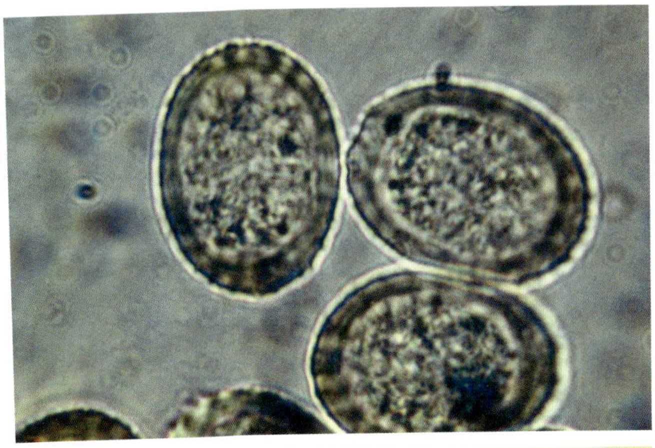

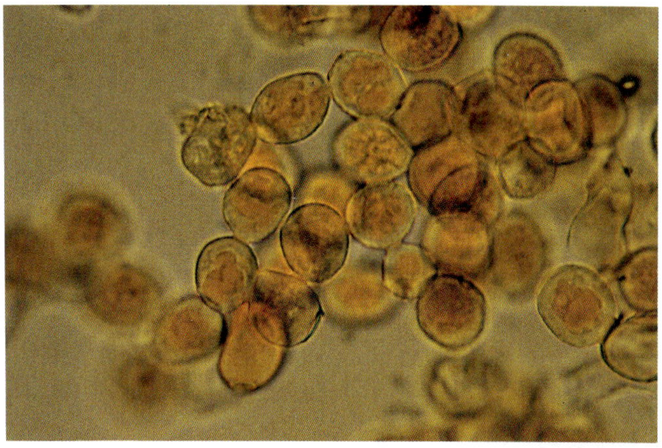

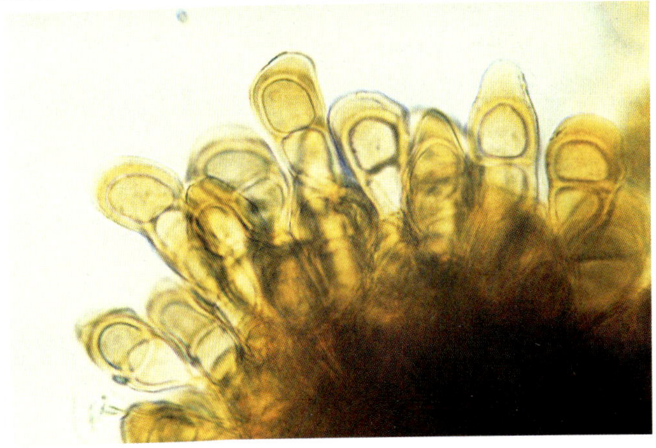

Fig. 114. Urediniospores (top) and teliospores (bottom) of *Puccinia graminis* (left) and *P. coronata* var. *coronata* (right). (Courtesy R. W. Smiley—© APS)

Rust fungi are initially grouped according to uredinial characteristics as follows: (1) the presence or absence of paraphyses (sterile, hairlike appendages within or at the periphery of the uredinium); (2) the ornamentation of the urediniospore wall (Fig. 114), which is either echinulate (spiny) or verrucose (having flat or rounded warts); and (3) the arrangement of germ pores in the urediniospore wall, either equatorial (aligned around the equator of the spore) or scattered (not in a single line at the equator).

The most common rusts of *Lolium* and *Poa* spp. in temperate climates can be readily distinguished into stem rust (caused by *Puccinia graminis*), stripe rust (caused by *P. striiformis*), crown rust (caused by *P. coronata*), and other leaf rusts (*P. recondita* and *P. brachypodii*). The stripe rust fungus has linear uredinia, whereas the others have rounded to oblong uredinia. Only one or a few species cause rust of *Axonopus, Buchloë, Cynodon, Puccinellia, Stenotaphrum,* and *Zoysia* spp. (Table 3). Stem and crown rusts were formerly thought to be specific to Kentucky bluegrass and perennial ryegrass, respectively, but research in which molecular techniques were used has demonstrated that stem rust and crown rust can occur on multiple turfgrass species. In addition, multiple rust diseases can be observed on the same turfgrass plant.

Physiologic forms (races) of rust fungi also occur. The races differ in their ability to attack cultivars within a grass genus or species, and certain cultivars such as Midnight Kentucky bluegrass are known to be highly susceptible to rust. Specific genes in rust fungi determine their virulence, and the interaction of these genes with specific host genes determines the infection type on a given cultivar. Infection types are used to distinguish races of rust fungi.

Rust fungi can produce as many as five different types of spores, but the urediniospore is the most important in the rust disease cycle on most turfgrasses. Urediniospores are usually round to oval, 15–40 μm long, echinulate (possessing fine spines), and colorless to yellow, orange, or cinnamon brown. Millions of urediniospores can be produced in just a square foot of rust-infected turfgrass, which results in dissemination of the fungus to new areas, thinning of the turfgrass stand, and staining of shoes and clothes.

Disease Cycle

Infected turfgrass foliage serves as the overwintering site for mycelia and urediniospores of rust fungi in areas with year-round growing seasons (Fig. 115). When the weather is conducive for spore germination or mycelial growth, new foliar

TABLE 3. Rust Fungi on Turfgrasses[a]

Rust Fungus	Host Genera	Distribution
Puccinia		
agrostidicola Tai	*Agrostis*	East Asia
brachypodii G. H. Otth var. *arrhenatheri* (Kleb.) Cummins & H. C. Greene	*Agropyron, Festuca, Poa, Agrostis, Phleum*	Worldwide
brachypodii G. H. Otth var. *poae-nemoralis* (G. H. Otth) Cummins & H. C. Greene	*Agrostis, Festuca, Lolium, Phleum, Poa*	Worldwide
chaetochloae Arthur	*Paspalum, Pennisetum*	North and South America
cockerelliana Bethel ex Arthur	*Festuca*	North America
coronata Corda var. *coronata*	*Agropyron, Agrostis, Festuca, Lolium, Paspalum, Phleum, Poa, Puccinellia*	Worldwide
coronata Corda var. *himalensis* Barclay	*Poa, Agrostis*	Worldwide
crandallii Pammel & H. H. Hume	*Festuca, Poa*	Western North America
cynodontis Lacr. ex Desm.	*Cynodon*	Worldwide
dolosa Arthur & Fromme	*Paspalum*	North and South America
dolosoides Cummins	*Paspalum*	Africa
festucae Plowr.	*Festuca*	Europe, Asia, Alaska
festucae-ovinae F. L. Tai	*Festuca*	China
graminis Pers.:Pers.	*Agropyron, Agrostis, Cynodon, Festuca, Lolium, Phleum, Poa*	Worldwide
graminis Pers.:Pers. subsp. *graminicola* Z. Urb.	*Festuca, Lolium, Phleum*	Worldwide
gymnothrichis P. Henn.	*Pennisetum*	Central and South America
hordei G. H. Otth	*Lolium*	Worldwide
kansensis Ellis & Barthol.	*Buchloë*	North America
levis (Sacc. & Bizz.) Magnus	*Axonopus, Paspalum*	North and South America
levis (Sacc. & Bizz.) Magnus var. *panici-sanguinalis* (Rangel) Ramachar & Cummins	*Paspalum, Pennisetum*	Worldwide
montanensis Ellis	*Agropyron, Lolium*	North America
paspalina Cummins	*Paspalum*	Africa, East Asia, Australia
pattersoniana Arthur	*Agropyron*	North America
piperi Ricker subsp. *scillae-rubrae* (Cruchet) Cummins	*Festuca*	Europe
poarum P. Nielsen	*Agrostis, Festuca, Phleum, Poa*	Worldwide
pygmaea Erikss.	*Agrostis, Festuca, Poa*	High elevations
recondita Roberge ex Desm.	*Agropyron, Agrostis, Festuca, Lolium, Poa, Cynodon*	Worldwide
sessilis W. G. Schneid.	*Festuca*	Northern Hemisphere
stenotaphricola J. Walker	*Pennisetum, Stenotaphrum*	Worldwide
striiformis Westend.	*Agropyron, Agrostis, Festuca, Lolium, Phleum, Poa*	Worldwide
subcentripora Arthur & Cummins	*Pennisetum*	Philippine Islands
substriata Ellis & Barthol.	*Paspalum, Pennisetum*	Worldwide
zoysia Dietel	*Zoysia*	East Asia, North America
Physopella compressa (Mains) Cummins & Ramachar	*Axonopus, Paspalum*	North and South America
Uromyces		
dactylidis G. H. Otth	*Festuca, Poa, Agrostis*	Worldwide
dactylidis G. H. Otth var. *poae* (Rabenh.) Cummins	*Agrostis, Poa*	Worldwide
fragilipes Tranzschel	*Agropyron, Agrostis, Festuca*	Northern Hemisphere
setariae-italicae Yoshino	*Pennisetum, Stenotaphrum, Paspalum*	Worldwide

[a] Prepared by R. W. Smiley, P. H. Dernoeden, and B. B. Clarke—© APS. Adapted from Cummins (1971). This reference contains a detailed key to the genera and species of rust fungi, arranged according to host genera.

infections are initiated and new uredinia are formed that produce new urediniospores. Urediniospores can be transported over long distances by wind, and these windblown spores from warm overwintering regions usually serve as the primary inoculum in colder regions where mycelium and urediniospores are not able to survive the winter. Once a urediniospore lands on a susceptible host, it can infect, form a uredinia, and produce thousands of new urediniospores in a matter of weeks when infection conditions are optimal. The urediniospore constitutes the repeating stage of rust fungi, and new cycles can produce exponentially larger numbers of urediniospores that spread to new areas and initiate new infections.

Most rust fungi also produce teliospores when plant foliage matures or growth begins to slow in the fall. Teliospores can overwinter in cold climates and then germinate to produce a third type of spore, the basidiospore. Basidiospores can become airborne, and if they land on a susceptible nongrass host, known as the alternate host, they can germinate and cause new infections. For example, the alternate host of stem rust is common barberry (*Berberis canadensis* and *B. vulgaris*), and when *P. graminis* basidiospores germinate on and infect barberry, they produce two more spore types, pycniospores and aeciospores. Aeciospores can infect only grasses, giving rise to urediniospores and thus completing the stem rust life cycle. Buckthorn is the alternate host of *P. coronata*.

Epidemiology

Optimal temperatures for growth and sporulation of rust fungi vary by species, but temperatures of 20–30°C (68–86°F) are generally optimal for stem rust development, while temperatures of 10–20°C (50–68°F) are optimal for crown and stripe rust development. Leaf moisture is required for rust spores to germinate and infect turfgrass leaves, but rust symptoms are typically most severe under dry conditions.

Rust diseases usually become most severe in grasses that are growing slowly under stressful conditions. Typical stresses include drought, nutrient deficiency, low mowing height, shade, and soil compaction. On cool-season turfgrasses in temperate climates, rust severity tends to be greatest in late summer and early fall when these stresses are most common.

Management

Optimal turfgrass growth conditions should be maintained to minimize rust development. This includes proper irrigation and fertilization to avoid water- and nutrient-related stresses that impede grass growth. Other cultural practices beneficial in

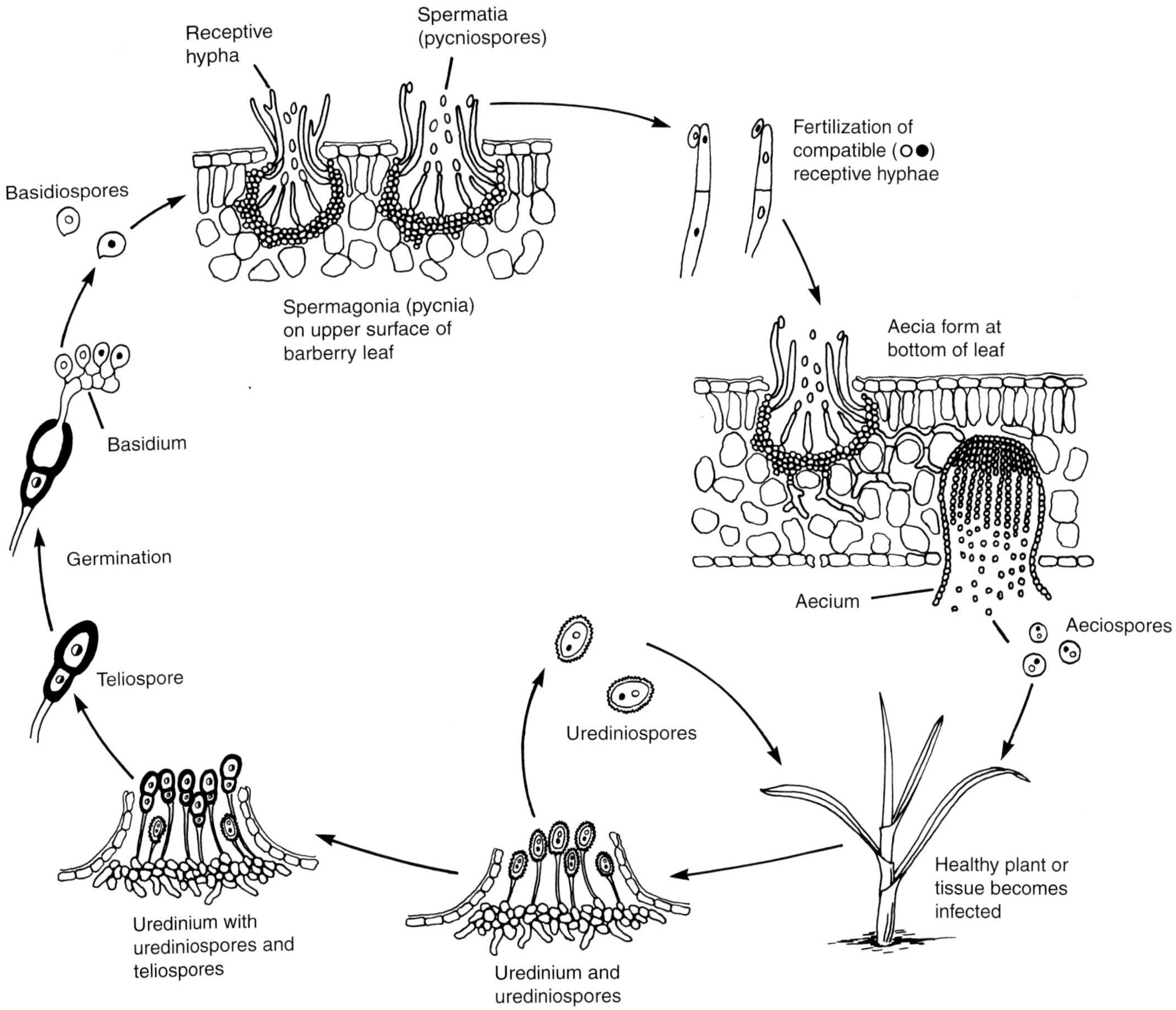

Fig. 115. Disease cycle of stem rust, caused by *Puccinia graminis*. (Courtesy C. R. Whittaker—© APS. Adapted from G. N. Agrios, 1978. Plant Pathology, 2nd ed., Academic Press, New York.)

reducing rust include aerification to alleviate soil compaction, traffic management to avoid excessive wear, adequate sunlight to encourage turf growth and leaf drying, and proper mowing height.

There are significant differences in rust resistance among turf species and cultivars (Fig. 112). Tall fescue and fine fescues are generally less susceptible to rust than perennial ryegrass and Kentucky bluegrass. Within Kentucky bluegrass, cultivars such as Midnight and those genetically related to it are more susceptible to rust than other bluegrass cultivars. Plant species mixtures or cultivar blends that include one or more resistant cultivars reduce overall rust severity in the turfgrass stand.

Fungicides are not typically recommended for control, since rusts do not usually kill the turf. Fungicides may be warranted in severe outbreaks, however. Products from the demethylation inhibitor class are generally the most effective for rust control.

Selected References

Agrios, G. N. 2005. Plant Pathology. 5th ed. Academic Press, San Diego, CA.

Beirn, L. A., Moy, M., Meyer, W. A., Clarke, B. B., and Crouch, J. A. 2011. Molecular analysis of turfgrass rusts reveals the widespread distribution of *Puccinia coronata* as a pathogen of Kentucky bluegrass in the United States. Plant Dis. 95:1547-1557.

Cummins, G. B. 1971. The Rust Fungi of Cereals, Grasses and Bamboos. Springer-Verlag, New York.

Cummins, G. B., and Hiratsuka, Y. 2003. Illustrated Genera of Rust Fungi. 3rd ed. American Phytopathological Society, St. Paul, MN.

Karakkat, B. B., Jackson, V. L., and Koch, P. L. 2018. Incidence and distribution of *Puccinia coronata* and *P. graminis* on turfgrass in the Midwestern United States. Plant Dis. 102:955-963.

Roderick, H. W., and Thomas, B. J. 1997. Infection of ryegrass by three rust fungi (*Puccinia coronata, P. graminis* and *P. loliina*) and some effects of temperature on the establishment of the disease and sporulation. Plant Pathol. 46:751-761.

(Prepared by P. L. Koch)

Southern Blight

Southern blight has been observed in *Agrostis, Cynodon, Festuca, Lolium, Poa,* and *Zoysia* spp. and in broadleaf turf (*Dichondra* spp.) in warm temperate climates throughout the world.

Symptoms

The initial symptoms usually are yellow or reddish brown, circular to crescent-shaped patches or rings up to 30 cm in diameter (Fig. 116). The grass may become chlorotic and thin. Weeds such as clover (*Trifolium* spp.) are also blighted. As the disease develops, a portion of the grass dies in the patch or at the periphery of a ring, usually with green grass remaining in the center, producing frog-eyes and arcs (Fig. 117). In *Agrostis* spp. and *Poa annua,* the disease initially appears as yellow rings, arcs, and patches, and the grass turns reddish brown as it dies. In *P. pratensis,* the rings are initially reddish brown or gray and fade to tan. The symptoms are similar to those of necrotic ring spot, but southern blight is more likely to appear during hot weather. Rings of dead grass may enlarge rapidly in hot, humid weather, growing as much as 20 cm per week. Patches up to 1 m in diameter are common, and some are as much as 3 m across. When moisture and humidity levels are high, white aerial mycelium may form on dying grass leaves at the periphery of the ring (Fig. 118). White sclerotia, 1–3 mm in diameter, develop from mycelium on dead grass or thatch in the ring. Sclerotia become dark brown when mature. Patches in *Agrostis* and *Poa* spp. are usually recolonized by these grasses during the autumn.

Fig. 116. Severe damage from southern blight on an *Agrostis stolonifera* golf green during the summer. (Courtesy Z. K. Punja—© APS)

Causal Agent

Athelia rolfsii produces abundant white mycelium and sclerotia, which initially are white and then turn brown (Fig. 119).

Fig. 117. Southern blight of *Festuca arundinacea.* (Courtesy M. Reiter—© APS)

Fig. 118. Southern blight of an *Agrostis stolonifera* putting green. (Courtesy A. Windham—© APS)

The sclerotia are round (1–3 mm in diameter), and, as revealed by dissection, each consists of a distinct medulla and cortex surrounded by a pigmented rind. Hyphae grow very rapidly, up to 2.5 cm per day. Primary hyphae are 4–9 μm in diameter and bear clamp connections at widely spaced septa (about 240 μm apart). Secondary hyphae are about 2 μm wide and usually have no clamp connections. *A. rolfsii* produces basidiospores, but their importance in southern blight of turfgrasses is unknown.

Disease Cycle

A. rolfsii produces abundant sclerotia, which serve as survival structures in soil and thatch. They germinate mostly in warm to hot weather (24°C [75°F]) when moisture is abundant. *A. rolfsii* is a facultative parasite, capable of existing as a saprophyte or as a pathogen, and attacks more than 500 plant species. It can move rapidly and indiscriminately through thatch or soil and into the plant canopy. The pathogen is easily spread from infected turf to surrounding areas, where it causes secondary infections, resulting in satellite rings around older patches.

Infected plants quickly lose their cellular integrity. Once the translocation of nutrients and water has been inhibited, tillers turn chlorotic or gray and then reddish brown as they die. Sclerotia are formed on dead grass and in thatch and may persist for long periods of time (Fig. 120).

Fig. 119. Sclerotia of *Athelia rolfsii* produced in a thatch layer of *Festuca arundinacea*. (Courtesy M. Reiter—© APS)

Fig. 120. Dark brown, mature sclerotia of *Athelia rolfsii* at the surface of a core from a putting green. (Courtesy R. W. Smiley—© APS)

Epidemiology

The growth of *A. rolfsii* is favored by hot, moisture-saturated conditions and an abundance of organic matter. The pathogen is most active when air temperatures are 30–35°C (86–95°F) and the relative humidity is above 90% for an extended period. Cool weather and poorly aerated or neutral to alkaline soils tend to restrict the growth of *A. rolfsii*. Germination of sclerotia increases after several days of dry conditions, and a period of high moisture following drought is particularly conducive to extensive plant damage. Sclerotia are formed most abundantly in the light when soil is moist, hot (above 30°C [86°F]), acidic (pH less than 7), and well aerated.

Management

Southern blight may be suppressed by reducing thatch accumulation through aerification or vertical mowing. Application of lime to increase soil pH may reduce the severity of southern blight but may also favor the development of other diseases such as summer patch. Southern blight is less apparent on vigorously growing, well-fertilized grass. Certain quinone outside inhibitor, demethylation inhibitor, and succinate dehydrogenase inhibitor fungicides are labeled for control of southern blight in turfgrasses.

Selected References

Endo, R. M., Ohr, H. D., and Wilbur, W. D. 1982. Sclerotium blight of cool-season turfgrasses in southern California and its control. Calif. Turfgrass Cult. 32:6.

Lucas, L. T. 1976. *Sclerotium rolfsii* on bentgrass in North Carolina. Plant Dis. Rep. 60:820-822.

Punja, Z. K. 1985. The biology, ecology, and control of *Sclerotium rolfsii*. Annu. Rev. Phytopathol. 23:97-127.

(Prepared by R. W. Smiley, P. H. Dernoeden, and B. B. Clarke; revised by M. Tomaso-Peterson)

Yellow Tuft (Downy Mildew)

Yellow tuft occurs in most turfgrass species. In *Stenotaphrum secundatum* and *Zoysia japonica,* the disease is also known as downy mildew.

Symptoms and Signs

The name "yellow tuft" is more descriptive than "downy mildew" because the disease does not always exhibit the same foliar symptoms (white growth) as those associated with downy mildew. Early symptoms include slightly stunted growth and slightly thickened or broadened leaf blades without any discoloration. When the disease is severe, small, yellow spots or patches, 0.6–10 cm in diameter, appear in the turf (Fig. 121). The spots are typically less than 2.5 cm across on close-cut *Agrostis* spp., *Poa* spp., and fine-leaf *Festuca* spp. but are larger in higher-cut *Lolium* and *Poa* spp. Each spot is a single, excessively tillered plant that contains a dense cluster of yellow shoots with shortened roots (Fig. 122). These symptomatic plants are easily detached from the turf.

Sporangia (Fig. 123) form a white, downy growth on leaf surfaces during cool, wet periods. Microscopic examination of leaves may reveal the presence of oospores, which are similar to those of *Pythium* spp. but much larger.

Symptoms are most prominent during the late spring and autumn and appear first in poorly drained areas. The yellow tuft symptom may persist throughout the summer in irrigated turf. Infected plants may be winter-killed or may wither and die during hot, dry periods, but often many infected plants survive stressful periods and appear healthy.

Symptoms in *S. secundatum* differ from those in cool-season grasses described above. Infected plants do not appear to be adversely affected but exhibit symptoms suggestive of viral diseases. Linear white streaks develop parallel to leaf veins, and the epidermis over the streaks is slightly raised. The pathogen produces white, downy growth on *S. secundatum* in wet weather.

In *Z. japonica,* the yellow tuft symptom is most prominent, but the disease may also cause a diffuse yellowing.

Causal Agent

Sclerophthora macrospora is a widely distributed, obligate parasite that causes downy mildew in a large number of grasses. It produces coenocytic mycelium of widely different diameters (3–6 μm), oospores 50–75 μm in diameter (larger than those of *Pythium* spp.), and ellipsoidal sporangia (about 40–60 × 70–100 μm), which release ovoid to pyriform zoospores (5–10 × 14–20 μm). *S. macrospora* resembles *Phytophthora* spp.

A similar pathogen that also causes downy mildew in some grasses is *Sclerospora graminicola*. These organisms are taxonomically classified within the kingdom Chromista (see Diseases and Their Causal Agents in the Introduction section).

Disease Cycle

S. macrospora mycelium spreads systemically into stems, sheaths, leaves, culms, and inflorescences. It can invade seed and become seedborne. *S. macrospora* persists as mycelium in intercellular spaces for indefinite periods. In moist weather, it produces sporangiophores (fruiting branches), which protrude through leaf stomata. Pearly white, pyriform sporangia are borne on sporangiophores protruding from stomata on wet leaves from late spring to autumn. The sporangia remain turgid only as long as the leaf surface is wet. When the leaves dry, the sporangia collapse and appear as a dirty white residue on leaves.

Sporangia germinate to release numerous zoospores, which swim actively in films of moisture on leaves and in surface water. Depending on the temperature, a zoospore encysts within a few minutes or hours of being released and then germinates to produce a hyphal strand, which can infect the juvenile meristem of another plant. Newly germinated seedlings are most susceptible to infection.

Oospores may be produced in infected leaves of perennial grasses, but they are more common and more numerous in annuals, such as *Digitaria* spp. Oospores can persist for several months or longer; and when conditions are suitable, they germinate to produce sporangia or germ tubes.

Epidemiology

S. macrospora requires free water for local movement of infectious spores. Therefore, the disease is associated mostly with low-lying, poorly drained areas. The pathogen lives in close harmony with the infected plant, and it grows best under conditions that are also optimal for the growth of the grass.

Management

Adequate drainage should be maintained to remove excess water from the surface during irrigation and rainfall. The turf should be managed in a manner that promotes active but not lush growth. Yellow-tufted plants on golf and bowling greens

Fig. 121. Yellow tuft in an *Agrostis stolonifera* putting green. (Courtesy L. P. Tredway—© APS)

Fig. 122. Dense clusters of chlorotic tillers characteristic of yellow tuft. (Courtesy P. H. Dernoeden—© APS)

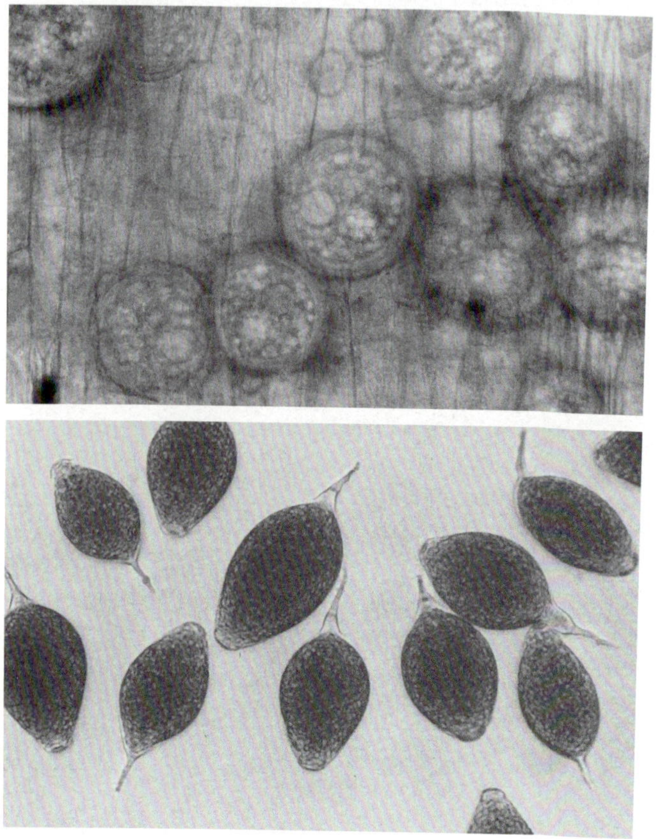

Fig. 123. Oospores of *Sclerophthora macrospora* in an *Agrostis stolonifera* leaf (top) and sporangia from *Poa pratensis* (bottom). (Courtesy N. Jackson—© APS)

can be detached by vertical cutting. No cultivars are known to be resistant to yellow tuft. Mefenoxam is the most effective fungicide when applied preventively. Fungicides may provide little or no curative control.

Selected References

Bruton, B. D., Toler, R. W., and Grisham, M. P. 1986. Prevention and curative control of downy mildew of St. Augustinegrass by metalaxyl. Plant Dis. 70:413-415.

Cheon, C. W., Han, J. J., Kim, D. S., and Kwak, Y. S. 2016. Morphological characteristics and occurrence of yellow tuft on zoysiagrass (*Zoysia japonica*) in cultivation fields. Weed Turfgrass Sci. 5:17-22.

Dernoeden, P. H., and Jackson, N. 1980. Infection and mycelial colonization of gramineous hosts by *Sclerophthora macrospora*. Phytopathology 70:1009-1013.

Dernoeden, P. H., and Jackson, N. 1980. Managing yellow tuft disease. J. Sports Turf Res. Inst. 56:9-17.

Jackson, N. 1980. Yellow tuft. Pages 135-137 in: Advances in Turfgrass Pathology. P. O. Larsen and B. G. Joyner, eds. Harcourt Brace Jovanovich, Duluth, MN.

Jones, B. L., and Amador, J. 1969. Downy mildew, a new disease of St. Augustine grass. Plant Dis. Rep. 53:852-854.

(Prepared by R. W. Smiley, P. H. Dernoeden, and B. B. Clarke; revised by J. P. Kerns and E. L. Butler)

Rhizoctonia and *Waitea* Diseases

Diseases caused by *Rhizoctonia* spp. are extremely important in turfgrass systems. Species of *Rhizoctonia* and related genera cause multiple turfgrass diseases including brown patch, large patch, brown ring patch, yellow patch, and leaf and sheath spot (Table 4). The pathogens that cause these diseases were previously classified as species of *Rhizoctonia,* but recent molecular evidence has differentiated some of them into the genus *Waitea*.

Rhizoctonia and *Waitea* spp. produce similar hyphal forms. They seldom produce sexual spores. Hyphae of *R. solani,* however, can break apart to form asexual spores called "arthroconidia." Young hyphae may have acute angular branching and few septa, with branch hyphae leaving the parent hypha at angles of about 45° (Fig. 124). Mature hyphal branches more commonly leave the parent hypha at angles of about 90°. Other characteristics used for diagnostic purposes to identify these pathogens include (1) branching near the distal septum of cells in young vegetative hyphae; (2) formation of septa near the origin of hyphal branches; (3) constriction of hyphal branches near their origin; (4) formation of dolipore septa; (5) absence of conidia, clamp connections, and rhizomorphs; and (6) formation of sclerotia.

TABLE 4. Diagnostic Features of Turfgrass Diseases Caused by *Rhizoctonia* and *Waitea* Species[a]

	Turfgrass Disease					
Character	**Brown Patch/ Large Patch**	**Brown Ring Patch**	**Leaf and Sheath Spot**	**Leaf and Sheath Spot**	**Basal Leaf Blight**	**Yellow Patch**
Fungal species	*Rhizoctonia solani*	*Waitea circinata*[b]	*Waitea oryzae*[b]	*Waitea zeae*[b]	*Waitea prodiga*[b]	*Rhizoctonia cerealis*
Hosts[c]	Many C3 (brown patch) and C4 (large patch) hosts	As, L, Pa, Pp, Pt	As, C, E, Fa, L, Pa, Pp, Pv, S, Z	As, C, E, Fa, L, Pa, Pp, Pv, S, Z	Pv[d]	As, C, Fa, L, Pa, Pp, Z
Symptoms	Circular to irregular, brown to straw-colored patches (C3 ≤ 1 m wide, C4 up to 8 m wide); extensive thinning; border often with black smoke ring (C3), orange firing (C4)	Circular to irregular yellow rings, often scalloped, that may fade to light yellow or reddish brown; infected area 5–100 cm wide	C3 hosts: similar to brown patch; C4 hosts: chlorotic spots (5–10 cm wide) turning into frog-eye patches (10–30 cm wide) with bronze to tan rings	C3 hosts: similar to brown patch; C4 hosts: chlorotic spots (5–10 cm wide) turning into frog-eye patches (10–30 cm wide) with bronze to tan rings	Irregularly shaped patches of blighted foliage (15–150 × 5–90 cm); progressing on older, basal leaves to form diffuse necrotic patches in turf swards	Circular to irregular yellow rings that may fade to brown; infected area 5–100 cm in diameter
Optimal temperature range for development of symptoms	21–32°C (70–90°F)	18–29°C (64–84°F)	28–36°C (82–97°F)	28–36°C (82–97°F)	25–35°C (77–95°F)	10–20°C (50–68°F)
Colony						
Color[e]	White to buff to gray, maturing to dark brown	Pinkish white, maturing to dark brown	White to cream buff to salmon pink	White to pink, maturing to pinkish orange	Yellow to pink	White to buff
Nuclei per cell	3–10	3–9	3–8	3–10	6–7	2
Hyphal diameter (μm)	4–13	4–10	5–11	5–11	4.5–5.5	2–6
Sclerotia						
Diameter (mm)	>1–10	1–5	1–3	0.5–1	3–>7	0.25–2
Shape	Irregular to spherical	Irregular	Irregular	Subspherical	Irregular to spherical	Irregular
Color	White, pale buff to dark brown	Pinkish orange to dark brown	Salmon pink to orange	Orange to dark orange	Salmon to yellow to light brown	Light to dark to brown

[a] Prepared by B. B. Clarke—© APS.

[b] Although these species of *Waitea* have morphological characteristics similar to those of *Rhizoctonia* spp., they do not share the same descent as *Ceratobasidium* and *Rhizoctonia* spp., despite having been assigned *Rhizoctonia* anamorphs in the past.

[c] As = *Agrostis stolonifera*, C = *Cynodon dactylon*, C3 = cool-season turfgrasses, C4 = warm-season turfgrasses, E = *Eremochloa ophiuroides*, Fa = *Festuca arundinacea*, L = *Lolium perenne*, Pa = *Poa annua*, Pp = *Poa pratensis*, Pt = *Poa trivialis*, Pv = *Paspalum vaginatum*, S = *Stenotaphrum secundatum*, and Z = *Zoysia* spp.

[d] Pathogenicity confirmed on Pv, A, C, and Pt in artificial inoculation but has been reported only on Pv in nature.

[e] Color of mycelial colony on potato dextrose agar at 25°C (77°F).

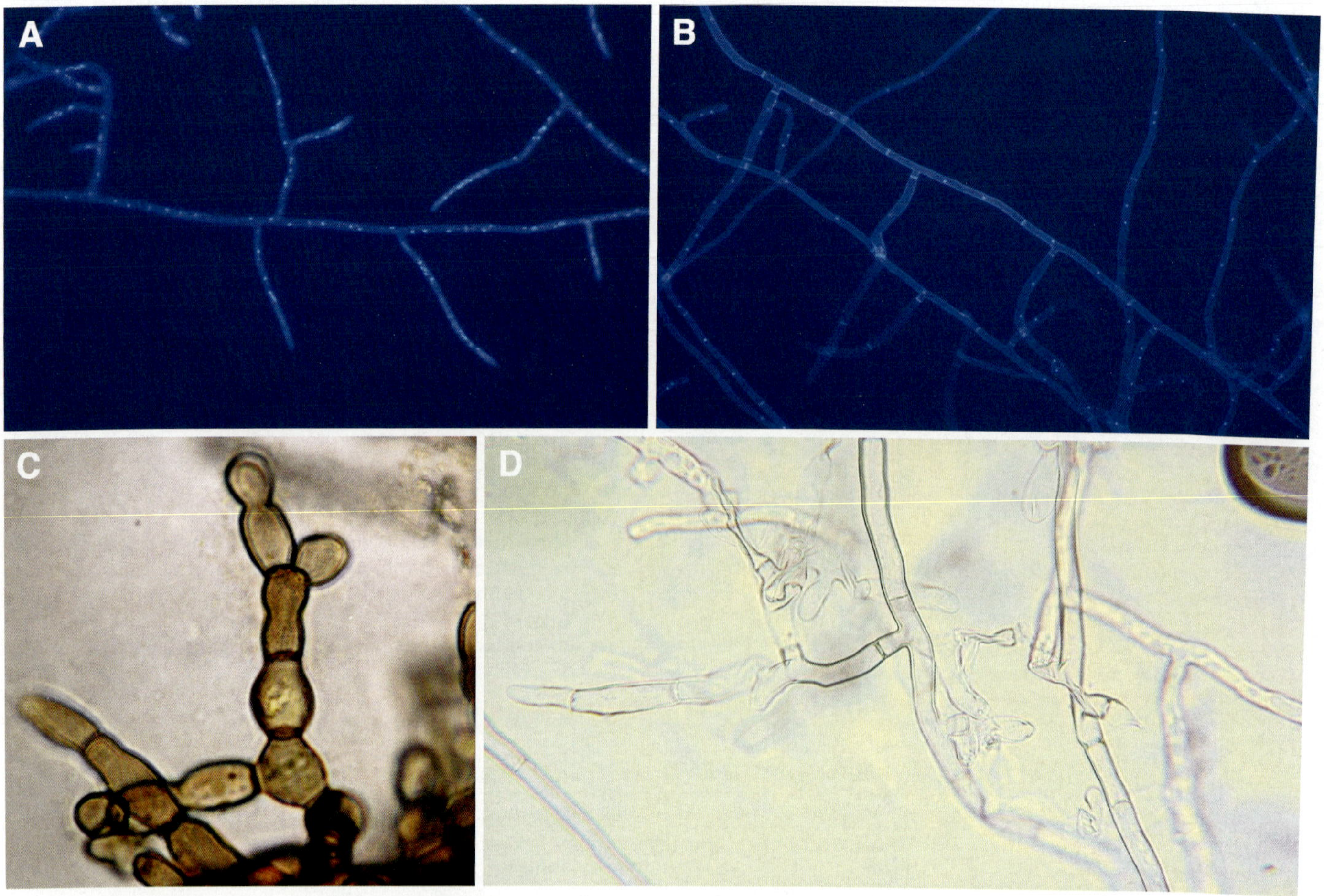

Fig. 124. Hyphal morphology of *Rhizoctonia* and *Ceratobasidium* spp. **A,** Binucleate hyphae of *C. cereale* and **B,** multinucleate hyphae of *R. solani* as revealed by DAPI staining; **C,** hyphae of *R. solani*; and **D,** hyphae of *R. solani* with initiation of monilioid cells. (A–C, Courtesy L. P. Tredway—© APS; and D, courtesy M. Tomaso-Peterson—© APS)

Although the mycelial branching pattern described is used by experienced diagnosticians as a preliminary diagnostic character, it is easy for the inexperienced microscopist to confuse it with branching patterns of other fungi.

Sclerotia (formerly referred to as "bulbils") may be produced in culture and vary in color and morphology. They may be white, buff, salmon, brown, red, red brown, orange, or black and range in size from 0.1 mm in diameter to aggregates of 5 mm or more in diameter. Sclerotia are initially white or nearly white. When mature, sclerotia of *W. zeae* are normally orange; *W. oryzae,* salmon; *R. cerealis,* light to dark brown; and *R. solani,* dark brown.

Distinguishing among species and strains of *Rhizoctonia* and *Waitea* is difficult. Molecular techniques and immunochemical assays have been developed to aid in the identification of *Rhizoctonia* species and anastomosis groups and subgroups. Mycelial cells of *R. solani* and *W. zeae* and immature cells of *W. oryzae* are multinucleate; mature cells of *W. oryzae* usually contain four nuclei per cell. Mycelial cells of *R. cerealis* are binucleate (Fig. 124). When grown on laboratory media, *R. solani* mycelium is usually brown, *R. cerealis* is white to buff, and *W. zeae* is white to salmon or pink. *W. oryzae* is similar to *W. zeae* in the color of its mycelium in culture. Primary hyphae are about 2–6 μm in diameter in *R. cerealis* and 5–11 μm in diameter in the other species.

Basidiospores (Fig. 125) are occasionally produced on a hymenium on the surface of infected leaf and sheath tissue, but their importance in turfgrasses is unknown.

Within species of *Rhizoctonia* and *Waitea,* groups of isolates have an affinity for hyphal fusion (anastomosis) and are thus assigned to anastomosis groups (AGs). Most isolates of *R. solani* from turfgrasses are in AG 1, AG 2-2, AG 4, AG 5, and AG 6, with AG 1, AG 2-2, and AG 4 the most common.

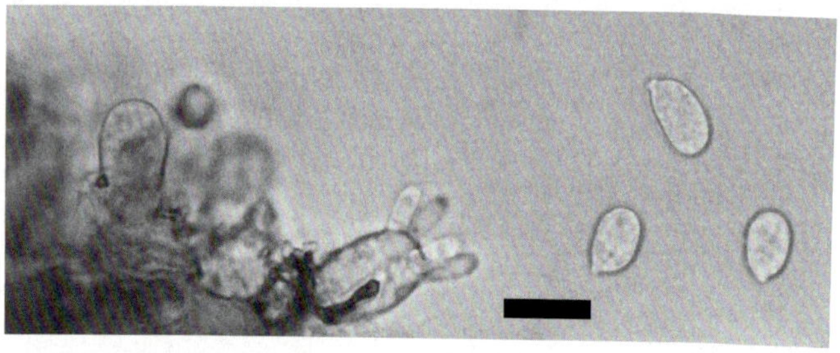

Fig. 125. Basidium and basidiospores of *Rhizoctonia solani* (bar = 10 μm). (Courtesy J. A. Frank)

Anastomosis groups of *R. solani* are genetically distinct and exhibit degrees of host specialization in turfgrasses and other crops. Anastomosis groups can be further subdivided according to the frequency of anastomosis or host origin. There is only one anastomosis group in *W. oryzae* (WAG-O), one in *W. zeae* (WAG-Z), and one in *R. cerealis* (AG-D).

Selected References

Burpee, L. L., and Martin, S. B. 1992. Biology of *Rhizoctonia* species associated with turfgrasses. Plant Dis. 76:112-117.

Stalpers, J., Redhead, S., Rossman, A. Y., Crouch, J. A., Cubeta, M. A., Kirschner, R., Langer, G. J., Larsson, K., Marvanová, L., May, T. W., Norvell, L., Oberwinkler, F., Roberts, P., Rajchenberg, M., Seifert, K. A., and Thorn, G. 2021. Competing sexual-asexual generic names in *Agaricomycotina* (*Basidiomycota*) with recommendations for use. IMA Fungus 12:22. doi:10.1186/s43008-021-00061-3

(Prepared by L. P. Tredway)

Brown Patch

Brown patch is an important disease of all cool-season grasses. It is a foliar disease that does not affect crowns, stolons, or rhizomes of turfgrass plants. Brown patch was first described in the late 1920s and has plagued turfgrass culture ever since. Even moderate outbreaks of the disease can result in thin, poor-quality turf that can lead to weed, moss, or algae encroachment. The disease is most severe on *Lolium perenne* and *Festuca arundinacea* in the Mid-Atlantic and southeastern United States. However, *Poa annua* and *Agrostis stolonifera* planted for golf course turf are also prone to brown patch.

Symptoms and Signs

The symptoms of brown patch vary according to mowing height. Where mowing height is greater than 2.5 cm, brown patch appears as roughly circular patches that are brown, tan, or yellow and range from 15 cm to more than 90 cm in diameter (Figs. 126 and 127). Symptomatic leaves typically remain upright with lesions that are irregular in shape and tan with dark brown borders (Fig. 128). When the leaves are wet or humidity is high, mycelium may be seen growing among affected leaves (Fig. 129). In close-cut turfgrasses (2.5 cm or less), brown patch develops in roughly circular patches, ranging from 7.6 cm to more than 90 cm in diameter, that are brown or orange (Figs. 130 and 131). Distinct foliar lesions are not visible and mycelium is typically not present, but a black or dark gray ring, called a "smoke ring," may surround the brown patches (Fig. 132). The smoke ring is evidence of active disease development and is present only when the turfgrass leaves are wet or humidity is near 100%.

Causal Agents

Brown patch traditionally has been associated with *Rhizoctonia solani,* but research demonstrates that multiple

Fig. 126. Brown patch symptoms in a *Festuca arundinacea* landscape. (Courtesy L. P. Tredway—© APS)

Fig. 127. Brown patch of *Festuca arundinacea* with smoke rings evident along the outer edges. (Courtesy P. H. Dernoeden—© APS)

Fig. 128. Foliar lesions characteristic of brown patch on *Festuca arundinacea*. (Courtesy L. P. Tredway—© APS)

Fig. 129. *Rhizoctonia solani* hyphae spreading from leaf to leaf. (Courtesy L. P. Tredway—© APS)

Rhizoctonia-like fungi can be isolated from typical brown patch symptoms, especially in tall fescue stands. *R. cerealis* and *Waitea zeae* (formerly known as *R. zeae*) can also be isolated, particularly during June, July, and August. Since these species induce similar symptoms, visual diagnosis is not a reliable tool to determine the causal agent.

Disease Cycle

Rhizoctonia spp. are capable of saprophytic growth in soil and survive periods unfavorable for growth in the form of sclerotia, monilioid cells, or thick-walled mycelia in plant debris (Fig. 124). These structures are highly tolerant of environmental extremes and pesticides. Sclerotia of *R. solani, W. zeae,* and *W. oryzae* can germinate over a wide range of temperatures (8–40°C [46–104°F]), with an optimum of 31°C (88°F). The optimal temperatures for infection and disease development, however, vary for different species and strains, ranging from 21 to 32°C (70 to 90°F).

When sclerotia germinate or hyphae emanate from an infected plant, the fungus spreads radially in the upper soil or thatch to form a roughly circular colony. Under humid, warm conditions, hyphae grow onto and infect moist turfgrass leaves and sheaths. Infected plant tissues become water soaked and dark. When exposed to sunlight or drying wind, infected spots or entire leaves turn tan or brown, shrivel, and die. Sclerotia may be formed on or in infected tissues and are liberated into thatch when tissues decompose.

Fig. 130. Smoke ring symptom of brown patch of *Agrostis capillaris* on a golf fairway. (Courtesy J. E. Kaminski—© APS)

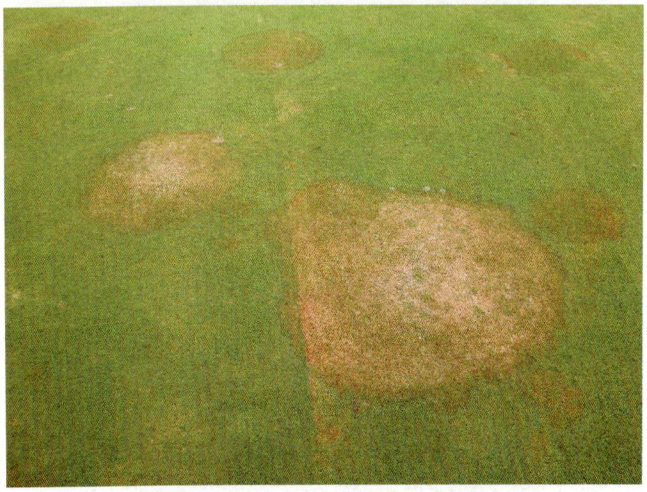

Fig. 131. Symptoms of brown patch on *Agrostis stolonifera* evident in the absence of dew or leaf wetness. (Courtesy L. P. Tredway—© APS)

Fig. 132. Brown patch, with smoke ring, in *Agrostis stolonifera*. (Courtesy B. B. Clarke—© APS)

Epidemiology

Disease caused by *R. solani* in cool-season grasses is often most severe when leaf wetness or humidity is high (above 95%) and night temperatures are above 20°C (68°F). Large areas of turf can succumb to blight quickly in hot, rainy, or humid weather. Prolonged periods of leaf wetness (10–12 hours) are necessary for extensive brown patch damage. Smaller patches or diffuse thinning of the stand generally result from infections that occur under warm, wet conditions. Models have been developed to assist in predicting outbreaks of disease caused by *R. solani* in cool-season turfgrasses.

Very humid or cloudy weather, poorly drained soil or thatch, and poor air circulation increase the duration of leaf wetness (dew) during the early morning. Dense, highly fertilized (greater than 6 lb N/ft^2/year), and frequently watered turf generally can promote infection by *Rhizoctonia* spp. In tall fescue, brown patch is most severe when growth is limited by low fertility and heat stress. Inadequate levels of phosphorous and potassium also contribute to injury from this disease.

Management

When environmental conditions favor brown patch, an abundance of readily available nitrogen may enhance the severity of the disease. On the other hand, insufficient nitrogen that limits turf growth may also intensify the disease. Slow-release sources of nitrogen are less likely to enhance brown patch, but applications of quick-release nitrogen sources at less than 0.25 lb N/1,000 ft^2 (12 kg N/ha) do not affect the severity of brown patch. Early-morning irrigation reduces brown patch severity, whereas evening irrigation intensifies the disease.

Good surface and subsurface water drainage are necessary to reduce excessive soil wetness and humidity in the turf canopy. Leaf surface wetness can be reduced by manual removal of dew in the early morning, which can be accomplished by mowing, dragging a garden hose or rope laterally across the grass, sweeping a long, flexible pole across the grass, or applying wetting agents. Additional drying may be promoted by fans or selective pruning of trees and shrubs to increase the circulation of air and the penetration of light to the turf canopy. Similarly, removal of hedges or barriers that impede the movement of air across the turf is helpful. Soil surfactants, products designed to promote infiltration of water into the soil, can limit brown patch development as well.

If thatch accumulates to a depth of more than 1.3 cm, it should be mechanically removed when the turf is actively growing.

Numerous fungicides are available for control of brown patch and other diseases caused by *Rhizoctonia* spp. In particular, the quinone outside inhibitor fungicides provide exceptional preventive control, and the sensitivity of *R. solani* to these fungicides has not decreased over time. Curative fungicide applications can be effective as long as cool-season turfgrass growth is not limited by heat stress. Weather-based forecasting models are available for predicting brown patch outbreaks in cool-season grasses.

Selected References

Amaradasa, B. S., Lakshman, D., Horvath, B. J., and Amundsen, K. L. 2014. Development of SCAR markers and UP-PCR cross hybridization method for specific detection of four major subgroups of *Rhizoctonia* from infected turfgrasses. Mycologia 106:163-172.

Burpee, L. L. 1985. Interactions among mowing height, nitrogen fertility, and cultivar affect the severity of Rhizoctonia blight of tall fescue. Plant Dis. 79:721-726.

Burpee, L. L., and Martin, S. B. 1992. Biology of *Rhizoctonia* species associated with turfgrasses. Plant Dis. 76:112-117.

Butler, E. L., Galle, G. H., and Kerns, J. P. 2019. Influence of nitrogen rate and timing, fungicide application method, and simulated rainfall after fungicide application on brown patch severity in tall fescue. Crop, Forage Turfgrass Manage. 5. doi:10.2134/cftm2019.03.0018

Fidanza, M. A., and Dernoeden, P. H. 1997. A review of brown patch forecasting, pathogen detection, and management strategies for turfgrasses. Int. Turfgrass Soc. Res. J. 8:863-874.

Fidanza, M. A., Dernoeden, P. H., and Grybauskas, A. P. 1996. Development and field validation of a brown patch warning model for perennial ryegrass turf. Phytopathology 86:385-390.

Green, D. E., and Burpee, L. L. 1997. Modelling the effects of host resistance on the progress of Rhizoctonia blight in tall fescue (*Festuca arundinacea*). J. Int. Turf. Res. Soc. 8:883-892.

Green, D. E., Burpee, L. L., and Stevenson, K. L. 1999. Components of resistance to *Rhizoctonia solani* associated with two tall fescue cultivars. Plant Dis. 83:834-838.

Green, D. E., Burpee, L. L., and Stevenson, K. L. 1999. Integrated effects of host resistance and fungicide concentration on the progress of Rhizoctonia blight in tall fescue turf. Crop Prot. 18:131-138.

Gross, M. K., Santini, J. B., Tikhonova, I., and Latin, R. 1998. The influence of temperature and leaf wetness duration on infection of perennial ryegrass by *Rhizoctonia solani*. Plant Dis. 82:1012-1016.

Koehler, A. A., and Shew, H. D. 2017. Seasonal dynamics and fungicide sensitivity of organisms causing brown patch of tall fescue in North Carolina. Mycologia 109:667-675.

Martin, S. B., and Lucas, L. T. 1984. Characterization and pathogenicity of *Rhizoctonia* spp. and binucleate Rhizoctonia-like fungi from turfgrasses in North Carolina. Phytopathology 74:170-175.

Parmeter, J. R., ed. 1970. *Rhizoctonia solani,* Biology and Pathology. University of California Press, Berkeley.

Schumann, G. L., Clarke, B. B., Rowley, L. V., and Burpee, L. L. 1994. Use of environmental parameters and immunoassays to predict Rhizoctonia blight and schedule fungicide applications on creeping bentgrass. Crop Prot. 13:211-218.

Sneh, B., Burpee, L., and Ogoshi, A. 1991. Identification of *Rhizoctonia* Species. American Phytopathological Society, St. Paul, MN.

Zhang, M., and Dernoeden, P. H. 1997. *Rhizoctonia solani* anastomosis groups and other fungi associated with brown patch-affected turfgrasses in Maryland. Int. Turfgrass Soc. Res. J. 8:959-969.

(Prepared by J. P. Kerns and E. L. Butler)

Brown Ring Patch

Brown ring patch, sometimes referred to as "Waitea patch," is an emerging disease of cool-season turfgrasses. It was first identified on *Agrostis stolonifera* in Japan in 2005 and has become a widespread disease of *Poa annua* putting greens in the United States since 2007. It has also been reported on *P. trivialis* and *Lolium perenne* used for overseeding in the southern United States; on *A. stolonifera* and *P. pratensis* in China; and on *A. stolonifera* in British Columbia, Canada, the U.S. state of Arizona, and Spain.

Symptoms and Signs

Brown ring patch is most common on putting greens, but it can occur on turf maintained at a higher cut (e.g., collars, fairways, and roughs). On *P. annua* and *P. trivialis*, symptoms typically appear in the spring as thin, yellow rings or crescents (semicircles) (Fig. 133). Infected leaves may retain their yellow color or become water soaked and fade to light brown to reddish brown with age (Fig. 134). Patches range from 5–10 cm to 1 m in diameter. Rings are circular to irregular and often appear "scalloped" (Fig. 135), a diagnostic feature that can be

Fig. 133. Typical symptoms of brown ring patch on a *Poa annua* golf green. (Courtesy P. Gradoville—© APS)

Fig. 134. Brown ring patch of *Poa annua* on a golf green. Note the lack of symptoms in aeration holes (lower left). (Courtesy S. J. McDonald—© APS)

Fig. 135. Brown ring patch of *Poa annua* on a golf green. (Courtesy J. E. Kaminski—© APS)

used in the field to differentiate brown ring patch from yellow patch (a similar disease caused by *Rhizoctonia cerealis* that occurs during cool weather). Symptoms can also be confused with those of fairy ring, summer patch, or southern blight. In severe cases, rings become sunken or depressed as the thatch is degraded, which can reduce the playability of putting greens (Figs. 136 and 137). Abundant white mycelium can be seen on the foliage, thatch, or upper soil layer when the turf is incubated in high humidity (Fig. 138). Outbreaks often appear in the spring, disappear in the summer, and then reappear in the autumn. Although the pathogen can infect both *A. stolonifera* and *P. annua,* typically only one host exhibits symptoms of brown ring patch at a time (Fig. 139).

On *A. stolonifera,* the disease appears as circular to irregularly shaped rings of tan to yellow brown turf. Affected leaves and stems become water soaked and fade to tan or brown as the disease progresses. Patches are 10–50 cm in diameter. Turf in the centers of patches often recovers and turns green.

Causal Agent

Waitea circinata is the causal agent of brown ring patch. In the past, *W. circinata* was referred to by five different varietal names: var. *agrostis,* var. *circinata,* var. *oryzae,* var. *prodigus,* and var. *zeae*. However, these varieties have now been elevated to the taxonomic rank of species: *W. agrostidis, W. oryzae, W. prodiga,* and *W. zeae. W. circinata* can be distinguished from *W. oryzae* and *W. zeae,* which cause leaf and sheath spot disease on turf, by comparing differences in hyphal and bulbil morphology (Table 4) as well as by molecular techniques and immunochemical assays.

Although *W. circinata* was not confirmed as the incitant of brown ring patch until 2005, it likely has been associated with turfgrasses for a long time and may be a more widespread pathogen of grasses than previously recognized. *W. circinata* was isolated from turf in California in the mid-1980s and has been identified on *A. stolonifera* putting greens in Japan since 1994. It has also been frequently isolated from agricultural soils since the 1980s. Recent sequencing of the intergenic spacer (ITS) region of *W. circinata* isolates collected from *P. annua* from throughout the United States has confirmed at least 17 unique ITS variants unrelated to geographic origin. This suggests that the pathogen is not a new arrival and that the recent appearance of brown ring patch on golf courses throughout the world may result from changes in management practices (reductions in fertility, mowing height, and soil water on putting greens since the 1980s to increase ball roll), fungicide use (the discontinuation of heavy metal-based products), and/or environmental conditions. The disease also appears to be more prevalent on sand-based putting greens.

In culture, colonies are floccose to appressed and white to buff and frequently fade to dark brown with age. Hyphal diam-

Fig. 136. Sunken ring symptom on a *Poa annua* golf green resulting from degradation of thatch by *Waitea circinata.* (Courtesy S. J. McDonald—© APS)

Fig. 137. Cross section of a soil profile from a *Poa annua* golf green with degraded yellow thatch (left) caused by *Waitea circinata.* (Courtesy M. Wolpoff—© APS)

Fig. 138. Water-soaked plant tissue and abundant mycelium of *Waitea circinata* from a *Poa annua* golf green sample incubated overnight in a humidity chamber. (Courtesy J. E. Kaminski—© APS)

Fig. 139. Brown ring patch of *Poa annua* on a golf green. Note the lack of symptoms on *Agrostis stolonifera* sod used to repair a hydraulic leak (blue green strip in center). (Courtesy S. J. McDonald—© APS)

eter is 4–10 μm, and there are three to nine nuclei per vegetative cell. Mycelium is septate with right-angle branches that are constricted at the base, similar to that of other *Rhizoctonia* species, and lacks clamp connections. Sclerotia are irregular in shape, are both submerged and appear on the agar surface, and 1–5 mm in diameter. They are initially white to orange and either remain orange or turn brown to dark brown with age. Sclerotia are rarely seen in the field but develop readily within 3–14 days on potato dextrose agar or in the thatch layer when the turf is incubated in high humidity (Fig. 140). The sexual state is not common, but basidiospores can be produced on the surface of V8 juice agar. Basidia are 8–13 × 6–9 μm and form on a hymenial layer. They have four sterigmata that are 2–5 μm long. Basidiospores are 6–10 × 4–5 μm. The importance of basidiospores in the disease cycle of brown ring patch is unknown.

Disease Cycle

W. circinata can survive unfavorable conditions as sclerotia or mycelium in foliage, crowns, roots, and thatch (Fig. 140). Sclerotia are tolerant of environmental extremes and can germinate over a wide range of temperatures. The fungus is believed to be transported in infected plant debris by equipment, people, animals, water, or wind. *W. circinata* is a pathogen primarily of leaves and stems. It has been isolated from roots close to the surface, but its role as a pathogen of roots is unknown. When conditions favor disease (warm, humid weather), sclerotia germinate and mycelium colonizes leaves and leaf sheaths. Infected tissue becomes chlorotic and water soaked and may turn tan or dark brown as the plant dies. Mycelia can grow at 10–35°C (50–95°F), but the optimal range is 25–30°C (77–86°F). The pathogen spreads laterally through the foliage, crown, thatch, and upper soil to form circular to irregularly shaped patches. Sclerotia form in the crown and thatch under warm, humid conditions (Fig. 141).

Fig. 140. Sclerotia and mycelium of *Waitea circinata* in thatch and leaves of a *Poa annua* golf green sample incubated in a humidity chamber. (Courtesy F. P. Wong—© APS)

Fig. 141. Mycelium and mycelial mats of *Waitea circinata* (precursors to the formation of sclerotia) in thatch and leaves of a *Poa annua* golf green sample incubated in a humidity chamber. (Courtesy S. J. McDonald—© APS)

Epidemiology

Symptoms typically occur in the spring and autumn but may also be present during the summer months if conditions are conducive for infection. Brown ring patch is favored by warm, cloudy, wet weather. On *P. annua* and *P. trivialis,* symptoms have been observed over a wide range of air temperatures (15–35°C [59–95°F]), but most reports of the disease have been associated with maximum daily air temperatures of 18–29°C (64–84°F). The disease has been observed on *A. stolonifera* over a similar temperature range; however, it has been reported at temperatures as low as 5°C (41°F). High humidity, poorly drained soils, and poor air circulation that increase the duration of leaf wetness can enhance the severity of disease outbreaks. Thatch often decomposes faster in the rings, resulting in a sunken appearance (Fig. 136). The pathogen infects primarily leaves, but in severe cases, when conditions are ideal for disease development, it can infect stems, crowns, and roots, causing plant death. Symptoms may disappear in the summer during hot weather only to reappear in the autumn with the return of warm, humid conditions.

Management

Unlike many turfgrass diseases caused by *Rhizoctonia* and *Waitea* species, the severity of brown ring patch is enhanced by low fertility. Adequate nitrogen levels reduce disease severity and encourage recovery. Water-soluble, quick-release forms of nitrogen (ammonium sulfate, calcium nitrate, and urea) should be applied to suppress brown ring patch and enhance fungicide effectiveness. Turf should be fertilized to avoid deficiencies of nitrogen, phosphorus, and potassium, and good thatch and organic matter management (aerification, vertical mowing, and sand topdressing) should be practiced. Dew and guttation fluids should be removed from leaf surfaces by mowing, rolling, or poling greens early in the morning to reduce the duration of leaf wetness. Good air circulation can be promoted by pruning or removing trees, shrubs, and other barriers to air flow. Application of growth regulators at high rates or short intervals may overregulate turf and should be avoided because this practice has been shown to enhance symptom expression and delay recovery when nitrogen fertility is low.

Fungicides such as azoxystrobin, chlorothalonil, metconazole, flutolanil, fludioxonil, polyoxin-D, propiconazole, tebuconazole, and triticonazole, applied alone or in tank mixtures, can provide effective control of brown ring patch. Similar to *W. zeae, W. circinata* appears to be naturally resistant to thiophanate-methyl. Preventive applications of fungicides are recommended where the disease is a chronic problem. Multiple applications may be required, and symptoms may persist for 2–3 weeks when fungicides are applied on a curative basis. Adequate spray volume and pressure should be maintained to ensure that fungicides penetrate the bases of the plants, and the addition of a surfactant to enhance fungicide efficacy should be considered. Posttreatment irrigation (3.8 mm) may enhance the efficacy of some fungicides (e.g., triticonazole). A turfgrass extension specialist from a local university may be consulted for current fungicide recommendations.

Selected References

Chen, C. M., de la Cerda, K. A., Kaminski, J. E., Douhan, G. W., and Wong, F. P. 2009. Geographic distribution and rDNA-ITS region sequence diversity of *Waitea circinata* var. *circinata* isolated from annual bluegrass in the United States. Plant Dis. 93:906-911.

Chen, C. M., Douhan, G. W., and Wong, F. P. 2007. First report of *Waitea circinata* var. *circinata* causing brown ring patch on *Poa trivialis* in California. Plant Dis. 91:1687.

de la Cerda, K. A., Douhan, G. W., and Wong, F. P. 2007. Discovery and characterization of *Waitea circinata* var. *circinata* affecting annual bluegrass from the western United States. Plant Dis. 91:791-797.

de la Cerda, K., Hsiang, T., and Joshi, V. 2010. First report of *Waitea circinata* from turfgrass in British Columbia, Canada. Plant Dis. 94:277

Gómez de Barreda, D., De Luca, V., Ramón-Albalat, A., León, M., and Armengol, J. 2019. First report of dollar spot caused by *Clarireedia jacksonii* and brown ring patch caused by *Waitea circinata* var. *circinata* on *Agrostis stolonifera* in Spain. Plant Dis. 103:1771.

Graf-Grachet, N., and Walker, N. R. 2016. First report of brown ring patch caused by *Waitea circinata* on perennial ryegrass (*Lolium perenne*) in Oklahoma. Plant Dis. 100:855.

Hyder, N., Chen, C. M., Towers, G., and Wong, F. P. 2011. First report of brown ring patch caused by *Waitea circinata* var. *circinata* on creeping bentgrass in Arizona. Plant Dis. 95:492.

Leiner, R. H., and Carling, D. E. 1994. Characterization of *Waitea circinata* (*Rhizoctonia*) isolated from agricultural soils in Alaska. Plant Dis. 78:385-388.

McDonald, S. J., Grala, R., and Clarke, B. B. 2014. Reducing brown ring patch severity on *Poa annua* greens. Golf Course Manage. 82:86-94.

McDonald, S. J., Settle, D., Stowell, L., Chen, C. M., and Wong, F. P. 2009. Chemical control of brown ring patch. Golf Course Manage. 77:74-80.

Ni, X. X., Li, B. T., Cai, M., and Liu, X. L. 2012. First report of brown ring patch caused by *Waitea circinata* var. *circinata* on *Agrostis stolonifera* and *Poa pratensis* in China. Plant Dis. 96:1821.

Pennycook, S. R. 1989. Plant Diseases Recorded in New Zealand, vol. 3. Plant Disease Division, Department of Scientific and Industrial Research, Auckland, New Zealand.

Toda, T., Mushika, T., Hayakawa, T., Tanaka, A., Tani, T., and Hyakumachi, M. 2005. Brown ring patch: A new disease on bentgrass caused by *Waitea circinata* var. *circinata*. Plant Dis. 89:536-542.

Wong, F. P., Chen, C., and Stowell, L. 2009. Effects of nitrogen and Primo MAXX on brown ring patch development. Golf Course Manage. 77:117-121.

(Prepared by B. B. Clarke and J. A. Crouch)

Large Patch

Large patch, caused by *Rhizoctonia solani,* was formerly termed "zoysia patch" because of its association with *Zoysia* spp. Disease symptoms also occur and may be severe on other warm-season turfgrass species including *Buchloë dactyloides, Cynodon dactylon* and *Cynodon* hybrids, *Eremochloa ophiuroides, Pennisetum clandestinum, Stenotaphrum secundatum,* and *Paspalum vaginatum.*

Symptoms and Signs

Large patch symptoms occur under cool conditions when warm-season grass metabolism is slowed and plants may be either going into or coming out of winter dormancy in autumn or spring. Patches can enlarge to a diameter of 8 m or more if left unchecked (Fig. 142). During active infection periods in the fall and spring, affected tillers at patch margins may turn yellow, orange, or reddish brown; the color is particularly noticeable in the early morning hours when dew is present (Figs. 143 and 144). In autumn, symptoms appear as rings or patches in which affected tillers are interspersed among living tillers and may remain until the grass becomes dormant (Fig. 145). In spring, symptoms become apparent 2–8 weeks after the beginning of green-up as patchy areas with fewer living tillers and a reduced rate of leaf growth (Fig. 146). Living tillers tend to spread through affected patches, and symptoms may slowly disappear during the summer depending on the recuperative potential of the turfgrass species. Patches typically appear in the same locations in successive years.

Leaf lesions are uncommon, but tip dieback can occur as the result of sheath infection. Roots are discolored but do not appear rotted. Water-soaked, reddish brown or black lesions appear on lower sheath tissue (Figs. 147 and 148) and later de-

Fig. 142. Large patch, caused by *Rhizoctonia solani,* in *Stenotaphrum secundatum.* (Courtesy T. K. Danneberger—© APS)

Fig. 143. Orange firing at the margins of large patch on *Zoysia japonica* in spring. (Courtesy G. L. Miller, Jr.—© APS)

Fig. 144. Large patch symptoms on *Eremochloa ophiuroides.* (Courtesy L. P. Tredway—© APS)

velop straw brown interiors surrounded by dark brown or purple borders. Infections also commonly occur on stolons or basal leaf sheaths, resulting in a basal rot rather than a leaf blight. Affected shoots can be easily pulled from their sheaths or points of attachment where infection has occurred.

Causal Agent

R. solani anastomosis group (AG) 2-2 LP causes large patch on warm-season turfgrasses. Anastomosis groups of *R. solani* are separated by their affinity for hyphal fusion; different groups are genetically distinct and exhibit degrees of host specialization in turfgrass (Table 4). These intraspecific groups infect different hosts at different times and differ genetically in culture morphology, growth rate, and optimal growth temperature. For example, *R. solani* AG 2-2 LP in culture has the slowest growth rate compared with intraspecific groups IIIB and IV and grows optimally at a lower temperature (23°C [73°F]).

On infected turfgrass sheath bases and stolons, *R. solani* AG 2-2 LP produces a dark brown, pigmented network of mycelium. Like those of other *Rhizoctonia* spp., the hyphae of *R. solani* AG 2-2 LP isolates are 5–17 µm in diameter without clamp connections, are separated by dolipore septa, produce monilioid cells, and have constricted right-angle branching near the septa. On potato dextrose agar, *R. solani* AG 2-2 LP cultures are dark brown with sparse aerial mycelium and irregular zonation. Irregular clusters of mycelia are produced, but sclerotia typically do not develop.

Disease Cycle

Rhizoctonia spp. survive periods unfavorable for growth in the form of sclerotia, monilioid cells, or thick-walled mycelia in plant debris. Isolates of *R. solani* AG 2-2 LP have been recovered over all seasons in zoysiagrass with a history of disease, most frequently from sheath tissues during disease development. Isolates have also been recovered from healthy zoysiagrass plants and rhizosphere soil, indicating that *R. solani* AG 2-2 LP may be ubiquitous in the phyllosphere and a capable soil saprophyte. Little information exists on root infection by *R. solani* AG 2-2 LP on warm-season turfgrasses.

R. solani most often infects through direct penetration of plant tissue, but entry through wounds, lenticels, and stomata has also been observed. *R. solani* AG 2-2 LP likely infects primarily through direct penetration of leaf sheaths and stem

Fig. 145. Large patch symptoms on a *Zoysia japonica* field in autumn. (Courtesy G. L. Miller, Jr.—© APS)

Fig. 146. Large patch, caused by *Rhizoctonia solani,* on a *Cynodon dactylon* athletic field. (Courtesy R. Webb—© APS)

Fig. 147. Water-soaked lesions on *Stenotaphrum secundatum* leaf sheaths characteristic of large patch development. (Courtesy M. Tomaso-Peterson—© APS)

Fig. 148. Lesions, a typical symptom of large patch, on leaf sheaths of *Stenotaphrum secundatum.* (Courtesy M. Tomaso-Peterson—© APS)

bases, since natural openings or wounds are sparse in this area. Hyphae accumulate along leaf sheaths and stem bases, leading to the formation of infection cushions with penetration pegs that exert hydrostatic pressure and penetrate the plant cells. Numerous secondary metabolites, including cutinase, endopolygalacturonase, and endopectinylase, are produced by *R. solani,* a necrotrophic pathogen that actively macerates tissue during the infection process.

Epidemiology

Although the disease may occur at any time during persistent rainfall, large patch symptoms are rare in summer when warm-season grasses are growing vigorously. Temperatures in spring and fall, particularly 20–25°C (68–77°F), are optimal for pathogen infection and sheath blighting in *Zoysia* spp. Patches form quickly in this temperature range, especially when combined with high relative humidity (above 90%), sustained cloud cover, abundant precipitation, or frequent irrigation. Extensive damage is most likely to occur on poorly drained soils or in areas of poor air circulation. When conditions are ideal for disease development, patches may coalesce to form large, irregularly shaped areas of blighted turf.

Management

Nitrogen fertilization during the spring and fall when large patch is actively developing does not increase disease severity on zoysiagrass. In the spring, a small application of nitrogen fertilizer (0.5 lb/1,000 ft^2 or 24 kg/ha), regardless of nitrogen source, may stimulate zoysiagrass growth and reduce disease severity. Nitrogen impact on the severity of large patch on other warm-season turfgrasses is unknown, as is the impact of slow-release or organic fertilizer types applied during the spring and fall infection period. Nitrogen source and application rate during the summer period do not influence large patch severity. Application of other nutrients (e.g., phosphorous or potassium) should be guided by soil test results.

Effective surface and subsurface water drainage are necessary to reduce large patch symptoms and remove excessive soil wetness and humidity in the turf canopy. Leaf surface wetness can be reduced by manual removal of dew in the early morning by mowing, dragging a garden hose or rope laterally across the grass, sweeping a long, flexible pole across the grass, or applying wetting agents. Additional drying may be promoted by selective pruning of trees and shrubs to increase the circulation of air and the penetration of light through the turf canopy. Similarly, the removal of hedges or barriers that impede the movement of air across the turf is helpful.

Large patch is more severe at low mowing heights (1.2–2.5 cm), so mowing heights should be raised above this level, if possible. If thatch accumulates to a depth of more than 1.2 cm, it should be mechanically removed. Aerification and verticutting should be performed when warm-season grasses are actively growing in summer and the potential for large patch infection through wounding is low.

Currently, cultivars of warm-season turfgrasses with resistance or improved tolerance to large patch are not commercially available.

Several fungicides are available for large patch control; quinone outside inhibitor, demethylation inhibitor, and succinate dehydrogenase inhibitor active ingredients are generally most effective. In high amenity stands, two to three applications may be required in the fall and spring to completely prevent disease occurrence. If a single application is planned on zoysiagrass, an early spring application when 5-day soil temperature averages 10°C (50°F) is more effective in preventing large patch outbreaks than a single application made the previous fall or a later spring application. Increased spray rate volumes of 2–4 gallons of water per 1,000 ft^2 (748–1,496 L/ha) enhance control efficacy by depositing fungicide spray on the leaf sheath and stem bases where most infections occur.

Selected References

Aoyagi, T., Kageyama, K., and Hyakumachi, M. 1998. Characterization and survival of *Rhizoctonia solani* AG2-2 LP associated with large patch disease of zoysia grass. Plant Dis. 8:857-863.

Benelli, J. J., Horvath, B. J., Womac, A. R., Ownley, B. H., Windham, A. S., and Sorochan, J. C. 2018. Large patch (*Rhizoctonia solani* AG2-2 LP) severity on Japanese lawngrass (*Zoysia japonica*) influenced by fungicide and application target site. Crop Prot. 106:23-28.

Burpee, L. L., and Martin, S. B. 1992. Biology of *Rhizoctonia* species associated with turfgrasses. Plant Dis. 76:112-117.

Green, D. E., II, Fry, J. D., Pair, J. C., and Tisserat, N. A. 1993. Influence of management practices on *Rhizoctonia* large patch disease in zoysiagrass. HortScience 29:186-188.

Haygood, R. A., and Martin, S. B. 1990. Characterization and pathogenicity of species of *Rhizoctonia* associated with centipedegrass and St. Augustinegrass in South Carolina. Plant Dis. 74:510-514.

Hurd, B., and Grisham, M. P. 1983. *Rhizoctonia* spp. associated with brown patch of Saint Augustinegrass. Phytopathology 73:1661-1665.

Hyakumuchi, M., Mushika, T., Ogiso, Y., Toda, T., Kageyama, K., and Tsuge, T. 1998. Characterization of a new cultural type (LP) of *Rhizoctonia solani* AG2-2 isolated from warm-season turfgrasses, and its genetic differentiation from other cultural types. Plant Pathol. 47:1-9.

Koehler, J. F., and Miller, G. L. 2019. Impact of nitrogen application timing, nitrogen source, and a single fungicide application on large patch of zoysiagrass. Crop Sci. 59:1285-1293.

Miller, G. L., Earlywine, D. T., Braun, R., Fry, J. D., and Kennelly, M. M. 2016. Influence of nitrogen source and application timing on large patch of zoysiagrass. Crop, Forage Turfgrass Manage. 2. doi:10.2134/cftm2015.0189

Obasa, K., Fry, J., St. John, R., and Kennelly, M. 2013. Effect of cultivation and timing of nitrogen fertilization on large patch disease of zoysiagrass. Plant Dis. 97:1075-1081.

Sneh, B., Burpee, L., and Ogoshi, A. 1991. Identification of *Rhizoctonia* Species. American Phytopathological Society, St. Paul, MN.

Tredway, L. P., and Burpee, L. L. 2001. Rhizoctonia diseases of turfgrass. Plant Health Instructor. doi:10.1094/PHI-I-2001-1109-01

(Prepared by G. L. Miller, Jr.)

Leaf and Sheath Spot

"Leaf and sheath spot" is a term broadly used to describe symptoms in turfgrass caused by *Waitea zeae* and *W. oryzae.* Although *W. oryzae* has been associated with leaf and sheath spot symptoms, *W. zeae* is more commonly isolated from symptomatic turfgrass. Leaf and sheath spot commonly occurs in cool-season grasses, especially during extreme heat, but it rarely causes widespread damage. On warm-season turfgrasses, particularly hybrid *Cynodon* spp. cultured as putting greens, this disease can be devastating. The disease is also known as "mini-ring," because stand symptoms typically develop as small rings on putting greens consisting of *Cynodon* spp. *W. zeae* has been documented as a pathogen of nearly all economically significant warm-season turfgrasses including *C. dactylon* and hybrids, *Eremochloa ophiuroides, Stenotaphrum secundatum,* and *Paspalum vaginatum.*

Symptoms and Signs

Warm-season grasses. Leaf and sheath spot symptoms in *Cynodon* spp. are distinctly different from those of large patch, caused by *Rhizoctonia solani.* Initially, the disease may appear as small, chlorotic or tan spots 5–10 cm in diameter (Fig. 149). Early in the disease cycle, leaf and sheath spot can be confused with other diseases such as take-all root rot or root-knot nematode damage. As the disease progresses, frog-eye patches 10–30 cm in diameter are produced (Figs. 150 and 151). The outer ring of the patch appears bronze, tan, or bleached, and the center is generally unaffected. Diseased leaf blades of *Cynodon* spp. do not have lesions but are often entirely blighted. The symptoms described above are most often associated with the disease on hybrid *Cynodon* golf course putting greens, but the disease can occur at higher cutting heights as well.

Leaf and sheath spot symptoms in *Eremochloa* and *Stenotaphrum* spp. consist of both foliar lesions and basal sheath rot. These symptoms may be difficult to discern from lesions characteristic of large patch caused by *R. solani*. Symptoms of leaf and sheath spot in *Paspalum* spp. are not well described but generally present as chlorosis and necrosis of lower leaf blades in the absence of foliar lesions. In addition to *W. zeae* and *W. oryzae*, *W. prodiga* appears to play some role in the development of leaf and sheath spot in *Paspalum* spp.

Cool-season grasses. In cool-season turf, primarily *Agrostis stolonifera* and *Festuca arundinacea*, symptoms of leaf and sheath spot can be similar to those of brown patch (*R. solani*), and diagnosis requires laboratory isolation of the causal fungus. In *A. stolonifera*, *W. zeae* may incite rings or patches that are yellow, orange, brown, or gray (Fig. 152). A smoke ring typical of brown patch does not form. In *Lolium* and *Festuca* spp., symptoms are indistinguishable from those caused by *R. solani*.

Signs of leaf and sheath spot are rarely visible to the unaided eye. However, white to cream-colored mycelium has been observed in cool-season turfgrasses when environmental conditions are favorable. With the use of a dissecting microscope, mycelia deep in the turfgrass canopy or thatch layer are often visible. Although the disease is called leaf and sheath spot, most of the pathogen activity in *Cynodon* spp. is observed on stolons and rhizomes (Fig. 153). Careful examination of

Fig. 149. Early symptoms of leaf and sheath spot (mini-ring) on a hybrid *Cynodon* putting green. (Courtesy L. A. Dant—© APS)

Fig. 150. Preliminary symptoms of leaf and sheath spot (mini-ring) on a hybrid *Cynodon* putting green. (Courtesy L. P. Tredway—© APS)

Fig. 151. Advanced frog-eye symptoms of leaf and sheath spot (mini-ring) on a hybrid *Cynodon* putting green. (Courtesy L. A. Dant—© APS)

Fig. 152. Symptoms of leaf and sheath spot on an *Agrostis stolonifera* putting green. (Courtesy J. P. Kerns—© APS)

Fig. 153. Hyphae of *Waitea zeae* in the rhizome tissue of a *Cynodon* sp. exhibiting symptoms of leaf and sheath spot. (Courtesy J. P. Kerns—© APS)

numerous turfgrass tissues is required to make a confident diagnosis of this disease. Sclerotia are rarely found in turfgrass grown under field conditions. When a diseased *Cynodon* sp. is incubated or grown under controlled conditions optimal for *W. zeae,* sclerotia have been observed on leaves, sheaths, and roots and are often embedded near the crown of the plant.

Causal Agents

W. zeae, W. oryzae, and *W. prodiga* are documented causes of leaf and sheath spot in turfgrasses. These fungi rarely produce asexual or sexual spores naturally. It appears that reproduction is nearly always via vegetative hyphae. Hyphae are 3–10 µm in diameter and multinucleate and form right-angle branches. Hyphal branches are commonly constricted, and a septum is present near the origin of each branch. Microscopic characteristics of *W. zeae, W. oryzae,* and *W. prodiga* hyphae are similar; therefore, macroscopic characteristics of isolate cultures and sclerotia are often used to identify isolates to the species level (Table 4). Mycelia of *W. zeae* and *W. oryzae* tend to grow at or below the agar surface and are white to buff to salmon colored on potato dextrose agar. In comparison, *W. prodiga* commonly forms yellow to pink aerial mycelium on potato dextrose agar. Mature sclerotia of *W. zeae* are orange to dark orange or dark brown, 0.5–1.0 mm in diameter, and uniformly spherical. In contrast, sclerotia of *W. oryzae* are not uniformly spherical but somewhat angular, vary in diameter from less than 1 mm to greater than 3 mm, and are salmon colored. Like those of *W. oryzae,* mature sclerotia of *W. prodiga* vary widely in size and shape and range in color from yellow to salmon to brown.

Disease Cycle

The disease cycle of leaf and sheath spot has not been extensively researched. The causal fungi likely survive unfavorable environmental conditions as sclerotia embedded in plant tissue or in the soil and as mycelia within plant tissue or organic matter. Germinating sclerotia or surviving mycelia are most likely the sources of initial infection. In growth chamber studies, the initial infection in *Cynodon* spp. appears to take place in the root or at the soil surface near the crown. This coincides with diagnostic observation of diseased *Cynodon* spp. turf, where *W. zeae* mycelium is generally present in the soil–thatch interface or deep within the plant canopy.

Epidemiology

Leaf and sheath spot in cool- and warm-season turfgrasses most often occurs at temperatures of 28–36°C (82–97°F) (optimal 30–32°C [86–90°F]). This temperature range is above optimal for cool-season turf and is generally when symptoms become visible. In warm-season turfgrass, the ideal temperature for leaf and sheath spot development is similar to the temperature range favorable for turfgrass growth; consequently, factors other than high temperatures often trigger the onset of disease symptoms. Mechanical damage from agronomic practices, low light conditions, and the onset of cool conditions appear to encourage symptom expression, especially in *Cynodon* spp. It is unclear whether leaf wetness or humidity plays a significant role in the development of leaf and sheath spot.

Management

Leaf and sheath spot control in cool-season turfgrass has not been extensively studied. It is assumed that cultural controls similar to those employed to combat brown patch, such as reducing turfgrass stress, avoiding excessive nitrogen fertilization, managing thatch, and eliminating excess moisture during favorable environmental conditions, also suppress leaf and sheath spot. Leaf and sheath spot is generally considered a minor disease in cool-season turfgrass; therefore, little is known about resistant cultivars. Limited research has shown that endophytic fungi may suppress *W. zeae* in *F. arundinacea*. Fungicides applied to control brown patch also appear to suppress leaf and sheath spot, with the exception of thiophanate-methyl, which is not active against these pathogens.

In warm-season turf, specifically *Cynodon* spp., leaf and sheath spot has proved difficult to manage. Symptoms often appear when the turf is not actively growing; therefore, recovery is hampered even when the causal fungi are suppressed. For this reason, preventive strategies must be employed. Fungicides should be applied when environmental conditions favor the causal fungi prior to the appearance of leaf and sheath spot symptoms. Balanced fertility, including adequate nitrogen to promote active growth, is required to manage leaf and sheath spot in *Cynodon* spp. Research has shown that ammonium sulfate increases leaf and sheath spot severity, although the mechanisms are not well understood. Alternative nitrogen sources, such as urea, should be utilized when leaf and sheath spot is problematic. Cultivation practices and abiotic stress should be minimized when turfgrass is not actively growing.

Selected References

Burpee, L. L., and Martin, S. B. 1992. Biology of *Rhizoctonia* species associated with turfgrasses. Plant Dis. 76:112-117.

Dant, L. A., Martin, S. B., Kerns, J. P., and McCarty, L. B. 2021. Nitrogen source impacts *Rhizoctonia* leaf and sheath spot severity in ultradwarf bermudagrass. Int. Turfgrass Soc. Res. J. doi:10.1002/its2.15

Elliott, M. L. 1999. Comparison of *Rhizoctonia zeae* isolates from Florida and Ohio turfgrasses. HortScience 34:298-300.

Haygood, R. A., and Martin, S. B. 1990. Characterization and pathogenicity of species of *Rhizoctonia* associated with centipedegrass and St. Augustinegrass in South Carolina. Plant Dis. 74:510-514.

Kammerer, S. J. 2011. *Waitea* and *Rhizoctonia* pathogens of seashore paspalum—The role of salinity in disease expression and characterization of a new *Waitea circinata* variety causing basal leaf blight. Ph.D. diss. University of Florida, Gainesville.

Kammerer, S. J., Burpee, L. L., and Harmon, P. F. 2011. Identification of a new *Waitea circinata* variety causing basal leaf blight of seashore paspalum. Plant Dis. 95:515-522.

Kerns, J., Yan, Y., and Butler, E. 2017. Characterization, pathogenicity, and in vitro sensitivity of *Rhizoctonia* spp. associated with leaf and sheath spot of bermudagrass putting greens in North Carolina and Alabama. Int. Turfgrass Soc. Res. J. 13:203-212.

Koehler, A. M., and Shew, D. H. 2017. Seasonal dynamics and fungicide sensitivity of organisms causing brown patch of tall fescue in North Carolina. Mycologia 109:667-675.

Martin, B. 2009. Identification, pathogenicity, and control of *Rhizoctonia* leaf and sheath spot of bermudagrass putting greens. USGA Turfgrass Environ. Res. 22:1-10.

Martin, S. B., Campbell, C. L., and Lucas, L. T. 1983. Horizontal distribution and characterization of *Rhizoctonia* spp. in tall fescue turf. Phytopathology 73:1064-1068.

Martin, S. B., and Lucas, L. T. 1983. Pathogenicity of *Rhizoctonia zeae* on tall fescue and other turfgrasses. Plant Dis. 67:676-678.

Martin, S. B., and Lucas, L. T. 1984. Characterization and pathogenicity of *Rhizoctonia* spp. and binucleate Rhizoctonia-like fungi from turfgrasses in North Carolina. Phytopathology 74:170-175.

Pańka, D., West, C. P., Guerber, C. A., and Richardson, M. D. 2013. Susceptibility of tall fescue to *Rhizoctonia zeae* infection as affected by endophyte symbiosis. Ann. Appl. Biol. 163:257-268.

Stalpers, J., Redhead, S., Rossman, A. Y., Crouch, J. A., Cubeta, M. A., Kirschner, R., Langer, G. J., Larsson, K., Marvanová, L., May, T. W., Norvell, L., Oberwinkler, F., Roberts, P., Rajchenberg, M., Seifert, K. A., and Thorn, G. 2021. Competing sexual-asexual generic names in *Agaricomycotina* (*Basidiomycota*) with recommendations for use. IMA Fungus 12. doi:10.1186/s43008-021-00061-3

Tomaso-Peterson, M., and Trevathan, L. E. 2007. Characterization of *Rhizoctonia*-like fungi isolated from agronomic crops and turfgrasses in Mississippi. Plant Dis. 91:260-265.

(Prepared by L. A. Dant and S. B. Martin)

Yellow Patch

Yellow patch, also called "cool-weather brown patch" or "winter brown patch," is a cool- and wet-weather disease incited by *Rhizoctonia cerealis*. The disease is most common on cool-season grasses including *Agrostis, Festuca, Lolium,* and *Poa* spp., but it is also observed in *Cynodon* and *Zoysia* spp.

Symptoms

Symptoms of yellow patch appear as yellow or reddish brown rings, arcs, and patches (Figs. 154–156). Affected patches are circular or irregularly shaped and up 1 m in diameter. Multiple rings often intersect to create larger patches with scalloped edges. With the advent of lower temperatures, which suppress the pathogen, the outer edges of patches or rings develop a brown, winter dormant color. Conversely, symptoms may dissipate rapidly during warm, sunny, dry periods. Foliar lesions may occur, but they often are absent. The disease tends to be most severe on wet, shaded, and poorly drained sites.

Causal Agent

R. cerealis produces hyphae with right-angle branches that are constricted at the bases, similar to other *Rhizoctonia* spp. Hyphae are binucleate (Fig. 124) and narrower than those of *R. solani* (2–6 µm in diameter compared with 4–13 µm) (Table 4). Mature colonies on laboratory media are typically white to buff with light to dark brown sclerotia.

Many other binucleate *Rhizoctonia* spp. are associated with turfgrasses and are either nonpathogenic or weak pathogens. Some may even have some utility as biological control agents. *R. cerealis* isolates form a single anastomosis group identified as CAG1 or AG-D.

The sexual stage of *R. cerealis* has been observed in culture and identified as *Ceratobasidium cereale,* but it is not frequently observed in nature.

Disease Cycle

The disease cycle of *R. cerealis* has not been studied in detail, but it is assumed to be similar to that of *R. solani.* Under conditions favorable for disease development, *R. cerealis* mycelia apparently spread over leaf sheaths and onto the lower, older leaves. The pathogen behaves mostly as a foliar blighter, but under ideal conditions, it can invade the crown and cause plant death.

Epidemiology

Yellow patch is favored by extended periods of overcast, rainy, cool weather. Optimal temperatures for disease development are 10–18°C (50–64°F). Disease development is interrupted when temperatures drop below 5°C (41°F) or exceed 24°C (75°F). Extensive foliar blighting occurs in rainy weather, but symptoms may disappear abruptly with a shift to sunny, drier weather.

Fig. 154. Mild symptoms of yellow patch, caused by *Rhizoctonia cerealis,* in *Agrostis stolonifera.* (Courtesy M. A. Fidanza—© APS)

Fig. 155. Severe symptoms of yellow patch, caused by *Rhizoctonia cerealis,* in *Agrostis stolonifera.* (Courtesy P. M. Halisky—© APS)

Fig. 156. Reddish brown rings of yellow patch, caused by *Rhizoctonia cerealis,* in *Agrostis stolonifera.* (Courtesy P. H. Dernoeden—© APS)

Management

Yellow patch severity can be reduced by improving surface and subsurface drainage. Removing trees or other obstructions to increase sunlight penetration and air movement is also likely beneficial. Preventive fungicide applications for snow mold control in late autumn have been shown to suppress yellow patch in the spring. Curative applications of fungicides may have little impact on the expression of symptoms as long as weather conditions are conducive to disease development. In general, fungicide-treated turf recovers more rapidly than untreated turf with the advent of warm weather in the spring. Application of nitrogen fertilizer helps speed recovery once winter dormancy breaks and the turf resumes active growth.

Selected References

Burpee, L. L. 1980. *Rhizoctonia cerealis* causes yellow patch of turfgrasses. Plant Dis. 64:1114-1116.

Burpee, L. L., and Martin, S. B. 1992. Biology of *Rhizoctonia* species associated with turfgrasses. Plant Dis. 76:112-117.

Hamada, M. S., Yin, Y., Chen, H., and Ma, Z. 2011. The escalating threat of *Rhizoctonia cerealis,* the causal agent of sharp eyespot in wheat. Pest Manage. Sci. 67:1411-1419.

Martin, S. B. 1987. Rapid tentative identification of *Rhizoctonia* spp. associated with diseased turfgrasses. Plant Dis. 71:47-49.
Martin, S. B., and Lucas, L. T. 1984. Characterization and pathogenicity of *Rhizoctonia* spp. and binucleate Rhizoctonia-like fungi from turfgrasses in North Carolina. Phytopathology 74:170-175.
Murray, D. I. L., and Burpee, L. L. 1984. *Ceratobasidium cereale* sp. nov., the teleomorph of *Rhizoctonia cerealis*. Trans. Br. Mycol. Soc. 82:170-172.
Sneh, B., Burpee, L., and Ogoshi, A. 1991. Identification of *Rhizoctonia* Species. American Phytopathological Society, St. Paul, MN.
Tredway, L. P., and Burpee, L. L. 2001. Rhizoctonia diseases of turfgrass. Plant Health Instructor. doi:10.1094/PHI-I-2001-1109-01

(Prepared by L. P. Tredway)

Snow Mold and Other Low-Temperature Diseases

"Snow mold" is an umbrella term that encompasses several different diseases caused by psychrophilic (cold-loving) fungi. Some snow molds require deep snow cover to cause disease, while others merely require prolonged periods of cool and wet conditions. The most common and destructive snow mold diseases are Microdochium patch, gray snow mold, and speckled snow mold and are described in more detail in the following sections. Less common snow molds include snow scald, Coprinus snow mold, and Pythium snow blight. The pathogens causing these diseases are described in Key to Common Snow Molds.

Two additional snow molds of geographically limited occurrence are caused by *Nectria tuberculariformis* and *Ceratorhiza rhizodes*. *N. tuberculariformis* produces cream-colored to white sclerotia, which mature to orange buff when dry and bright carrot orange when wet (Fig. 157). The shapes of these sclerotia vary widely. The hyphae do not have clamp connections. *N. tuberculariformis* is mostly saprophytic, living on organic debris. *C. rhizodes* causes frost scorch or string of pearls disease. This fungus produces rows of beadlike sclerotia, 1–5 mm in diameter, on bleached, rigid leaf blades. The sclerotia are initially white and turn black as they mature and dry. The hyphae do not have clamp connections.

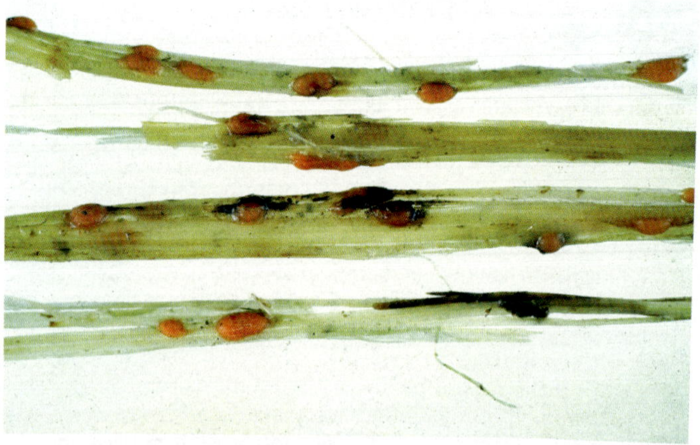

Fig. 157. Sclerotia of *Nectria tuberculariformis* on grass leaves. (Courtesy J. D. Smith—© APS)

Selected References

Hoshino, T., Xiao, N., Hoshino, T., and Tkachenko, O. B. 2009. Cold adaptation in the phytopathogenic fungi causing snow molds. Mycoscience 50:26-38.
Matsumoto, N., and Hsiang, T. 2016. Snow Mold: The Battle Under Snow Between Fungal Pathogens and Their Plant Hosts. Springer, Singapore.
McBeath, J. H. 2002. Snow mold-plant-antagonist interactions: Survival of the fittest under the snow. Plant Health Instructor. doi:10.1094/PHI-I-2002-1010-01
Smith, J. D. 1980. Snow molds of turfgrasses: Identification, biology and control. Pages 75-80 in: Advances in Turfgrass Pathology. P. O. Larsen and B. G. Joyner, eds. Harcourt Brace Jovanovich, Duluth, MN.
Smith, J. D. 1981. Snow molds of winter cereals: Guide for diagnosis, culture, and pathogenicity. Can. J. Plant Pathol. 3:15-25 (erratum, 3:185).
Smith, J. D. 1987. Winter-hardiness and overwintering diseases of amenity turfgrasses with special references to the Canadian prairies. Agric. Can. Tech. Bull. 1987-12E.

(Prepared by R. W. Smiley, P. H. Dernoeden, and B. B. Clarke; revised by P. L. Koch)

Key to Common Snow Molds[a]

I. Mycelium with clamp connections
- A. Sclerotia present, mycelium white to gray
  - 1. Sclerotia globular, ovate, or flattened spherical .... *Typhula* spp.
    - a. Sclerotia up to 5 mm in diameter, orange or red to brown .... *T. incarnata*
    - b. Sclerotia up to 2 mm in diameter, amber to brown to black, never pink to red .... *T. ishikariensis*
  - 2. Sclerotia irregularly shaped, flattened, 1–3 mm in diameter, white to tan to brownish black .... *Coprinopsis psychromorbida*
- B. Sclerotia absent, mycelium white to gray .... *C. psychromorbida*

II. Mycelium without clamp connections
- A. Mycelium coenocytic (aseptate), white; sporangia or oospores may be present in leaves .... *Globisporangium iwayamae*
- B. Mycelium septate
  - 1. Sclerotia present
    - a. Sclerotia beadlike, 1–5 mm in diameter, white to black .... *Ceratorhiza rhizodes*
    - b. Sclerotia variably shaped
      - 1) Sclerotia white to orange .... *Nectria tuberculariformis*
      - 2) Sclerotia dull black .... *Sclerotinia borealis*
  - 2. Sclerotia absent; mycelium white to pink; abundant fusoid conidia, pink en masse .... *Monographella nivalis*

[a] Prepared R. W. Smiley, P. H. Dernoeden, and B. B. Clarke; revised by P. L. Koch—© APS.

Coprinus Snow Mold

Coprinus snow mold, sometimes called "cottony snow mold," is a relatively rare disease caused by two strains of *Coprinopsis psychromorbida,* a sclerotium-forming strain and a nonsclerotial strain.

Symptoms and Signs

Receding snow reveals the presence of circular or irregularly shaped patches in which foliage is covered by a white to gray, woolly or appressed mycelium (Figs. 158 and 159). Leaf lesions or entire leaves of infected plants become rotted and water soaked. The lesions turn pale brown when dry and have dark reddish brown margins.

The sclerotial and nonsclerotial strains of the pathogen cause slightly different symptoms. The nonsclerotial strain tends to grow faster and produce more cottony growth than the sclerotial strain. Thus, patches formed by the nonsclerotial strain are often somewhat larger (15–100 cm in diameter) and whiter than those formed by the sclerotial strain. Patches associated with the sclerotial strain tend to be more superficial.

In infections by the sclerotial strain, sclerotia are conspicuous on leaf surfaces. They initially appear as white hyphal knots 0.25–3 mm across but turn light brown and then brownish black as they mature. Mature sclerotia are ellipsoidal or flattened, irregularly shaped, and 1–3 mm in diameter.

Causal Agent

C. psychromorbida is a basidiomycete fungus. It produces white to gray mycelium with clamp connections at the septa. Some strains of the fungus form sclerotia and are referred to as sclerotial low-temperature basidiomycete strains (see Symptoms), while others do not and are known as low-temperature basidiomycete strains.

Fig. 158. Coprinus snow mold of *Poa pratensis*. (Courtesy J. D. Smith—© APS)

Fig. 159. Coprinus snow mold. (Courtesy J. D. Smith—© APS)

Disease Cycle and Epidemiology

A prolonged period of deep snow cover favors the disease. Optimal temperatures for the growth of the pathogen are 5–15.5°C (41–60°F). Growth can continue at temperatures slightly below 0°C (32°F). Extracellular enzymes produced by the pathogen degrade leaves and contribute to disease development. When present, sclerotia can be found on the leaf surface or embedded within the leaf sheath. Sclerotia serve as the survival structures for the sclerotial strains. The means of persistence of nonsclerotial strains are not known but are assumed to be affected tissues.

Management

Heavy applications of nitrogen fertilizers in late autumn should be avoided. Removing snow drifts or selectively placing snow fencing promotes early melting of snow on high-value turfs. Two or three applications of fungicides are required from autumn into winter; the last application should be made just before the first persistent snow cover is expected. Several resistant cultivars of *Poa pratensis* and *Festuca rubra* have been identified. *Bromus, Elymus,* and *Agropyron* spp. used for low-maintenance lawns in Canada have been found to be very resistant, whereas *Poa trivialis* and most cultivars of *Agrostis* spp. are susceptible.

Selected References

Gaudet, D. A., and Laroche, A. 1999. Snow mold crop-environment interactions. Page 191 in: Biotechnological Applications of Cold-Adapted Organisms. R. Margesin and R. Schinner, eds. Springer-Verlag, New York.

Gaudet, D. A., and Sholberg, P. L. 1990. Comparative pathogenicity of *Coprinus psychromorbidus* monokaryons and dikaryons on winter wheat, alfalfa, grass, and pome fruit. Can. J. Plant Pathol. 12:31-37.

Smith, J. D. 1987. Winter-hardiness and overwintering diseases of amenity turfgrasses with special references to the Canadian prairies. Agric. Can. Tech. Bull. 1987-12E.

Traquair, J. A., and Smith, J. D. 1982. Sclerotial strains of *Coprinus psychromorbidus,* a snow mold basidiomycete. Can. J. Plant Pathol. 4:27-36.

(Prepared by R. W. Smiley, P. H. Dernoeden, and B. B. Clarke; revised by P. L. Koch)

Gray Snow Mold

Gray snow mold, caused by *Typhula incarnata,* can be a damaging disease of turf in areas where a deep snow cover persists for longer than 2 months. In the United States, this includes the Northeast, the southern Great Lakes region, and the Rocky Mountain West. Outside the United States, this includes southern Canada, northern Europe, Scandinavia, and northern Japan. Gray snow mold is sometimes combined with speckled snow mold (*T. ishikariensis*) and referred to as "Typhula blight," but there are distinct differences in the identification, epidemiology, and control of these two pathogens. Gray snow mold can sometimes be found in the same stand with other snow mold diseases including speckled snow mold and Microdochium patch. Preventive control measures are often required in regions that experience long durations of snow cover because little can be done to impact disease development once snow has fallen. All cool-season turfgrasses are susceptible to speckled snow mold development, although annual bluegrass and creeping bentgrass are more susceptible than other species.

Symptoms and Signs

The symptoms of gray snow mold are first apparent when the snow melts as circular areas of light yellow, straw-colored, or grayish brown turf 15 cm to 1 m in diameter (Fig. 160). The leaves in affected areas are matted together and may be covered with white to gray mycelium within 1–2 days after the snow melts (Fig. 161). As the grass dries, the mycelium desiccates and the leaves become grayish to silvery white, brittle, and matted (Fig. 162). Damage from gray snow mold is normally confined to the leaves, which recover once turf growth resumes, but severe cases may lead to thin turf canopies that are susceptible to weed invasion. Visual symptoms of gray snow mold are similar in appearance to those of speckled snow mold (Fig. 163).

Sclerotia are long-term survival structures produced by *T. incarnata* after the snow melts that act as key identifying tools to help separate gray snow mold from speckled snow mold. Sclerotia of *T. incarnata* are irregular in shape to roughly spherical and 0.5–5 mm in diameter. The sclerotia are typically found on or embedded within affected leaf tissue (Fig. 164). When young, they tend to be a light pink or cream color and darken to reddish brown upon maturity. By comparison, sclerotia produced by *T. ishikariensis* (speckled snow mold) are smaller, spherical, and black. *Monographella nivalis,* the causal agent of Microdochium patch, does not produce sclerotia.

Causal Agent

Gray snow mold is caused by *T. incarnata*. The presence of clamp connections on the mycelium of this fungus characterizes it as a basidiomycete (Fig. 73). The cortical cells of the sclerotium interlock to form rind patterns that resemble pieces of a jigsaw puzzle. These rind patterns aid in distinguishing *Typhula* spp. from other sclerotium-bearing fungi on grasses. Sporocarps (fruiting bodies) are sometimes observed in au-

Fig. 162. Stand symptom of gray snow mold caused by *Typhula incarnata* on *Agrostis stolonifera*. (Courtesy P. L. Koch—© APS)

Fig. 160. Gray snow mold symptoms on *Agrostis stolonifera*. (Courtesy B. B. Clarke—© APS)

Fig. 163. Gray snow mold of *Agrostis stolonifera* managed as a golf course fairway. (Courtesy P. L. Koch—© APS)

Fig. 161. Mycelium of *Typhula incarnata* in *Agrostis stolonifera*. (Courtesy P. L. Koch—© APS)

Fig. 164. Reddish brown sclerotia of *Typhula incarnata* in *Agrostis stolonifera*. (Courtesy P. L. Koch—© APS)

tumn and early winter; they are slender, white to pinkish, club-shaped structures, 1–2 cm long.

Disease Cycle

T. incarnata survives as sclerotia during the summer (Fig. 165). In late autumn, sclerotia are exposed to wet and cool conditions that favor germination. Optimal temperatures for sclerotial germination are 10–18°C (50–64°F). Soon after the sclerotia germinate, sporocarps may form, depending on the exposure of sclerotia to light. Diffuse light high in the ultraviolet spectrum favors sporocarp formation, whereas direct light or an absence of light (on sclerotia buried in soil or covered by deep snow or mulch) prevents sporocarp formation. Mycelia from germinating sclerotia or basidiospores produced on the sporocarps may infect turfgrass plants under an insulating snow cover where conditions are relatively mild and moist. The fungus grows slowly, even under optimal conditions, necessitating long periods of snow cover to cause disease. As the snow melts in the spring, sclerotia are produced on infected leaves or within sheaths and fall into the thatch as infected leaves decompose. The disease, therefore, tends to recur in the same areas year after year if conditions are favorable for its development.

Epidemiology

Gray snow mold is most severe under deep snow on unfrozen soil. Optimal infection conditions include high soil moisture, high relative humidity, and temperatures just above freezing (1–2°C [34–36°F]) for 2 months or longer. Snow cover exceeding 3 months tends to favor the development of speckled snow mold over gray snow mold. Snow cover favors gray snow mold development by keeping temperatures in the turf canopy near or slightly above freezing and relative humidity above 65%. The fungus begins to grow in late fall under deep snow cover and slowly colonizes and infects the turf as long as the snow cover is sufficient to insulate the turf. Synthetic winter covers, tree leaves, or any other insulating material that traps heat and moisture can also lead to fungal growth and gray snow mold development. Applications of nitrogen fertilizer in the fall prior to winter dormancy may favor the disease by promoting the development of succulent leaf tissue that is more susceptible to fungal infection during the winter.

Management

Cultural control methods for gray and speckled snow mold are similar. Applications of nitrogen fertilizer should be avoided in late autumn prior to winter dormancy. Fall fertilizer should be applied in early fall so the plants can slow their growth before snow cover or as a dormant application after turf growth has stopped. Research has shown that increasing rates of potassium fertilizer applied throughout the year can increase the severity of gray snow mold; however, research has also demonstrated potassium to be an important component of winter survival of annual bluegrass. If creeping bentgrass is the predominant grass species present, limiting the amount of potassium applied should be considered to reduce gray snow mold. Grass should be mowed into late autumn to ensure that the turf does not lodge and trap heat and moisture at the soil surface. However, the grass should not be mowed abnormally

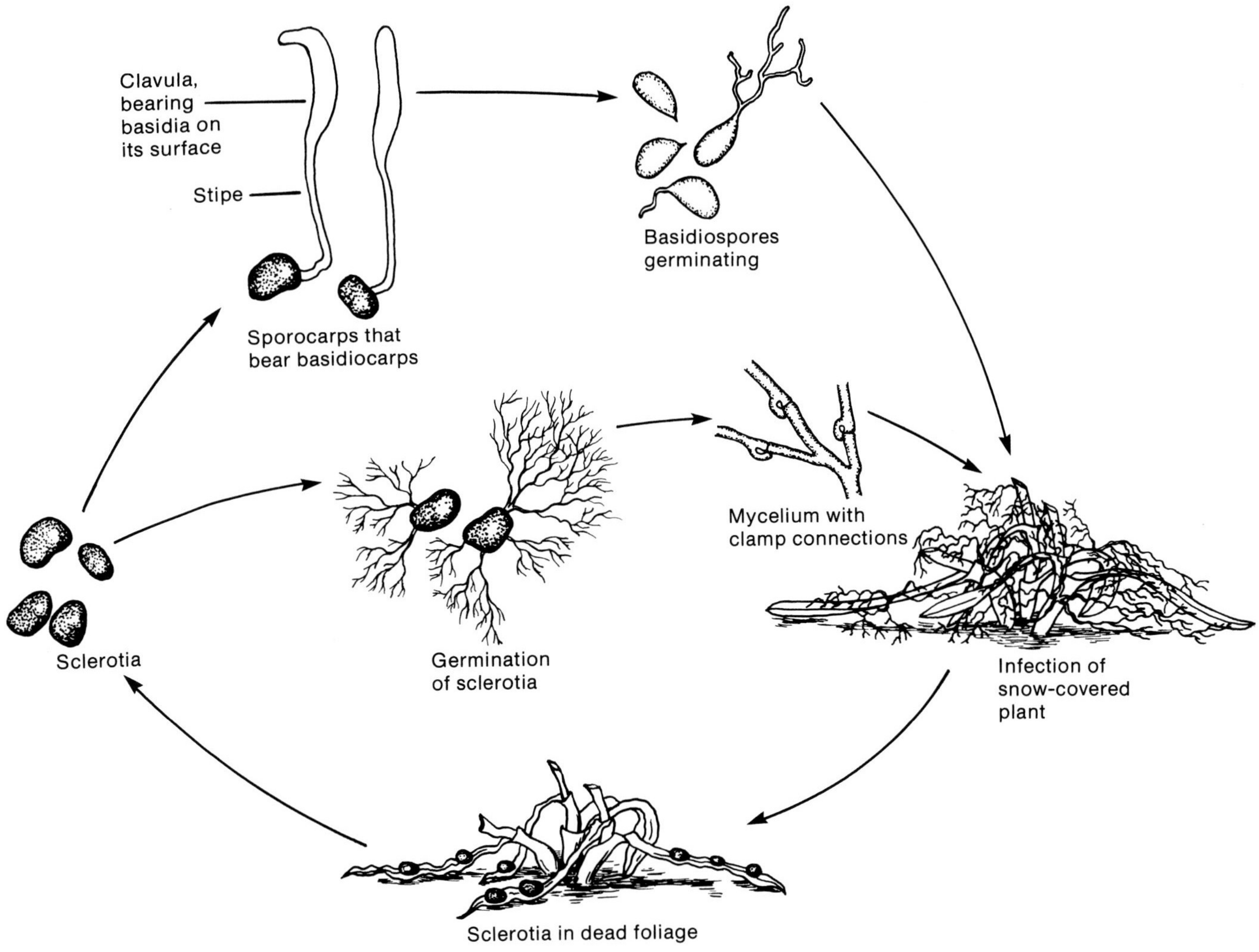

Fig. 165. Disease cycle of gray snow mold, caused by *Typhula incarnata*. (Courtesy C. R. Whittaker—© APS. Adapted from Jackson, N., and Fenstermacher, J. M. 1969. Typhula blight: Its cause, epidemiology and control. J. Sports Turf Res. Inst. 45:67-73.)

short (i.e., scalped), because doing so results in removal of stored plant carbohydrates and potentially increases the risk for winter injury and/or weed encroachment in the spring. Placing snow fences, windbreak plantings, or similar barriers prevents the formation of large snowdrifts on important turf areas. Removing snow and cutting channels in the snow and ice to facilitate drainage promotes rapid drying and warming of disease-prone areas in early spring. Recovery of damaged turf in the spring can be enhanced by lightly raking affected areas to promote sunlight and oxygen penetration and lightly fertilizing with quick-release nitrogen sources. In contrast to speckled snow mold, gray snow mold rarely results in plant death and reseeding is not normally required.

Fungicides may be required for acceptable gray snow mold control in areas that experience deep snow cover for 2 months or longer. Applications should be made in the fall within 1 month of expected snow cover. An additional application made 3–4 weeks prior to the final application can inhibit early fungal growth and may improve snow mold control. Fewer fungicide active ingredients are normally required to control gray snow mold relative to speckled snow mold, but high-value areas should still be protected with two or three active ingredients from different chemical classes. Fungicides from the demethylation inhibitor class are particularly effective against *Typhula* spp., so it is recommended that a product from this class be included in any application targeting gray snow mold. Research has shown that winter rainfall and melting snow can result in increased fungicide degradation. However, reapplication of fungicides to control snow mold is normally recommended only if this occurs early in the winter. Fungicide applications in the spring after the snow melts are not recommended for gray snow mold because the fungus is no longer actively infecting the turf. Green pigments are often added to snow mold fungicide mixtures and can provide a brief period of darker green turf the following spring that dissipates after the first or second mowing. Pigments have not been shown to improve snow mold control when mixed with fungicides. Biological control of gray snow mold has been achieved experimentally under field conditions, but at present no effective commercially available biological control product is available.

Selected References

Årsvoll, K. 1975. Fungi causing winter damage on cultivated grasses in Norway. Meld. Nor. Landbrukshoegsk. 54:1-49.

Årsvoll, K., and Smith, J. D. 1978. *Typhula ishikariensis* and its varieties, var. *idahoensis* comb. nov. and var. *canadensis* var. nov. Can. J. Bot. 56:348-364.

Berthier, J. 1976. Monographie des *Typhula* Fr., *Pistillaria* Fr. et genres voisins. Bull. Soc. Linn. Lyon Numéro Spec. 45:1-213.

Burpee, L. L., Kaye, L. M., Goulty, L. G., and Lawton, M. B. 1987. Suppression of gray snow mold on creeping bentgrass by an isolate of *Typhula phacorrhiza*. Plant Dis. 71:97-100.

Hsiang, T., Matsumoto, N., and Millett, S. M. 1999. Biology and management of Typhula snow molds of turfgrass. Plant Dis. 83:788-798.

Hsiang, T., Wu, C., and Cook, S. 1999. Residual efficacy of *Typhula phacorrhiza* as a biocontrol agent of grey snow mold on creeping bentgrass. Can. J. Plant Pathol. 21:382-387.

Jackson, N., and Fenstermacher, J. M. 1969. Typhula blight: Its cause, epidemiology and control. J. Sports Turf Res. Inst. 45:67-73.

Koch, P. 2017. Optimal fungicide timing for suppression of Typhula blight under winter covers. Agron. J. 109:1771-1776.

Koch, P. L., and Kerns, J. P. 2013. Impact of spring and fall fungicide timings on the development of Typhula blight and dollar spot on golf course turfgrass. Int. Turfgrass Soc. Res. J. 12:97-101.

Koch, P. L., Stier, J. C., and Kerns, J. P. 2015. Snow cover has variable effects on persistence of fungicides and their suppression of microdochium patch on amenity turfgrass. Plant Pathol. 64:1417-1428.

Smith, J. D. 1987. Winter-hardiness and overwintering diseases of amenity turfgrasses with special references to the Canadian prairies. Agric. Can. Tech. Bull. 1987-12E.

Woods, M. S., Ketterings, Q. M., Rossi, F. S., and Petrovic, A. M. 2006. Potassium availability indices and turfgrass performance in a calcareous sand putting green. Crop Sci. 46:381-389.

(Prepared by P. L. Koch)

Microdochium Patch

Microdochium patch occurs on both cool- and warm-season turfgrasses in temperate climates around the world. It is the most common turfgrass disease in areas with cool, humid climates such as the Pacific Northwest of the United States and Canada, the United Kingdom, northern Europe, and parts of Scandinavia. Snow cover is not required for Microdochium patch to occur, but in regions where snow cover persists for 2 months or longer, the disease can be found in the same locations as gray snow mold and speckled snow mold. All commonly used turfgrass species are believed to be susceptible to Microdochium patch development, but *Agrostis stolonifera* and *Poa annua* are the most susceptible.

Symptoms and Signs

Symptoms typically differ if the disease develops under deep snow cover versus in the absence of snow. Under snow cover, circular patches of tan or brown turf (15–30 cm in diameter) can develop (Fig. 166). In the absence of snow, the patches first appear as small, water-soaked spots less than 5 cm in diameter with a dark brown or rust-colored appearance (Fig. 167). In both instances, white, fluffy mycelium may be present when conditions are highly favorable for infection. A faint pinkish hue can also be observed around the outer edges of patches, although this is not always apparent. The pink ring is evidence of sporodochia formed on the surfaces of infected leaves, which occurs in response to sunlight exposure. Individual spots or patches may coalesce into large areas of blighted turf (Figs. 168 and 169). After the snow melts, Microdochium patch may resemble gray and speckled snow mold; however, the presence of a pink ring around the outer edge of the patch is a key identifying characteristic of Microdochium patch (Fig. 170). In addition, the causal agents of both gray snow mold (*Typhula incarnata*) and speckled snow mold (*T. ishikariensis*) produce sclerotia in

Fig. 166. Microdochium patch on a *Poa pratensis* lawn. (Courtesy of P. L. Koch—© APS)

infected leaf tissue, whereas the causal agent of Microdochium patch does not.

Causal Agent

Monographella nivalis causes Microdochium patch. The fungus produces septate mycelium without clamp connections. It does not form sclerotia. Salmon pink or white sporodochia (Fig. 171) are produced in abundance in cool weather. Conidia (Fig. 172) are lunate, 2.5–5 × 10–30 µm, and mostly one to three septate and lack the foot cells typical of *Fusarium* spp.

Other fungi sometimes found in association with *M. nivalis* include *Fusarium culmorum, F. equiseti, F. avenaceum,* and *F. incarnatum,* all of which can infect grasses at low temperatures.

Disease Cycle

M. nivalis survives periods of unfavorable conditions in infected plants and dead tissue. When conditions are conducive, mycelium grows from infected debris or germinating conidia and infects leaves. The fungus initially grows slowly, and infections may go undetected for several weeks or longer. The disease spreads rapidly under snow cover or wet and overcast conditions when temperatures are near freezing to around 16°C (61°F). The disease becomes inactive as the turf canopy dries during sunny and warm periods. Conidia and hyphal fragments are easily carried via mechanical means like mowers, foot traffic, or surface water flow to infect healthy turf.

Fig. 167. Microdochium patch of *Agrostis stolonifera*. Note the water-soaked zone around the perimeter of each well-defined circular patch. (Courtesy G. Jung—© APS)

Fig. 168. Microdochium patch on a *Cynodon dactylon* × *transvaalensis* (ultradwarf cultivar Champion) putting green. (Courtesy M. Tomaso-Peterson—© APS)

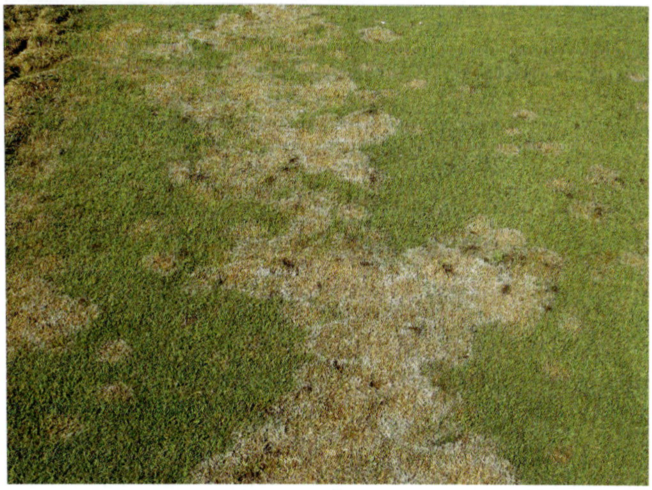

Fig. 169. Microdochium patch of *Agrostis stolonifera* on a putting green. (Courtesy of P. L. Koch—© APS)

Fig. 170. Microdochium patches with pink rings around the outer edges on a *Poa annua* putting green. (Courtesy B. B. Clarke—© APS)

Fig. 171. Sporodochia of *Monographella nivalis* on a leaf of *Agrostis stolonifera*. (Courtesy J. E. Kaminski—© APS)

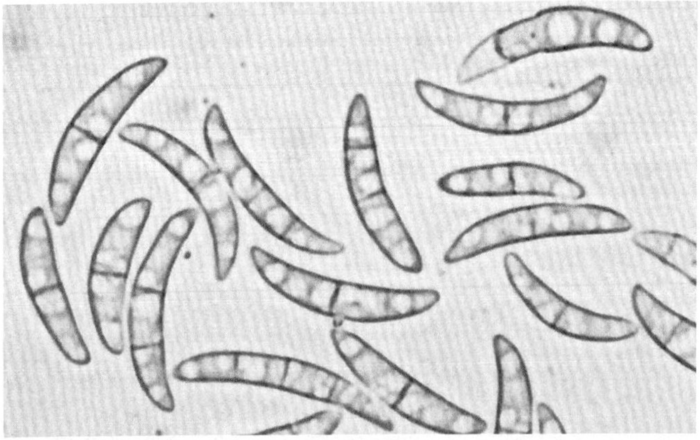

Fig. 172. Conidia of *Monographella nivalis.* (Courtesy T. A. Toussoun and P. E. Nelson—© APS)

Epidemiology

Microdochium patch is most severe on unfrozen turf under deep, insulating snow and during cool (0–8°C [32–46°F]), wet periods. Under prolonged wet conditions, symptoms can develop at temperatures up to 18°C (64°F). In most climates, fungal growth begins during the late fall as conditions cool and moisture becomes more consistent. High nitrogen fertility that promotes turf growth during the autumn or winter prevents the turf from hardening off before winter and increases susceptibility to disease. Potassium has been shown to increase Microdochium patch severity on *A. stolonifera,* although other research has indicated potassium may decrease the disease on *P. annua*. Poor drainage and limited airflow can also favor disease development.

M. nivalis, F. culmorum, F. avenaceum, and *F. equiseti* grow best at 21–25°C (70–77°F) in culture and are capable of growth at temperatures between –6°C (21°F) and 30°C (86°F). At 3°C (37°F), *M. nivalis* grows slightly better than *F. avenaceum* and only half as fast as *F. equiseti.* All three of these fungi are destructive snow mold pathogens at 3°C (37°F), and *F. culmorum* is only slightly less pathogenic. However, these fungi are considerably less tolerant of cold than *Typhula* spp. and *Sclerotinia borealis,* which grow optimally at 9°C (48°F) but retain nearly half their growth capacity at 0°C (32°F).

Management

Application of nitrogen fertilizer should be limited late in the fall or during extended periods of cool, wet weather. Potassium fertilizer should be limited on stands of predominantly *A. stolonifera,* unless soil tests indicate a potassium deficiency. Mowing should be continued until the turf is dormant to prevent lodging, which can trap heat and moisture at the soil surface and intensify disease development. The grass should not be mowed abnormally short (i.e., scalped), because doing so results in removal of stored plant carbohydrates and potentially increases the risk for winter injury. Research has shown that repeated applications of sulfur and phosphite products can significantly reduce Microdochium patch in the absence of snow cover; however, to date, these products have not been effective at reducing disease during extended periods of snow cover. Recovery of damaged turf can be promoted by lightly raking affected areas to promote sunlight and oxygen penetration into the canopy and lightly fertilizing with quick-release nitrogen sources. Mowing or traveling through infected areas should be avoided, because mycelium and conidia can be easily tracked to other locations and begin new infections.

Fungicides may be required for acceptable control of Microdochium patch if favorable conditions persist for prolonged periods of time. In areas that experience prolonged snow cover, control is typically achieved with a fungicide application made within 1 month of expected snow cover. A second fungicide application made 3–4 weeks earlier may improve control in areas with heavy disease pressure. In locations that do not experience prolonged snow cover, repeated fungicide applications may be needed every 2–4 weeks when conditions are optimal for infection. Fungicides effective against gray and speckled snow mold are not always effective against Microdochium patch, so a local university extension specialist may be consulted for the most effective products available in a specific region. Applications of mixtures consisting of multiple active ingredients provide control against Microdochium patch, gray snow mold, and speckled snow mold in areas that experience prolonged snow cover.

Selected References

Årsvoll, K. 1975. Fungi causing winter damage on cultivated grasses in Norway. Meld. Nor. Landbrukshoegsk. 54:1-49.

Arx, J. A. von. 1985. Notes on *Monographella* and *Microdochium.* Trans. Br. Mycol. Soc. 83:373-374.

Bier, P. V., Persche, M., Koch, P., and Soldat, D. J. 2018. A long term evaluation of differential potassium fertilization of a creeping bentgrass putting green. Plant Soil 431:303-316.

Kaminski, J. E., Dernoeden, P. H., and Bigelow, C. A. 2004. Soil amendments and fertilizer source effects on creeping bentgrass establishment, soil microbial activity, thatch, and disease. HortScience 39:620-626.

Koch, P. L., Stier, J. C., and Kerns, J. P. 2015. Snow cover has variable effects on persistence of fungicides and their suppression of microdochium patch on amenity turfgrass. Plant Pathol. 64:1417-1428.

Litschko, L., and Burpee, L. L. 1987. Variations among isolates of *Microdochium nivale* collected from wheat and turfgrasses. Trans. Br. Mycol. Soc. 89:252-256.

Mann, R. L., and Newell, A. J. 2005. A survey to determine the incidence and severity of pests and diseases on golf course putting greens in England, Ireland, Scotland, and Wales. Int. Turf Res. J. 10:224-229.

McDonald, B., Mattox, C., Gould, M., and Kowalewski, A. 2018. Effects of sulfur and calcium source on pH, anthracnose severity, and Microdochium patch management on annual bluegrass in western Oregon. Crop, Forage Turfgrass Manage. 4. doi:10.2134/cftm2018.03.0018

Sanders, P. L., and Cole, H., Jr. 1981. The *Fusarium* diseases of turfgrass. Pages 195-209 in: *Fusarium:* Diseases, Biology and Taxonomy. P. E. Nelson, T. A. Toussoun, and R. J. Cook, eds. Pennsylvania State University Press, University Park.

Simpson, D. R., Rezanoor, H. N., and Parry, D. W. 2000. Evidence for differential host preference in *Microdochium nivale* var. *majus* and *Microdochium nivale* var. *nivale.* Plant Pathol. 49:261-268.

Woods, M. S., Ketterings, Q. M., Rossi, F. S., and Petrovic, A. M. 2006. Potassium availability indices and turfgrass performance in a calcareous sand putting green. Crop Sci. 46:381-389.

(Prepared by P. L. Koch)

Pythium Snow Blight

Pythium snow blight of turfgrasses was first documented in Japan but has also been observed in North America. This disease also occurs in small grain crops and is most common in Japan where winter wheat or barley are grown in paddy fields. *Globisporangium iwayamae,* originally described as *Pythium iwayamae,* is the causal agent in turfgrasses. Several other species of *Pythium* and *Globisporangium* are associated with the disease in cereal crops.

Symptoms

Small, tan or orange spots may appear on the turf, or large areas of turf may be uniformly blighted. Foliar mycelium may or may not be present. Rotted leaf tissue appears brown or light tan and is filled with oospores of the pathogen. Roots are mostly unaffected, but crowns can be extensively rotted. Plants with infected crowns die quickly.

Causal Agent

G. iwayamae produces coenocytic hyphae, sporangia, and oospores similar to those of *Pythium* spp. Sporangia are terminal or intercalary, globose to subglobose, and 28–48 × 26–44 µm. Oogonia are also terminal or intercalary and 20–25 µm in diameter, with one or two antheridia per oogonium. Oospores are simple, aplerotic, and 17–20 µm in diameter.

Epidemiology

Pythium snow blight is favored by heavy snow accumulation on unfrozen, saturated soil. In laboratory growth studies at 0.5°C (33°F), *G. iwayamae* grew five times faster than *Typhula ishikariensis* and 10 times faster than *Microdochium nivale*. Zoospores were produced at 0–10°C (32–50°F), with optimal production at 5°C (41°F). The disease is encouraged by high soil fertility and poor drainage.

Management

Improved surface and subsurface drainage may reduce the severity of Pythium snow blight. Low to moderate fertility levels in late fall to winter are recommended. Application of a *Pythium*-active fungicide such as mefenoxam prior to snowfall may be warranted where the disease is a regular occurrence.

Selected References

Araki, T., Shimanuki, T., and Ozaki, M. 1974. Pythium snow rot on grasses in the eastern Hokkaido. Ann. Phytopathol. Soc. Jpn. 40:117.

Lipps, P. E., and Bruehl, G. W. 1978. Snow rot of winter wheat in Washington. Phytopathology 68:1120-1127.

Masumoto, S., Shigyo, T., and Motoaki, T. 2009. Pythium snow blight of Kentucky bluegrass turf in a golf course in Hokkaido, Japan. J. Jpn. Soc. Turfgrass Sci. 38:33-36.

Uzuhashi, S., Kakishima, M., and Tojo, M. 2010. Phylogeny of the genus *Pythium* and description of new genera. Mycoscience 51:337-365.

(Prepared by R. W. Smiley, P. H. Dernoeden, and B. B. Clarke; revised by L. P. Tredway)

Snow Scald

Snow scald is believed to occur on all cool-season turfgrasses but is most frequently observed on annual bluegrass and creeping bentgrass. Snow scald is not commonly observed, and much remains unknown about the disease and its causal agent. The disease is believed to be favored by deep snow lying on frozen soil and is most common in areas that experience cold, windswept winters such as western Minnesota and the northern Great Plains of the United States and Canada.

Symptoms and Signs

Visual symptoms resemble those of gray snow mold (*Typhula incarnata*) or speckled snow mold (*T. ishikariensis*) and appear as roughly circular patches 15–30 cm in diameter as the snow recedes (Fig. 173). Affected leaves are initially water soaked and covered with sparse, gray mycelium and tan sclerotia. The leaves then become bleached and turn tan or white. The sclerotia turn black as they mature and are considerably larger than those produced by *T. incarnata* or *T. ishikariensis*. Snow scald sclerotia resemble mouse droppings embedded in affected turf (Fig. 174).

Causal Agent

Sclerotinia borealis produces gray mycelium without clamp connections and dull black sclerotia of variable size. Sclerotia may be spherical, oval, or flakelike and are 0.5–7 mm long (Fig. 175). Apothecia of *S. borealis* are up to 5 mm in diameter and are borne on a stipe (stalk) up to 5 mm high.

Disease Cycle and Epidemiology

Very little research has been conducted on snow scald. From what is known, snow scald is most severe following a prolonged period of snow cover on frozen soil. Regions that experience bitterly cold winter conditions and are relatively open and windswept tend to experience the most severe snow scald symptoms. Additional observations suggest that low soil moisture going into winter can increase susceptibility to snow scald. The pathogen grows most rapidly in laboratory cultures

Fig. 173. Snow scald, caused by *Sclerotinia borealis*. (Courtesy W. C. Stienstra—© APS)

Fig. 174. Sclerotia of *Sclerotinia borealis* produced in the canopy of *Agrostis stolonifera*. (Courtesy P. L. Koch—©APS)

Fig. 175. Mature sclerotium of *Sclerotinia borealis*. (Courtesy W. C. Stienstra—© APS)

at 5–10°C (41–50°F) and is capable of growth at temperatures as low as –5°C (23°F). Sclerotia are long-term survival structures produced by *S. borealis* that allow the fungus to survive the summer. Fungal hyphae germinate from the sclerotia in the fall and initiate new infections the following winter.

Management

Use of wetting agents to improve soil moisture going into winter has successfully suppressed snow scald in certain situations. Preventive fungicide applications are recommended to suppress disease in areas that experience favorable conditions for snow scald. Fungicide mixtures that provide excellent control of speckled snow mold have been reported to provide poor control of snow scald, although it is unclear which fungicides are most effective against snow scald because of limited research. Recovery of damaged areas in the spring can be promoted by lightly raking affected patches to promote sunlight and oxygen penetration and lightly fertilizing with quick-release nitrogen sources. Snow scald does not normally result in death of the entire plant.

Selected References

Kohn, L. M. 1979. Delimitation of the economically important plant pathogenic *Sclerotinia* species. Phytopathology 69:881-886.

Smith, J. D. 1987. Winter-hardiness and overwintering diseases of amenity turfgrasses with special references to the Canadian prairies. Agric. Can. Tech. Bull. 1987-12E.

(Prepared by P. L. Koch)

Speckled Snow Mold

Speckled snow mold causes severe damage to turfgrass in areas where deep snow cover persists for 3 months or longer, including the far northern United States and much of Canada, the Rocky Mountain West, Scandinavia, and northern Japan. The disease may be present in affected stands with other snow mold diseases such as gray snow mold (*Typhula incarnata*) and Microdochium patch (*Monographella nivalis*). Accurate disease identification is important for effective snow mold control because snow mold pathogens respond differently to fungicides and other management practices. Preventive control measures are often required in regions that experience long durations of snow cover because little can be done to impact disease development once snow has fallen. All cool-season turfgrasses are susceptible to speckled snow mold development, although annual bluegrass and creeping bentgrass are generally more susceptible than other species.

Symptoms and Signs

Symptoms of speckled snow mold are first apparent when the snow melts as circular areas of light yellow, straw-colored, or grayish brown turf 15 cm to 1 m or more in diameter (Fig. 176). The leaves in affected areas are matted together and may be covered with white to gray mycelium within 1–2 days after the snow melts. As the grass dries, the mycelium desiccates and the leaves become grayish to silvery white, brittle, and matted. Large areas of turf may be killed when deep snow cover is present for several months (Fig. 177), although in mild cases the damage is confined to the leaves and the plants recover once growth begins in the spring. Visual symptoms of speckled snow mold are similar in appearance to those of gray snow mold.

Sclerotia are long-term survival structures produced by *Typhula ishikariensis* after the snow melts and are useful for differentiating speckled snow mold from gray snow mold. Sclerotia of *T. ishikariensis* are roughly spherical and less than 2 mm in diameter (Fig. 178). The sclerotia are typically found on or embedded within affected leaf tissue. When young, they tend to be dark amber or chestnut. They turn dark brown or

Fig. 176. Speckled snow mold on a mixed stand of *Agrostis stolonifera* and *Poa annua* managed as a golf course putting green. (Courtesy P. L. Koch—© APS)

Fig. 177. Speckled snow mold on a golf course fairway consisting of *Agrostis stolonifera* and *Poa annua*. Healthy plots were treated with preventive fungicides before snow cover. (Courtesy P. L. Koch—© APS)

Fig. 178. Sclerotia of *Typhula ishikariensis* in a mixed stand of cool-season grasses. (Courtesy P. L. Koch—© APS)

black upon maturity and look like pepper sprinkled on the turf. Sclerotia produced by *T. incarnata* (gray snow mold) are larger and dark red. *M. nivalis* (Microdochium patch) does not produce any sclerotia.

Causal Agents

Speckled snow mold is caused by three biotypes (varieties) of *T. ishikariensis:* var. *ishikariensis,* var. *idahoensis,* and var. *canadensis.* Recent molecular studies have shown extensive genetic diversity within varieties of *T. ishikariensis.*

The presence of clamp connections on the mycelium of these fungi characterizes them as basidiomycetes (Fig. 73). The cortical cells of the sclerotium interlock to form rind patterns that resemble pieces of a jigsaw puzzle. These rind patterns aid in distinguishing *Typhula* spp. from other sclerotium-bearing fungi on grasses. Sporocarps (fruiting bodies) are sometimes observed in autumn and early winter; they are slender, white to pinkish, club-shaped structures, 1–2 cm long.

Disease Cycle, Epidemiology, and Management

The disease cycle, epidemiology, and management of speckled snow mold and gray snow mold are very similar; see the section Gray Snow Mold for further details.

Selected References

Årsvoll, K. 1975. Fungi causing winter damage on cultivated grasses in Norway. Meld. Nor. Landbrukshoegsk. 54:1-49.

Årsvoll, K., and Smith, J. D. 1978. *Typhula ishikariensis* and its varieties, var. *idahoensis* comb. nov. and var. *canadensis* var. nov. Can. J. Bot. 56:348-364.

Berthier, J. 1976. Monographie des *Typhula* Fr., *Pistillaria* Fr. et genres voisins. Bull. Soc. Linn. Lyon Numéro Spec. 45:1-213.

Burpee, L. L., Kaye, L. M., Goulty, L. G., and Lawton, M. B. 1987. Suppression of gray snow mold on creeping bentgrass by an isolate of *Typhula phacorrhiza.* Plant Dis. 71:97-100.

Hsiang, T., Matsumoto, N., and Millett, S. M. 1999. Biology and management of Typhula snow molds of turfgrass. Plant Dis. 83:788-798.

Hsiang, T., Wu, C., and Cook, S. 1999. Residual efficacy of *Typhula phacorrhiza* as a biocontrol agent of grey snow mold on creeping bentgrass. Can. J. Plant Pathol. 21:382-387.

Jackson, N., and Fenstermacher, J. M. 1969. Typhula blight: Its cause, epidemiology and control. J. Sports Turf Res. Inst. 45:67-73.

Koch, P. L., Stier, J. C., and Kerns, J. P. 2015. Snow cover has variable effects on persistence of fungicides and their suppression of microdochium patch on amenity turfgrass. Plant Pathol. 64:1417-1428.

Smith, J. D. 1987. Winter-hardiness and overwintering diseases of amenity turfgrasses with special references to the Canadian prairies. Agric. Can. Tech. Bull. 1987-12E.

Smith, J. D., Jackson, N., and Woolhouse, A. R. 1989. Fungal Diseases of Amenity Turf Grasses. E. & F. N. Spon, London.

Woods, M. S., Ketterings, Q. M., Rossi, F. S., and Petrovic, A. M. 2006. Potassium availability indices and turfgrass performance in a calcareous sand putting green. Crop Sci. 46:381-389.

(Prepared by P. L. Koch)

Diseases of Roots, Rhizomes, and Stolons

Our understanding of turfgrass diseases caused by root-infecting fungi has increased markedly since the 1980s. Diseases commonly referred to as "patch diseases," characterized in part by circular patches of dying turf with annual or periodic recurrence of symptoms in the same locations, have been recognized for more than 70 years. However, take-all patch of creeping bentgrass, caused by *Gaeumannomyces avenae,* was the only turfgrass root disease that was well-defined prior to 1984. Since then, advanced research methods and diagnostic techniques have increased our understanding of these diseases, most recently through molecular techniques that permit more precise pathogen identification and taxonomic clarification. Today, root diseases are recognized as some of the most important challenges in the management of bentgrass, Bermudagrass, and bluegrass species in golf course, athletic field, and landscape turf.

Patch diseases, including take-all patch, summer patch, spring dead spot, and take-all root rot, are caused by a group of fungi collectively referred to as "ectotrophic root-infecting fungi" (Table 5). Although these fungi are not all closely related from an evolutionary standpoint, they colonize root surfaces with strands of darkly pigmented, ectotrophic runner hyphae (Fig. 179). These diseases pose a significant challenge for diagnosticians because similar ectotrophic hyphae can be produced by weakly pathogenic or nonpathogenic fungi. Careful consideration of field symptoms, weather conditions, and microscopic observations are essential to diagnose these diseases accurately.

Infections of turfgrass roots by *Pythium* species are commonly observed, particularly on golf course putting greens in warm, humid climates. Many species of *Pythium* and allied genera have been documented in turfgrass roots and vary widely in their pathogenicity. There is much to be learned about the impact of *Pythium* species on turfgrass roots and how to manage these pathogens in a sustainable way.

Plant-parasitic nematodes have been recognized as a significant turfgrass pest in the southeastern United States since the 1960s, but more recent surveys have demonstrated their widespread presence in turfgrass systems. Plant-parasitic nematodes reduce the efficiency of water and nutrient absorption by turfgrass roots, increasing fertilizer and irrigation requirements and the potential for negative impacts on the environment. Organochloride and organophosphate compounds traditionally used for nematode control have been banned in most areas. Although modern nematicides pose less risk to the environment and human health, these products are more narrow in spectrum and/or less persistent in the soil. A better understanding of nematode biology is needed to make the most efficient use of modern nematicidal compounds.

(Prepared by L. P. Tredway)

TABLE 5. Morphological Characteristics of Ectotrophic Root-Infecting Fungi of Cool- and Warm-Season Turfgrass[a]

Fungal Species	Hosts	Growth Rate (mm/day) on PDA[b] at 25°C (77°F)	Asexual Stage (conidia)	Sexual Stage	Vegetative (hyphopodia)	Colony Pigment on PDA at 25°C (77°F)
Candidacolonium cynodontis	*Cynodon* spp.	2.7	...	...	Olivaceous brown, chiroid or mitten-shaped, 11.6 × 8.8 µm	White, turning pale to dark grayish buff
Gaeumannomyces arxii	*Stenotaphrum, Pennisetum* spp.	10.0	Lunate, fusiform, hyaline, 4–10 × 1–2 µm	...		Grayish olivaceous or grayish sepia
G. avenae	*Agrostis* spp.	8.0	Straight to slightly curved, hyaline, 5–14 × 2–4 µm	Ascospores 107–141 µm long, five to 11 septa	Simple	White, turning gray or black
G. californicus	*Stenotaphrum* spp.	12.0	Lunate, allantoid or fusiform, hyaline, 4–11 × 1–1.5 µm	...	Brown when mature, lobed, 25–32.5 × 24–30 µm	White, submerged gray to olivaceous
G. floridanus	*Stenotaphrum* spp.	12.0	Lunate, curved, hyaline, 5–11 × 1–1.5 µm	...	Lobed, brown, 18–27 × 14.5–26.5 µm	White, submerged dark, margins effuse, rhizoid
G. graminicola	*Cynodon, Eremochloa, Stenotaphrum* spp.	10.5	Lunate, curved, hyaline, 5–11.5 × 1–2 µm	...	Lobed, brown, 16.5–24 × 15.5–23.5 µm	White, submerged dark, center hazel gray to olivaceous gray, rhizoid margin
G. graminis	*Cynodon, Paspalum, Pennisetum, Zoysia* spp.	10.0	Lunate, allantoid, hyaline, 4–10 × 1–2 µm	Ascospores 70–110 µm long, three to five septa, 17–27 × 20–30 µm	Lobed, brown	Olivaceous gray or buff, submerged darker
G. hyphopodioides	*Pennisetum* spp.	8.6	Lunate, allantoid, hyaline, 5.5–10.5 × 1–2 µm	...	Simple and lobed, hyaline to brown, 17–28 × 18–25 µm	Aerial cottony, white to gray; submerged hazel, olivaceous, dull green, rhizoid margins
G. nanograminis	*Cynodon* spp.	9.6	...	...	Deeply lobed, brown, 19.4–20 µm in diameter	Olivaceous gray or buff, submerged darker
G. oryzinus	*Cynodon* spp.	11.0	Lunate, allantoid to fusiform, hyaline, 5–11 × 1–2.5 µm	Asci cylindrical, 72 × 7–16 µm; ascospores 70–112 × 2–4.6 µm	Lobed, brown, 19–45 × 15.5–36 µm	White to pale gray, submerged olivaceous gray to smoke gray
G. walker	*Stenotaphrum* spp.	9.3	Fusiform to lunate, hyaline, 5–14 × 1–1.5 µm	...	Lobed, brown, 20–31 × 18.5–24.5 µm	Scarce, pale olivaceous in center, hyaline to margin
G. wongoonoo	*Stenotaphrum* spp.	8.8	Cylindrical, hyaline, 5–12.5 × 3–5 µm	Asci 80–140 × 10–14 µm; ascospores 12 septa, 36–75 × 3–5 µm	Simple or slightly lobed, 20 µm in diameter	Gray to olivaceous black, margins hyaline
Magnaporthiopsis agrostidis	*Agrostis* spp.	7.5	Hyaline, aseptate, filiform, 5–20 × 1.5 –3 µm	...	...	Aerial gray becoming olivaceous brown
M. cynodontis	*Agrostis, Cynodon, Festuca, Poa* spp.	6.3	Ovoid to ellipsoid, slightly curved, hyaline, 7–13 × 2–6.5 µm	...	...	Parrot green, surface floccose, aerial yellowish
M. incrustans	*Cynodon, Eremochloa, Festuca, Poa, Stenotaphrum, Zoysia* spp.	12.0	Subglobose or ovate, hyaline, 3–6 × 2–3 µm	Asci eight-spored, 65–107 µm long; ascospores fusoid, three to five septa, hyaline, 33–53 × 2–4 µm	Slightly lobed, 6–14 × 4–6 µm; crusts of black on surfaces of stolons, roots, and rhizomes	Initially hyaline, then gray to olivaceous black
M. meyeri-festucae	*Festuca* spp.	4.5	Ellipsoidal, straight to slightly curved, aseptate, hyaline, 3–5 × 1–2.5 µm	...	...	Warbler green, aerial yellowish to brownish
M. poae	*Agrostis, Poa* spp.	12.6	Slightly curved, hyaline, 3–8 × 1–3 µm	Asci 63–108 × 7–15 µm; ascospores fusoid, three septa, terminal cells hyaline, 23–42 × 4–6 µm	Globose, single or aggregated, 6–12 µm in diameter	Initially hyaline, then gray to olivaceous brown
Ophiosphaerella agrostidis	*Agrostis, Cynodon* spp.	...	...	Asci 100–230 × 10–15 µm; ascospores filiform, pale brown, 75–150 × 2–2.5 µm, six to 15 transverse septa	...	Pink to quartz rose
O. herpotricha	*Cynodon, Zoysia* spp.	4	...	Asci 150–190 × 7–9 µm; ascospores filiform, yellow to pale brown, 140–180 × 2–2.5 µm, eight to 20 septa	...	White, turning tan to brown
O. korrae	*Axonopus, Cynodon, Poa, Zoysia* spp.	5	...	Asci 145–200 × 10–15 µm; ascospores filiform, pale brown, 120–180 × 4–5 µm, one to 15 septa	...	White, turning dark gray
O. narmari	*Cynodon, Pennisetum, Stenotaphrum* spp.	5	...	Asci 100–140 × 12–17 µm; ascospores elliptical to filiform, pale brown, 40–50 × 5–6 µm, five septa	...	White to buff, darkening with age

[a] Prepared by M. Tomaso-Peterson—© APS.

[b] Potato dextrose agar.

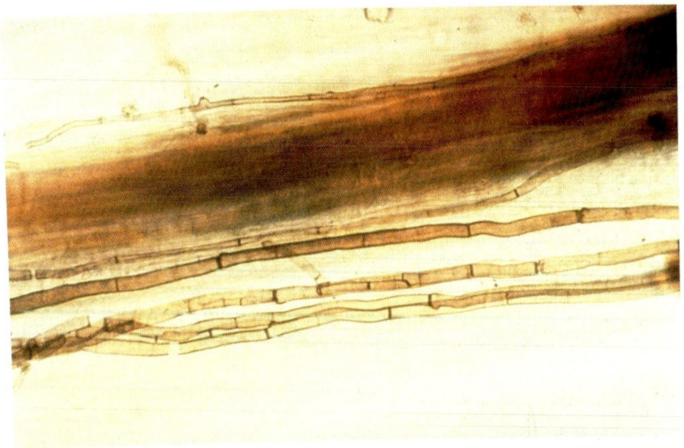

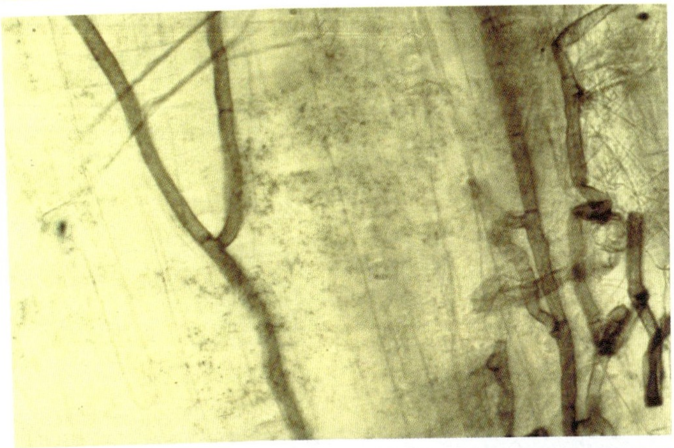

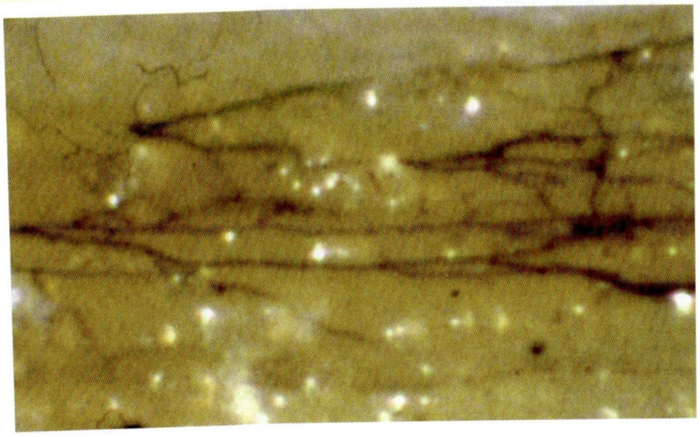

Fig. 179. Dark, ectotrophic hyphae characteristic of species of *Gaeumannomyces, Magnaporthiopsis,* and *Ophiosphaerella* and related root pathogens. (Courtesy P. J. Landschoot, M. L. Elliott, and R. W. Smiley–© APS)

Dead Spot

Dead spot, also known as bentgrass dead spot and Bermudagrass dead spot, is a disease of young stands of *Agrostis* and hybrid *Cynodon* spp. grown in soil mixes with a high sand content on golf putting greens and tees in the United States and Canada. The disease has also been observed in the spring on *Poa trivialis* that was overseeded into *Cynodon* spp. greens. Dead spot may develop when older greens are planted after fumigation of the soil. The disease is not known to affect turfs grown in native soils.

Symptoms and Signs

In closely mown *Agrostis* spp. putting greens and tees, dead spot initially appears as small, reddish brown spots (Figs. 180 and 181), approximately 1.5 cm in diameter, which can enlarge to approximately 9 cm in diameter. In the early stages of disease development, spots are reddish brown or copper colored and resemble ball-mark injury. As the disease progresses, grass in the center of a spot turns tan, while leaves in the outer edge of the spot appear reddish brown (Fig. 182). Rings of dead grass with living turf in the centers (frog-eyes) may occur but are uncommon. Roots and stem bases of dying plants are discolored or necrotic. Patches may be distributed throughout a green or

Fig. 180. Dead spot, caused by *Ophiosphaerella agrostidis,* on an *Agrostis stolonifera* golf green. (Courtesy J. E. Kaminski—© APS)

Fig. 181. Dead spot, caused by *Ophiosphaerella agrostidis,* on an *Agrostis stolonifera* golf green. (Courtesy J. E. Kaminski—© APS)

Fig. 182. Dead spot, caused by *Ophiosphaerella agrostidis,* on an *Agrostis stolonifera* golf green. Note the reddish brown appearance of leaf tissue on the outer edge of the spot. (Courtesy J. E. Kaminski—© APS)

may be localized. Spots generally do not coalesce, except in severe cases. They sometimes form depressions or pits in the putting surface. Spots recover very slowly, and stolon growth into dead patches is often restrained or inhibited. Foliar mycelium is not observed in the field, but a white to pale pink mycelium may develop on leaves of diseased plants incubated under high humidity for 3–5 days. Numerous pseudothecia (Fig. 183) can be found on necrotic leaf, sheath, and stolon tissues.

In closely mown *Cynodon* spp., patches are 2.5–10 cm in diameter, and infected leaves appear reddish brown or tan.

Dark brown ectotrophic hyphae (Fig. 179) may be observed on the nodes of stolons in *Agrostis* spp. but generally are not found on the roots. In *Cynodon* spp., dark brown hyphae may be observed within roots.

Causal Agent

Ophiosphaerella agrostidis causes dead spot. Colonies grown on potato dextrose agar at 25°C (77°F) are pink to rose quartz or rarely greenish gray, turning light brown to gray with age (Table 5). No asexual stage has been observed. *O. agrostidis* is a homothallic fungus and produces pseudothecia similar to those of other *Ophiosphaerella* species (Fig. 184), which are commonly found in nature. Pseudothecia are black and 130–350 μm in diameter, with necks 40–160 μm long. Ascospores (Fig. 185) are filiform (needle shaped), 70–190 × 2–2.5 μm, and pale brown, each with six to 15 transverse septa. Ascospores are spirally twisted in bitunicate asci and appear light brown en masse.

In *Agrostis* spp., dark brown to black ectotrophic hyphae appear on or near stolon nodes and stem bases. Ectotrophic hyphae are occasionally present in the internode region of the stolon and on roots emerging from stem bases or nodes.

In *Cynodon* spp., dark brown hyphae can be observed within roots but may not be evident on the roots or stolons.

Disease Cycle

O. agrostidis survives the winter as mycelium in crowns, stolons, and roots and as pseudothecia on colonized tissues. The fungus becomes active when temperatures reach or exceed 20°C (68°F) for a sustained period. It attacks all tissues, generally beginning with stolons or leaves of plants adjacent to previously infected plants. *O. agrostidis* produces large numbers of pseudothecia in the summer. Ascospores are physically discharged or exude en masse in the presence of moisture. They cause secondary infections, resulting in the development of small, reddish brown spots around the original patch within 3–6 days of ascospore discharge.

In *Agrostis* spp., the pathogen may be active from May to December, but disease is most severe during hot, dry periods in the summer.

Fig. 183. Pseudothecia of *Ophiosphaerella agrostidis* embedded in a stolon of *Agrostis stolonifera*. (Courtesy J. E. Kaminski—© APS)

In *Cynodon* spp., the disease appears in March or April during spring green-up. With the advent of higher temperatures, Bermudagrass rapidly recovers.

Dead spot is most severe in full sun and in areas prone to heat and drought stress, such as slopes and ridges. A few to several hundred dead spots can appear on a putting green. The disease is less severe in higher-cut collars and tees. Dead spot occurs in turf grown in sand mixes but has not been found in turf grown in native soils.

Epidemiology

The optimal temperatures for the growth of *O. agrostidis* are 25–30°C (77–86°F). Following winter, the disease reactivates in previously infected spots within 12–28 days after the advent of favorable temperatures. The optimal temperature for the

Fig. 184. Ascocarps of *Ophiosphaerella korrae* (top), *Magnaporthiopsis poae* (middle), and *Gaeumannomyces graminis* (bottom). (Courtesy P. H. Dernoeden, P. J. Landschoot, and N. R. O'Neill)

growth of *O. agrostidis* is higher than the temperatures favorable for the growth of cool-season grasses (16–24°C [61–75°F]). Hence, the pathogen has a competitive advantage over *Agrostis* spp. in the summer. Pseudothecia are rapidly produced as the day length increases in early summer. Ascospores are physically discharged or exuded from ostioles and are transported by wind or water. They can begin to germinate within 2 hours of being deposited on leaves or roots. Ascospores form germ tubes, each of which may form a terminal appressorium. The appressorium forms an infection peg, which directly penetrates host tissues. Germ tubes may also infect the host by entering open stomata on the leaf surface. Leaf and root exudates and light appear to be important factors in the germination of ascospores and subsequent infection of tissues during the morning. When conditions are ideal for growth of *O. agrostidis,* the pathogen can infect turf and produce a second generation of ascospores in as little as 12 days.

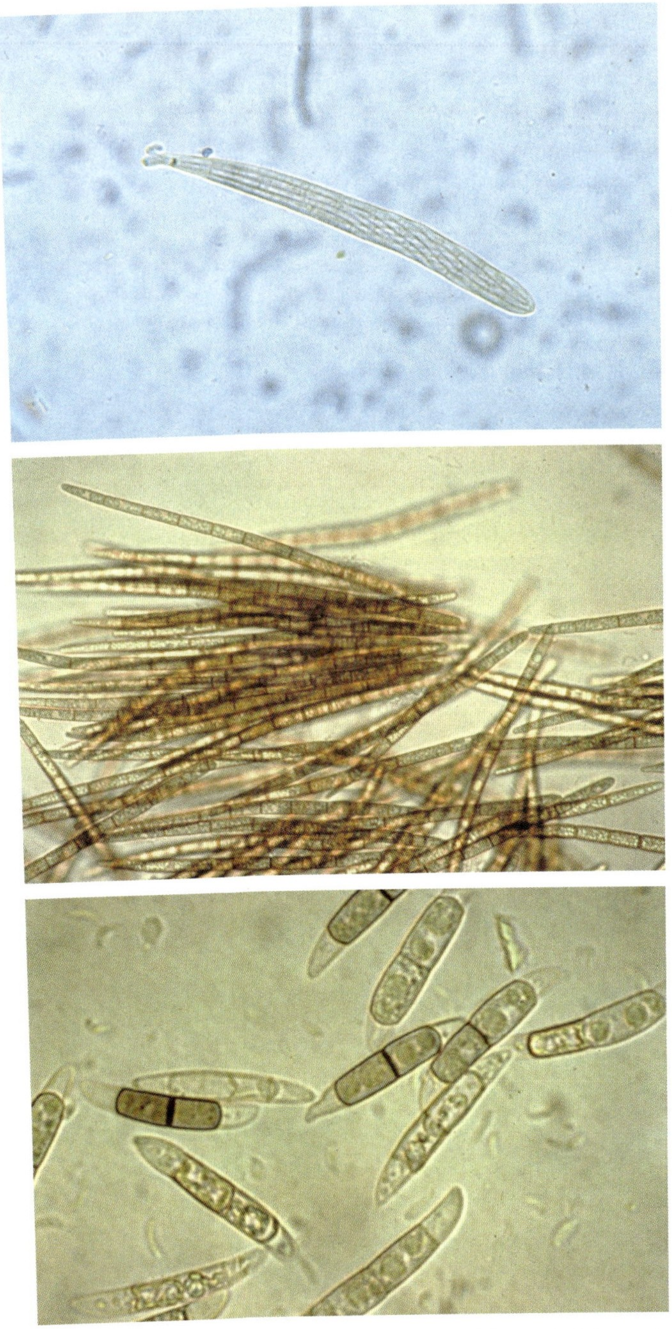

Fig. 185. Ascus of *Ophiosphaerella agrostidis* containing spirally twisted ascospores (top) and ascospores of *O. korrae* (middle) and *Magnaporthiopsis poae* (bottom). (Top, courtesy J. E. Kaminski and P. H. Dernoeden—© APS; middle, courtesy N. Jackson—© APS; and bottom, courtesy P. J. Landschoot—© APS)

Dead spot appears within 1–3 years of seeding or sprigging new greens and tees. It may also occur on older fumigated greens. In most cases, the disease naturally declines within 1–4 years after it first appears but may occur for longer periods in alkaline soils.

Management

Severe outbreaks of dead spot cannot be effectively controlled by cultural practices. When only a few dead spots are present, affected areas can be plugged out and the plugs should be discarded. The disease may be less severe where ammonium sulfate is used as a nitrogen source. The use of urea or nitrate nitrogen can result in a recurrence of dead spot in subsequent years. Like many diseases caused by ectotrophic root-infecting fungi on turfgrasses, the application of lime may intensify dead spot. An efficacious fungicide tank mixed with a low rate of nitrogen can arrest disease development and help speed recovery. Applications of fungicides may be needed every 7–10 days until the turf fully recovers. In *Agrostis* spp., it is important to control the disease and achieve complete recovery of the dead spots before winter. In *Cynodon* spp., turf recovers rapidly when temperatures become favorable for stolon and rhizome growth. An application of nitrogen at spring green-up is also likely to speed recovery.

Selected References

Câmara, M. P. S., O'Neill, N. R., van Berkum, P., Dernoeden, P. H., and Palm, M. E. 2000. *Ophiosphaerella agrostis* sp. nov. and its relationship to other species of *Ophiosphaerella.* Mycologia 92:317-325.

Dernoeden, P. H., O'Neill, N. R., Câmara, M. P. S., and Feng, Y. 1999. A new disease of *Agrostis palustris* incited by an undescribed species of *Ophiosphaerella.* Plant Dis. 83:397.

Kaminski, J. E. 2004. Biology of *Ophiosphaerella agrostis,* epidemiology of dead spot, and a molecular description of the pathogen. Ph.D. diss. Department of Natural Resource Sciences and Landscape Architecture, University of Maryland, College Park.

Kaminski, J. E., and Dernoeden, P. H. 2002. Geographic distribution, cultivar susceptibility, and field observations on bentgrass dead spot. Plant Dis. 86:1253-1259.

Kaminski, J. E., and Dernoeden, P. H. 2005. Dead spot of creeping bentgrass and hybrid bermudagrass. Appl. Turfgrass Sci. 2. doi:10.1094/ATS-2005-0419-01-DG

Kaminski, J. E., and Dernoeden, P. H. 2005. Nitrogen source impact on dead spot (*Ophiosphaerella agrostis*) recovery in creeping bentgrass. Int. Turfgrass Res. Soc. J. 10:214-223.

Kaminski, J. E., and Dernoeden, P. H. 2006. Dead spot severity, pseudothecia development, and overwintering of *Ophiosphaerella agrostis* in creeping bentgrass. Phytopathology 96:248-254.

Kaminski, J. E., Dernoeden, P. H., and Fidanza, M. A. 2007. Environmental monitoring and exploratory development of a predictive model for dead spot of creeping bentgrass. Plant Dis. 91:565-573.

Kaminski, J. E., Dernoeden, P. H., O'Neill, N. R., and Momen, B. 2002. Reactivation of bentgrass dead spot and growth, pseudothecia production, and ascospore germination of *Ophiosphaerella agrostis.* Plant Dis. 86:1290-1296.

Kaminski, J. E., and Hsiang, T. 2006. First report of *Ophiosphaerella agrostis* infecting creeping bentgrass in Canada. Plant Dis. 90:1114.

Krausz, J. P., White, R. H., Tisserat, N. A., and Dernoeden, P. H. 2001. Bermudagrass dead spot: A new disease of bermudagrass caused by *Ophiosphaerella agrostis.* Plant Dis. 85:1286.

Phookamsak, R., Liu, J.-K., McKenzie, E. H. C., Manamgoda, D. S., Ariyawansa, H., Thambugala, K. M., Dai, D.-Q., Camporesi, E., Chukeatirote, E., Wijayawardene, N. N., Bahkali, A. H., Mortimer, P. E., Xu, J.-C., and Hyde, K. D. 2014. Revision of Phaeosphaeriaceae. Fungal Diversity 68:159-238.

(Prepared by P. H. Dernoeden and J. E. Kaminski)

Necrotic Ring Spot

Necrotic ring spot is a disease of *Poa pratensis* and *P. annua* turf most frequently observed in northern and mountainous regions of North America. The pathogen, *Ophiosphaerella korrae,* has been reported in *Festuca rubra, Agrostis stolonifera,* and *P. trivialis* but is not recognized as a common problem in these turf species. This pathogen is also one of the principal incitants of spring dead spot of *Cynodon* and *Zoysia* spp. in North America.

Symptoms and Signs

Circular patches of infected turf first appear during periods of cool, wet weather. The first symptoms are small, light green spots 5–10 cm in diameter. They enlarge into patches that are typically less than 30 cm in diameter but can exceed 60–90 cm, particularly during periods of drought. As turf succumbs to infection, the leaves turn reddish brown to bronze and then fade to a light straw color. All the plants in a patch may die, leaving a sunken or craterlike depression about 2–3 cm deep (Fig. 186). Often, however, plants survive or recolonize infection centers, and the patch takes on a ringlike appearance (Fig. 187). Symptoms may also appear as yellow or brown foliage in diffuse patterns that coalesce into larger areas of blighted turf (Fig. 188). Leaf lesions are not associated with this disease.

Necrotic ring spot may be confused with summer patch, yellow patch, and Microdochium patch, since they cause similar patch symptoms. Unlike these diseases, however, necrotic ring spot can occur throughout the growing season. Like summer patch, necrotic ring spot is characterized by a blackening of roots and rhizomes (Fig. 189) and by the presence of dark brown ectotrophic hyphae on affected roots, rhizomes, and crowns (Fig. 179). Roots may become severely rotted (Fig. 190), and black pseudothecia (fruiting bodies) of the pathogen are occasionally found on these tissues (Fig. 184).

Fig. 186. Necrotic ring spot of *Poa pratensis* with no turf growth in the center of the patch. (Courtesy R. W. Smiley—© APS)

Fig. 187. Necrotic ring spot exhibiting "frog-eyes" as *Poa pratensis* recolonizes the centers of patches. (Courtesy P. L. Koch—© APS)

Causal Agent

Necrotic ring spot is caused by *O. korrae.* Colonies in vitro are white to light gray at first but often turn dark gray or black at maturity. In culture, mycelial growth is aerial and feltlike in appearance (Table 5). *O. korrae* attains maximum growth (3–6 mm per day) on potato dextrose agar at 20–28°C (68–82°F) and is inhibited at temperatures above 30°C (86°F) or below 10°C (50°F). The fungus forms brown, septate, ectotrophic hyphae on infected roots and crowns (Fig. 179). Thin, hyaline hyphae emerge from the ectotrophic hyphae to invade the underlying host tissue. Dark brown, flattened, septate mats (mycelial plaques), 50–400 µm in diameter, and pseudothecia later develop on infected plant parts. Pseudothecia (Fig. 184) are black, flask-shaped (300–500 × 400–600 µm), and thick walled. Asci are bitunicate and cylindrical to clavate (145–200 × 10–15 µm). Ascospores (Fig. 185) are filiform (needlelike), 4–5.5 × 120–180 µm, and pale brown, each with one to 15 septa. A conidial state has not been observed.

Molecular techniques have been developed for rapid identification of *O. korrae* and other ectotrophic root-infecting fungi isolated from turfgrass roots.

Disease Cycle

O. korrae is believed to survive unfavorable periods as septate mats or mycelium in plant debris. Little is known about its development in soil. The fungus is thought to move from plant to plant by growing along the surfaces of roots and rhizomes. Initially, ectotrophic hyphae colonize the surfaces of host roots. Slender, hyaline infection hyphae then infect the root cortex. Infection of roots, rhizomes, leaf sheaths, and crowns occurs during the spring and autumn. Aboveground symptoms typically appear during cool, wet weather in spring or fall. Patches may fade with the advent of warmer conditions in the summer, but in some regions they persist or intensify in response to heat and drought stress. Severely infected plants can be easily removed from the soil because of extensive rotting of roots, crowns, and rhizomes. Pseudothecia are occasionally produced on necrotic roots, rhizomes, leaf sheaths, and crowns, but the importance of ascospores in dissemination of the pathogen is unknown. The pathogen is most likely spread on infested sod and cultivation equipment.

Fig. 188. *Poa pratensis* severely affected with necrotic ring spot. (Courtesy B. B. Clarke—© APS)

Epidemiology

O. korrae infects and colonizes roots and rhizomes during periods of cool, wet weather in the spring and fall. Aboveground symptoms are most common during this time. In moderate climates, symptoms may recede during warmer and drier summer conditions. However, in transition zone environments, the symptoms often persist and intensify as a result of heat and drought stress. Necrotic ring spot can occur in soils with a wide pH range (5–8) and is most severe in compacted soils. The disease is most prevalent on 2- to 4-year-old lawns that were established with sod, but seeded areas and young turf may also be damaged. Like dead spot and take-all patch, necrotic ring spot often declines in severity after several years if left untreated.

Management

Cultural practices that increase root depth or density provide effective suppression of necrotic ring spot. Turf grown in excessively compacted soils is most prone to severe damage. Hollow-tine aerification or similar cultivation practices should be used as needed to alleviate soil compaction. Where possible, foot and vehicular traffic should be distributed to prevent excessive compaction of localized areas. Excessive thatch can be removed by verticutting or power raking. Topdressing or incorporation of compost or other organic amendments has been shown to reduce necrotic ring spot development.

Soil drainage should be improved and overirrigation avoided during spring and fall when the necrotic ring spot pathogen is most active. During periods of hot, dry weather, infected turf may require more-frequent irrigation to minimize heat and drought stress and reduce the expression of symptoms. Light applications of water should be made daily to reduce heat stress and stimulate recovery.

Necrotic ring spot may be less severe when fertilizers that contain slow-release or organic forms of nitrogen are used rather than quick-release (water-soluble) forms. Adequate and balanced fertility should be maintained, particularly nitrogen, phosphorus, and potassium fertility, as determined by soil and tissue tests.

Overseeding infected turf with *Lolium perenne, Festuca arundinacea,* or more-resistant cultivars of *P. pratensis* helps to reduce disease severity and limit symptom expression. Severely affected turfs may require complete renovation with new seed or sod. Although costly, conversion of *P. annua* turf to an *Agrostis* sp. also reduces the occurrence of necrotic ring spot on golf courses.

Several acropetal penetrant fungicides control necrotic ring spot when applied as a preventive measure in early to midspring. Contact fungicides have not provided adequate control in most tests. Local extension personnel may be consulted for a list of currently available fungicides and resistant cultivars.

Fig. 189. Rotting of roots and rhizomes of *Poa pratensis* caused by *Ophiosphaerella korrae.* (Courtesy N. Jackson—© APS)

Fig. 190. Severe root rot of *Poa pratensis* seedlings caused by *Ophiosphaerella korrae* (right) compared with noninoculated seedlings (left). (Courtesy R. W. Smiley—© APS)

Selected References

Brede, A. D. 2005. Necrotic ring spot and turf quality as affected by Kentucky bluegrass cultivar and species mixtures. Int. Turfgrass Res. Soc. J. 10:156-162.

Chastagner, G. A. 1997. Management of necrotic ring spot and take-all patch with DMI fungicides in the Pacific Northwest of the USA. Int. Turfgrass Res. Soc. J. 8:833-846.

Melvin, B. P., and Vargas, J. M., Jr. 1994. Irrigation frequency and fertilizer type influence necrotic ring spot of Kentucky bluegrass. HortScience 29:1028-1030.

Smiley, R. W., and Craven Fowler, M. 1984. *Leptosphaeria korrae* and *Phialophora graminicola* associated with Fusarium blight syndrome of *Poa pratensis* in New York. Plant Dis. 68:440-442.

Smiley, R. W., Craven Fowler, M., and Kane, R. T. 1985. Temperature and osmotic potential effects on *Phialophora graminicola* and other fungi associated with patch diseases of *Poa pratensis.* Phytopathology 75:1160-1167.

Wetzel, H. C., III, Dernoeden, P. H., and Millner, P. D. 1996. Identification of darkly pigmented fungi associated with turfgrass roots by mycelial characteristics and RAPD-PCR. Plant Dis. 80:359-364.

Worf, G. L., Stewart, J. S., and Avenius, R. C. 1986. Necrotic ring spot disease of turfgrass in Wisconsin. Plant Dis. 70:453-458.

(Prepared by R. W. Smiley, P. H. Dernoeden, and B. B. Clarke; revised by L. P. Tredway)

Nematodes

Plant-parasitic and saprophytic nematode species of diverse taxa are normal components of turfgrass ecosystems and are commonly present at high population densities in soil. Non-plant-parasitic nematodes are generally much more abundant than plant-parasitic species. However, in this compendium, the term "nematode" refers only to plant-parasitic species. The importance of these small animals in the spectrum of diseases varies among regions. Nematodes cause extensive losses in turfgrasses in warm temperate and subtropical regions and chronic debilitation of grasses in cooler regions.

Symptoms and Signs

Foliar symptoms (Figs. 191–195) include slight to severe chlorosis, declining growth and gradual thinning, wilting, and sometimes death. Symptoms may be uniform throughout a turf but more often appear in dispersed circular to irregularly shaped patches. Symptoms are most evident under stressful conditions, such as hot weather, drought, or low fertility, but may also appear when conditions are favorable for turf growth if nematode populations are very high. Turfgrasses injured by nematodes have increased susceptibility to attack by fungal pathogens. Nematodes may restrict the development of root systems (Figs. 196 and 197), and some cause distinctive symptoms on roots, such as lesions, swellings, or knots (Figs. 198 and 199).

The visible aboveground symptoms of nematode damage mimic symptoms of stress. Nematodes may be suspected as a possible cause of chlorosis and lack of turf vigor that does not respond to fungicide, irrigation, or fertilizer. Turf that does not respond to aeration of compacted or heavy soil is another indicator of nematode issues. The only way to determine whether plant-parasitic nematodes are involved is to perform an assay of soil and turf root samples. Diagnosis of nematode problems involves linking characteristic symptoms listed above with a nematode assay. Sampling procedures for nematode populations are critical to ascertaining a nematode problem. However, nematode diagnostic laboratories have certain procedures outlined for collecting samples that fit their specific nematode enumeration method. It is advised that turfgrass managers consult laboratory websites for the most current information on sampling techniques. Although sampling techniques vary among laboratories, there are methods that consistently provide accurate diagnoses: (1) Samples should be collected from healthy turf or from the perimeter of damaged turf. Nematodes are biotrophic and require a host to reproduce; therefore, collecting samples from severely damaged turf will likely result

Fig. 191. Sting nematode damage on *Cynodon dactylon* managed as a golf course fairway. (Courtesy L. P. Tredway—© APS)

Fig. 192. Sting and lance nematode damage on a hybrid *Cynodon* spp. putting green. (Courtesy L. P. Tredway—© APS)

Fig. 193. Chlorosis and thinning of a *Poa annua* putting green caused by the stem gall nematode, *Anguina pacificae*. (Courtesy D. K. Mosdell—© APS)

Fig. 194. *Eremochloa ophiuroides* damaged by ring nematodes (*Mesocriconema ornata*) present in high numbers in untreated turf (right) compared with nematicide-treated turf (left). (Courtesy W. T. Crow)

Fig. 195. Sting nematode damage on a tall fescue sward grown for sod production. (Courtesy L. P. Tredway—© APS)

in faulty analysis. (2) Given that laboratories vary with respect to preferred sampling techniques, enumeration methods, and threshold values, it is important to develop a relationship with a single laboratory rather than submitting samples to numerous laboratories. (3) Finally, nematode populations can vary dramatically across turf plantings. Therefore, turfgrass managers should consider breaking turfgrass swards, for example, a putting green, into smaller areas and consistently sampling the same area or areas.

Causal Agents

Plant-parasitic nematodes are transparent, microscopic animals that have complex internal organs and bodily functions (Fig. 200). These tiny roundworms, mostly 0.5–2 mm long, are biotrophs (obligate parasites) of plants. All plant-parasitic species have well-developed stylets (spearlike mouthparts, similar to hypodermic needles), with which they penetrate plant cells and withdraw cell contents. The parasitic genera most commonly reported on turfgrasses (not to be confused with those causing the most damage or those causing significant damage) are described in Key to Some Genera of Nematodes Parasitic on Turfgrasses. The common names of plant-parasitic nematodes refer to their shapes or anatomical characteristics or to symptoms of the damage they cause in plants (Fig. 201). Nematode taxonomy is becoming increasingly complex as modern diagnostic procedures are employed to distinguish certain genera and species. Current taxonomic keys are available on the Internet and in reference texts.

Nematodes that cause the most damage to turfgrasses in cool regions are dagger, spiral, stylet, stubby-root, ring, lesion, lance, and root-knot nematodes. Those that are most damaging in some warm regions include sting, lance, root-knot, stubby-root, ring, sheath, sheathoid, awl, spiral, and stunt nematodes.

Ectoparasitic nematodes feed along the root surface and in mature stages normally do not enter the root tissue (Fig. 201). Feeding by some of these nematodes can cause swellings and abnormal root function. Other ectoparasites can cause necrotic lesions that girdle the root and render it nonfunctional. Tiny, brown lesions, root girdling, and root pruning may result in a shallow or stubby root system.

Endoparasitic nematodes enter roots (Figs. 202 and 203) or permanently attach themselves to roots and feed on the cortical or vascular cells, causing discolored lesions or swellings. The endoparasites listed in the key are the cyst, lance, lesion, and root-knot nematodes. In larval stages, these nematodes can migrate out of roots and therefore act as both endo- and ectoparasites.

Life Cycle

Nematodes have distinctive life cycles, similar to those of certain insects. Females lay eggs, which hatch into juveniles. The juveniles mature through a series of four molts to become adults. Fertilization of eggs by males is required for reproduction in many but not all species. Nematodes typically survive adverse conditions as eggs in the soil. Species parasitic on turfgrasses are most active when environmental conditions favor turf growth. They are motile, but long-distance spread, over

Fig. 196. Damaged root system of a *Cynodon* sp. grown in soil containing sting nematodes (left) compared with healthy plants (right). (Courtesy W. T. Crow)

Fig. 197. *Stenotaphrum secundatum* roots damaged by lance nematodes (*Hoplolaimus* sp.) (right) compared with healthy roots (left). (Courtesy W. T. Crow)

Fig. 198. *Poa annua* roots damaged by the root-gall nematode *Subanguina radicicola*. The turfgrass canopy exhibited a general decline in response to applications of fertilizer and water. (Courtesy R. W. Smiley—© APS)

Fig. 199. *Cynodon dactylon* damaged by the root-knot nematode *Meloidogyne incognita*. (Courtesy J. L. Dale—© APS)

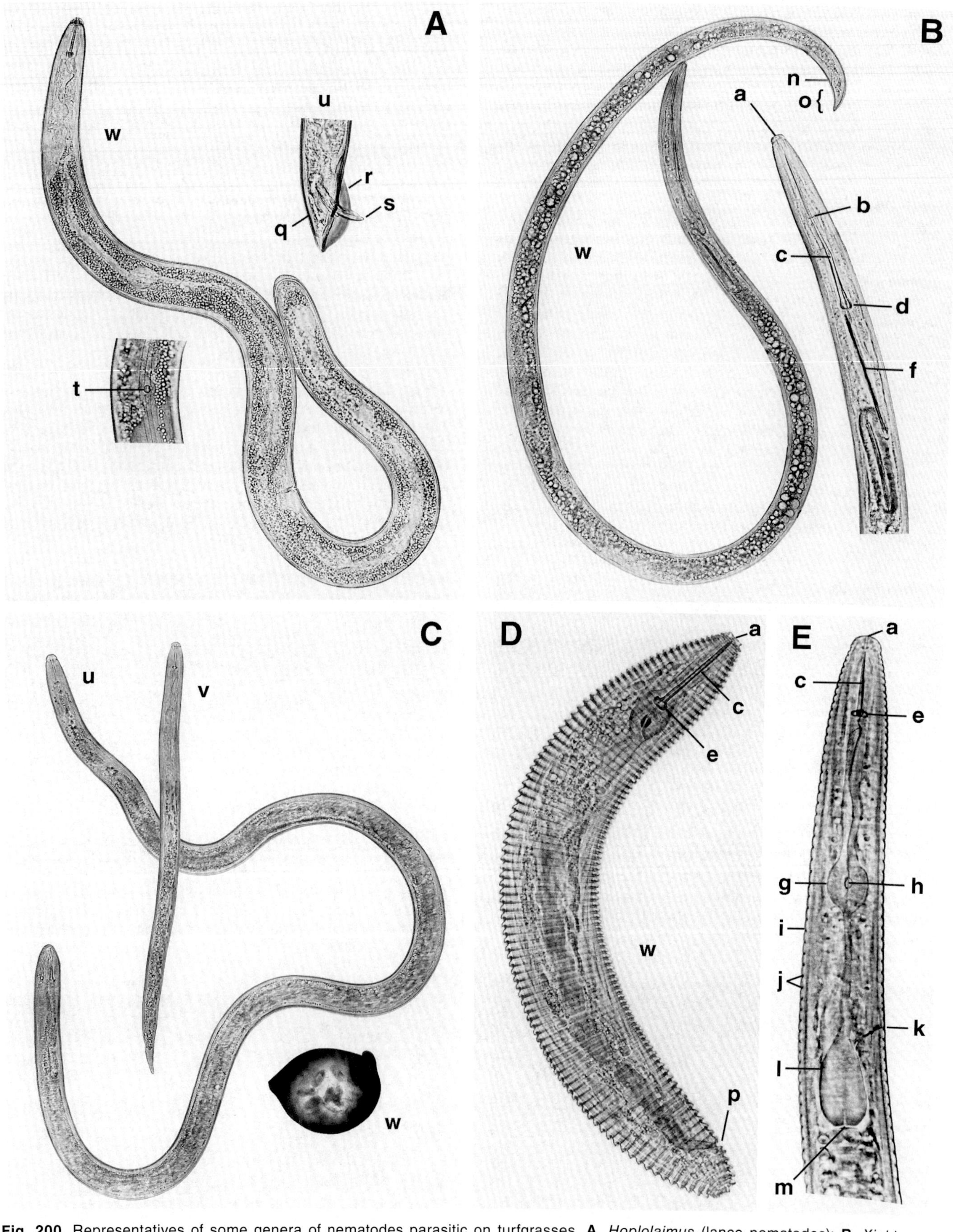

Fig. 200. Representatives of some genera of nematodes parasitic on turfgrasses. **A,** *Hoplolaimus* (lance nematodes); **B,** *Xiphinema* (dagger nematodes); **C,** *Heterodera* (cyst nematodes); **D,** *Macroposthonia* (ring nematodes); and **E,** *Tylenchorhynchus* (stylet or stunt nematodes). Anatomical features: (a) lips; (b) stylet guiding ring; (c) stylet; (d) stylet flanges; (e) basal bulbs of stylet; (f) developing stylet; (g) metacorpus; (h) valve of metacorpus; (i) striation; (j) annule; (k) excretory pore; (l) basal bulb of esophagus; (m) esophago-intestinal valve; (n) anus; (o) tail; (p) vulva (mature female); (q) gubernaculum (mature male); (r) bursa; (s) spiculum; and (t) phasmid. The stages pictured are (u) adult male, (v) larvae, and (w) adult female. (Courtesy W. F. Mai—© APS. Adapted from Mai, W. F., and Lyon, H. H. 1975. Pictorial Key to the Genera of Plant Parasitic Nematodes. Comstock Publishing, Cornell University Press, Ithaca, NY.)

Key to Some Genera of Nematodes Parasitic on Turfgrasses[a]

I. Stylet an odontostyle (spear or tooth) without basal knobs; two-part esophagus without valvulated apparatus (median bulb); male with two testes (Class Adenophorea)
- A. Odontostyle (tooth) short, curved, with dorsal groove; body short and thick, 0.45–1.5 mm long (Order Triplonchida) .... *Paratrichodorus* (stubby-root) or *Trichodorus* (stubby-root)
- B. Odontostyle (spear) long (50–200 μm), straight, hollow; body long (2–8 mm) and slender (Order Dorylaimida)
  - 1. Spear extensions with basal flanges; guiding ring near base of spear .... *Xiphinema* (dagger)
  - 2. Spear extensions without basal flanges; guiding ring near apex of spear .... *Longidorus* (needle)

II. Stylet a stomatostyle with basal knobs; three-part esophagus with valvulated metacorpus (median bulb); male with one testis (Class Secernentea, Order Tylenchida)
- A. Mature female greatly enlarged and pear or lemon shaped
  - 1. Female attached to root by neck only; body turning brown at maturity .... *Heterodera* (cyst)
  - 2. Female completely or nearly embedded in root; white body; distinct knots may form on roots .... *Meloidogyne* (root-knot) or *Hypsoperine* (pseudo-root-knot)
- B. Mature female vermiform (eel-like), not pear-shaped
  - 1. Tail[b] generally more than six times body diameter at anus; one ovary; thick, coarsely annulated cuticle .... *Caloosia*
  - 2. Tail generally less than six times body diameter at anus
    - a) One ovary; vulva located in posterior third of body
      - (1) Procorpus and metacorpus not swollen and combined into a large valvular bulb
        - (a) Stylet less than 15 μm long; tail acute; found in root galls .... *Subanguina* (root-gall)
        - (b) Stylet more than 15 μm long; tail rounded; $s < 1.5$;[c] esophagus overlapping intestine; not found in root galls .... *Pratylenchus* (lesion)
      - (2) Procorpus and metacorpus swollen and combined into a large valvular bulb
        - (a) Mature female without extra cuticle or sheath
          - (i) Cuticle with prominent retrorse annules .... *Criconemella* (ring), *Mesocriconema* (ring), *Criconemoides* (ring), or *Macroposthonia* (ring)
          - (ii) Cuticle without prominent retrorse annules .... *Paratylenchus* (pin)
        - (b) Mature female with extra cuticle or sheath
          - (i) Stylet knobs rounded, sloping anteriorly; cuticle usually with more than 200 annules .... *Hemicycliophora* (sheath)
          - (ii) Stylet knobs anchor-shaped with posterior projection; cuticle usually with fewer than 200 annules .... *Hemicriconemoides* (sheathoid)
    - b) Two ovaries; vulva located near center of body
      - (1) $s > 2.5$[c]
        - (a) Average body length usually more than 1.75 mm; stylet more than 80 μm long .... *Belonolaimus* (sting)
        - (b) Average body length usually less than 1.75 mm; stylet less than 80 μm long .... *Dolichodorus* (awl)
      - (2) Generally, $s < 2.5$[c]
        - (a) Tail usually less than 1.5 times body diameter at anus
          - (i) Both phasmids located posterior to vulva and anterior to anus .... *Peltamigratus*[d]
          - (ii) One phasmid located anterior to vulva and one posterior to vulva .... *Hoplolaimus* (lance)
        - (b) Tail usually more than 1.5 times body diameter at anus .... *Tylenchorhynchus* (stunt)

[a] Prepared by R. W. Smiley, P. H. Dernoeden, and B. B. Clarke—© APS. Modified from Mai and Mullin (1996), Siddiqi (2000), and University of Nebraska, Lincoln, Nematology Lab, Interactive diagnostic key to plant parasitic, freeliving and predaceous nematodes (https://nematode.unl.edu/key/nemakey.htm).

[b] The tail is measured from the anus to the posterior terminus.

[c] s = Stylet length divided by body diameter at the base of the stylet.

[d] *Peltamigratus* spp. on turfgrasses were formerly known as *Rotylenchus* spp.

distances of 1 m or more, usually occurs passively in surface water runoff or by the transport of infested soil or sod.

Nematodes can cause substantial impairment of root functions, but pathogenic fungi residing close to the site of nematode injury can enhance root damage. These pathogens become active in the presence of plant fluids leaking into the soil, and they can multiply rapidly and cause much more severe diseases than those caused by nematodes acting alone. These pathogens may eventually dominate the disease syndrome, and it can be difficult to identify the initial cause of the disease and the respective roles of the different pathogens acting in concert.

Epidemiology

Nematode activity is favored by soil temperatures of 20–30°C (68–86°F) and soil moisture in the range that is readily available to plants. Thatch appears to be an environment conducive to the proliferation of nematodes. Nematode activity is restricted in dry soils and in soils that are saturated long enough to become anaerobic. Highly compacted and heavy-textured soils generally are less favorable for nematodes than friable, light-textured soils. For instance, turfgrasses cultivated in sand root zones rather than in heavier soils are more vulnerable to nematode damage, even in cool temperate climates.

Nematodes are most active and most numerous on cool-season grasses in mid- to late spring and again in autumn and on warm-season grasses during the summer and early autumn. These times coincide with periods of active growth of turfgrass roots. Specifically, sting nematode populations primarily reside within the top 10 cm of soil when soil temperatures are 15–27°C (59–81°F). Populations migrate below the root zone as soil temperatures rise above 27°C (81°F). Root-knot nematode populations, however, remain in the upper 10 cm and counts continue to increase through the summer months. The effects of nematode feeding appear only when conditions become un-

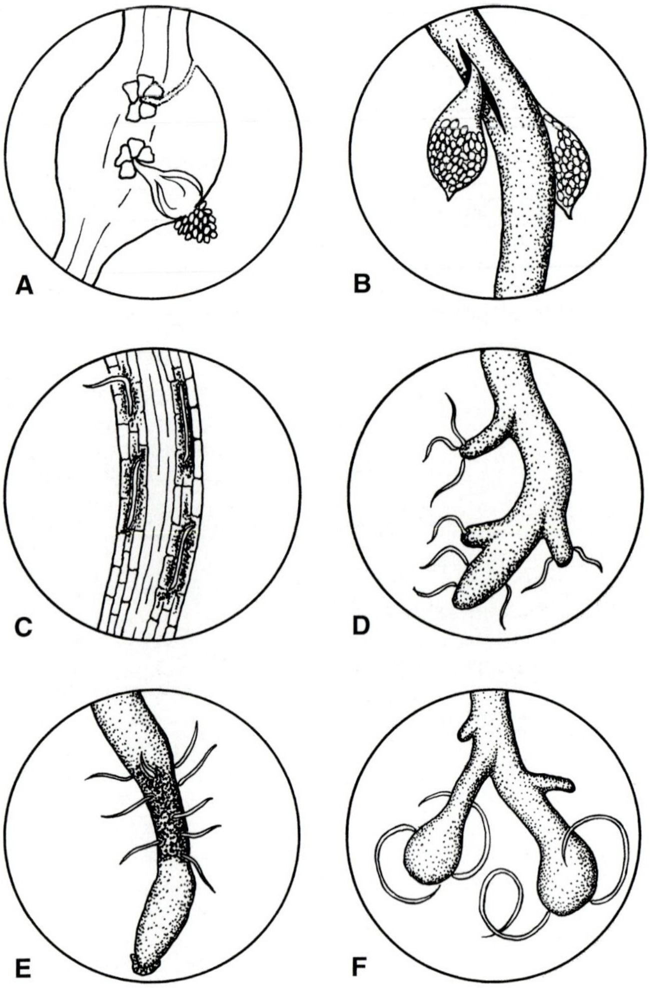

Fig. 201. Symptoms caused by nematodes parasitic on turfgrasses. **A,** *Meloidogyne* (root-knot nematodes); **B,** *Heterodera* (cyst nematodes); **C,** *Pratylenchus* (lesion nematodes); **D,** *Trichodorus* (stubby-root nematodes); **E,** *Macroposthonia* (ring nematodes); and **F,** *Xiphinema* (dagger nematodes). (Courtesy C. R. Whittaker—© APS. Adapted from Radewald, J. D. 1980. Nematode diseases of turfgrasses. Pages 40-42 in: Turfgrass Pests. W. R. Bowen and P. A. Davis, eds. University of California, Division of Agricultural Sciences, Berkeley.)

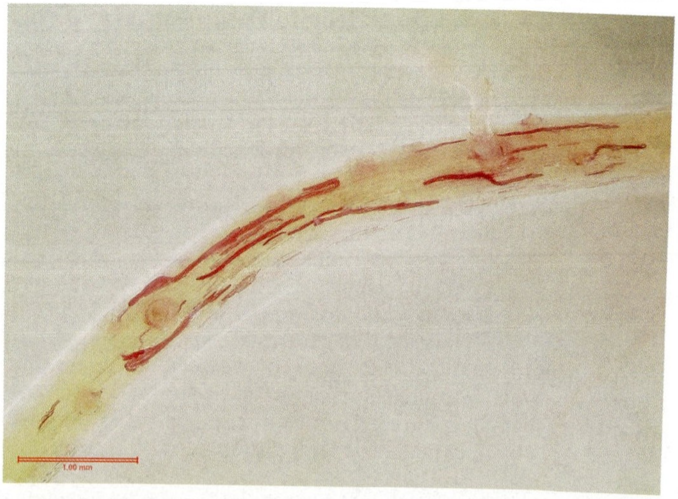

Fig. 202. Lance nematodes (*Hoplolaimus* sp.), stained red, in and on roots of a *Cynodon* sp. (bar = 1.00 mm). (Courtesy W. T. Crow)

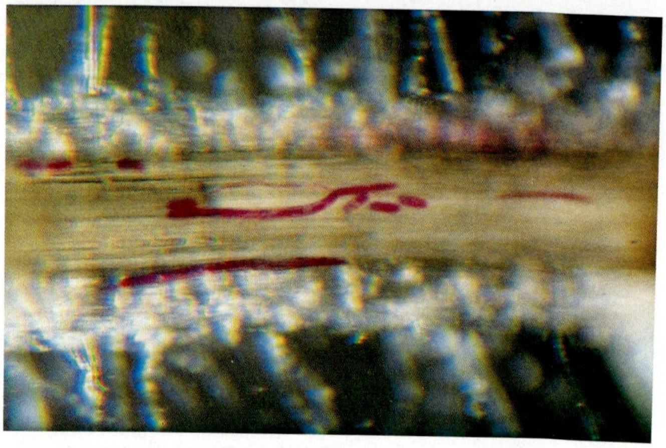

Fig. 203. Lesion nematode (*Pratylenchus neglectus*) eggs and adults, stained red, inside a root. (Courtesy R. W. Smiley—© APS)

favorable for the growth of the turf. Periods of stress, such as heat, drought, low fertility, and others, can result in symptoms of nematode feeding as well.

Management

No turfgrass cultivars are resistant to feeding by all nematode species. Different nematode species, however, exhibit preferential feeding on turfgrasses of certain species and cultivars. Cultural practices that promote vigorous turfgrass root growth can aid in suppressing visual symptoms of nematode damage. Nematodes feed on plant roots and predispose turf plants to various stresses including heat, drought, and nutrient stress. Fertilization and application of wetting agents and organic materials such as humates may assist in limiting nematode damage. Nematode management varies among nematode species and climatic regions; thus, it is prudent to consult local extension specialists on current nematode management strategies for specific nematode species in a particular climatic zone. For example, nematode management in tropical regions is likely to be more intensive than that in temperate regions.

Chemical fumigants or nematicides are advised when nematode counts exceed currently accepted thresholds in a particular region and when stand symptoms resemble nematode damage. Although nematode assays are essential for determining whether a chemical nematicide is required for management, periodic sampling after implementation of a chemical nematicide program may not document a decline in nematode populations. Most fumigants can be used only before seeding because they are highly toxic to plants. However, 1,3 dichloropropene can be injected 12 cm below the soil surface with specialized equipment. Chemical burn may occur in injection slits and can be visible for up to 14 days after the application. This active ingredient is currently labeled only for use on golf courses and athletic fields in Alabama, Florida, Georgia, North Carolina, and South Carolina. A few synthetic organic nematicides are currently registered for use on turf including abamectin, fluopyram, and furfural. Nematicides should be chosen for use on the basis of previous sampling for species, because none control all nematodes that may be present. Multiple nematicides may be necessary in a rotation to avoid selecting for certain species and for resistance management. Timing of applications usually coincides with turfgrass root growth and production, but the timing varies in different climatic zones. All nematicides require postapplication irrigation to be most effective. There are numerous biological materials available for use in turfgrass systems, but there is limited data supporting their efficacy. Nematicides and fumigants can be toxic to animals and should be used and handled only as directed on the labels.

Selected References

Bekal, S., and Becker, J. O. 2000. Population dynamics of the sting nematode in California turfgrass. Plant Dis. 84:1081-1084.

Chastagner, G. A., and McElroy, F. D. 1984. Distribution of plant-parasitic nematodes in putting green turfgrass in Washington. Plant Dis. 68:151-153.

Crow, W. T., and Davis, S. D. 2001. Survey of plant-parasitic nematodes in golf courses along the northern coast of the Gulf of Mexico. (Abstr.) Phytopathology 91(suppl.):S132.

Crow, W. T., and Walker, N. R. 2003. Diagnosis of *Peltamigratus christiei,* a plant-parasitic nematode associated with warm-season turfgrasses in the southern United States. Plant Health Prog. doi:10.1094/PHP-2003-0513-01-DG

Evans, K., Trudgill, D. L., and Webster, J. M. 1993. Plant Parasitic Nematodes in Temperate Agriculture. CAB International, Wallingford, U.K.

Heald, C. M., and Perry, V. G. 1969. Nematodes and other pests. Pages 360-369 in: Turfgrass Science. Agron. Monogr. 14. A. A. Hanson and F. V. Juska, eds. American Society of Agronomy, Madison, WI.

Mai, W. F., and Mullin, P. G. 1996. Plant Parasitic Nematodes: A Pictorial Key to Genera. Cornell University Press, Ithaca, NY.

Mitkowski, N. A., and Jackson, N. 2003. *Subanguina radicicola,* the root-gall nematode infecting *Poa annua* in New Brunswick, Canada. Plant Dis. 87:1263.

Radewald, J. D. 1980. Nematode diseases of turfgrasses. Pages 40-42 in: Turfgrass Pests. W. R. Bowen and P. A. Davis, eds. Publ. 4053. University of California, Division of Agricultural Sciences, Berkeley.

Reidel, R. M. 1980. Nematode problems on northern turfgrasses. Pages 59-62 in: Advances in Turfgrass Pathology. P. O. Larsen and B. G. Joyner, eds. Harcourt Brace Jovanovich, Duluth, MN.

Settle, D. M., Fry, J. D., Todd, T. C., and Tisserat, N. A. 2006. Population dynamics of the lance nematode (*Hoplolaimus galeatus*) in creeping bentgrass. Plant Dis. 90:44-50.

Siddiqi, M. R. 2000. Tylenchida: Parasites of Plants and Insects. 2nd ed. CAB International, Wallingford, U.K.

Trenholm, L. E., Lickfeldt, D. W., and Crow, W. T. 2005. Use of 1,3-dichloropropene to reproduce irrigation requirements of sting nematode-infested bermudagrass. HortScience 40:1543-1548.

Todd, T. C., and Tisserat, N. A. 1990. Occurrence, spatial distribution, and pathogenicity of some phytoparasitic nematodes on creeping bentgrass putting greens in Kansas. Plant Dis. 74:660-663.

Walker, N. R., Goad, C. L., Zhang, H., and Martin, D. L. 2002. Factors associated with populations of plant-parasitic nematodes in bentgrass putting greens in Oklahoma. Plant Dis. 86:764-768.

Yu, Q., Potter, J. W., and Gilby, G. 1998. Plant-parasitic nematodes associated with turfgrass in golf courses in southern Ontario. Can. J. Plant Pathol. 20:304-307.

Zeng, Y., Ye, W., Martin, S. B., Martin, M., and Tredway, L. 2012. Diversity and occurrence of plant-parasitic nematodes associated with golf course turfgrasses in North and South Carolina, USA. J. Nematol. 44:337-347.

Zeng, Y., Ye, W., Tredway, L., Martin, S., and Martin, M. 2012. Taxonomy and morphology of plant-parasitic nematodes associated with turfgrasses in North and South Carolina, USA. Zootaxa 3452:1-46.

(Prepared by R. W. Smiley, P. H. Dernoeden, and B. B. Clarke; revised by J. P. Kerns)

Pythium Root Rot

Pythium root rot is a devastating disease of fine amenity turfgrasses. All turfgrass species can develop Pythium root rot, but the disease is most prevalent and severe on golf course putting greens. Traditionally, Pythium root rot was problematic only in the transition zone of the United States, but as the global climate has warmed, Pythium root rot has become a major disease of golf course putting greens in many locations. It has been documented throughout the United States, Canada, Japan, and Australia. This disease is difficult to diagnose and manage and is frequently confused with Pythium root dysfunction. Although Pythium root rot and Pythium root dysfunction are related diseases, symptom expression occurs under different conditions and best management practices differ. Pythium root rot is far more widespread and common than Pythium root dysfunction. Pythium root rot, like many root diseases, can be exacerbated by nematode feeding or other stresses such as heat or traffic.

Symptoms and Signs

Numerous *Pythium* spp. are associated with crown and root diseases that cause a general decline of turfgrass stands. Symptoms of declining turfgrass health are typically nonspecific. Affected turf may appear thin, off color, and slow growing. The disease may occur in small patches (Fig. 204), or it may occur as a general decline affecting large areas and eventually causing widespread loss of turf.

Crown and root diseases caused by *Pythium* spp. are most often associated with highly maintained golf greens. Symptoms may become visible at any time during the growing season. Under cool, wet conditions in early spring and late autumn, symptoms on cool-season grasses first appear as small, diffuse, yellow patches 4–8 cm in diameter (Fig. 205). Initially, symptoms are most prominent in the periphery or "clean-up" areas of golf greens or in high-traffic areas. Infected plants may appear stunted, and their leaves are yellow and more narrow than

Fig. 204. Initial symptoms of Pythium root rot on an *Agrostis stolonifera* golf green. (Courtesy E. L. Butler—© APS)

Fig. 205. Pythium root rot on an *Agrostis stolonifera* golf green during cool, wet conditions in spring. (Courtesy J. P. Kerns—© APS)

healthy leaves. Patches of severely infected turf may coalesce into large areas in which the turf appears yellow or reddish brown and generally weakened (Fig. 206). As temperatures rise, large areas of turf may wilt, turn brown, and die. Under warm, wet conditions in midsummer, especially on shaded greens with excessively wet soil, the turf appears reddish brown or yellow and thins out in irregular patterns.

On putting greens planted with warm-season grasses, symptoms generally appear as off-white to yellow patches 10–30 cm in diameter (Fig. 207). Affected areas may coalesce and appear to track along drainage patterns. Symptoms of Pythium root rot on warm-season grasses can be easily confused with those of take-all root rot and root-knot nematode damage. Symptoms on warm-season grasses can occur at any time of the year, but symptom expression does not vary with the season as observed with cool-season turfgrasses. Algae invasion is very common in turfgrass stands thinned by Pythium root rot.

Unlike Pythium blight, crown and root diseases caused by *Pythium* spp. are not associated with the production of mycelium on leaves, and rarely can they be diagnosed from field symptoms alone. The crowns of severely infected plants may appear water soaked, and their root systems are discolored and greatly reduced in length and volume. Roots and crowns may be extensively discolored. Root depth is significantly impaired when Pythium root rot is severe. Microscopic examination of roots and crowns is required to determine whether the damage is associated with infection by *Pythium* spp. Laboratory diagnosis is often based on an abundance of oospores in root tissue (Fig. 208) and/or crown tissue. In severely damaged root systems, signs of *Pythium* spp. may not be present and diagnosis is based on symptomology. Typically, roots infected and killed by *Pythium* spp. are water soaked and slimy (Fig. 209). Often the cortical cells of the roots are sloughed off or significantly spread apart because *Pythium* spp. degrade pectin that is pres-

Fig. 206. Stand symptoms of severe Pythium root rot on an *Agrostis stolonifera* golf green. (Courtesy L. P. Tredway—© APS)

Fig. 207. Pythium root rot of an ultradwarf cultivar of hybrid *Cynodon* spp. (Courtesy J. P. Kerns—© APS)

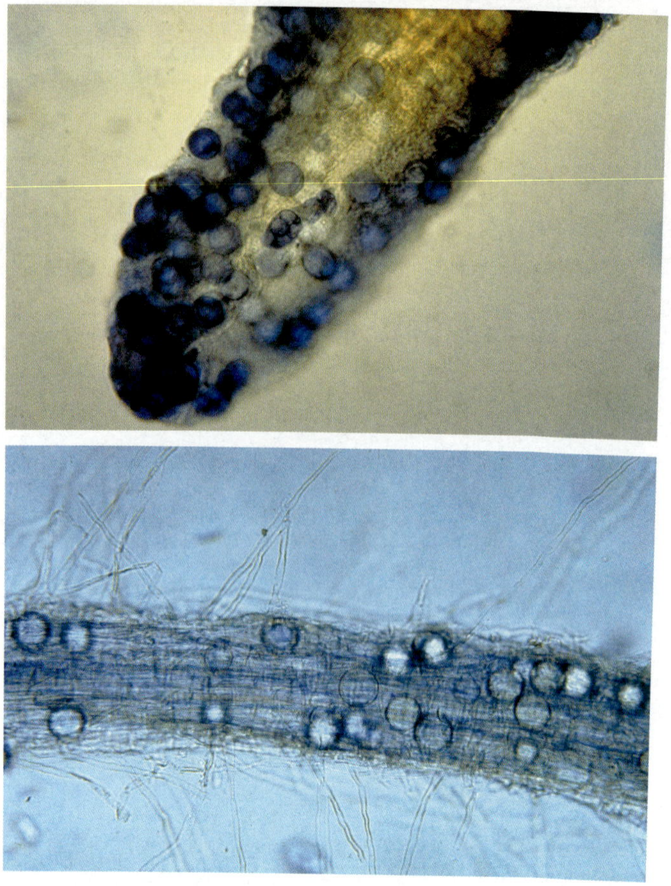

Fig. 208. Oospores of *Pythium* spp. in turfgrass roots. (Courtesy E. B. Nelson—© APS)

Fig. 209. Reduction in depth and necrotic, water-soaked appearance of *Agrostis stolonifera* roots caused by Pythium root rot. (Courtesy E. L. Butler—© APS)

ent in the middle lamella. Sporangia may also be seen in the cortex of infected roots during active periods of infection.

Causal Agents

Numerous *Pythium* spp. are associated with roots of turfgrass plants exhibiting symptoms of Pythium root rot. Isolation of these organisms into pure culture is difficult because once symptoms develop, the roots are often severely rotted and colonized with faster-growing saprophytes. Therefore, the etiology of this disease remains elusive.

The following species are frequently isolated from turf exhibiting symptoms of Pythium root rot: *P. aristosporum, P. arrhenomanes, P. catenulatum, P. graminicola, P. myriotylum, P. tardicrescens, P. vanterpoolii, P. volutum, Phytopythium vexans, Globisporangium irregulare,* and *G. ultimum*. These species are moderately or highly aggressive toward *Agrostis stolonifera* seedlings, but to date Pythium root rot has not been induced on mature turfgrass plants in a controlled environment.

Pythium torulosum is among the most common species associated with *A. stolonifera* roots, yet it routinely fails to kill seedlings in pathogenicity assays. In studies concerning the etiology of Pythium root dysfunction, *P. torulosum* was nonpathogenic to mature turfgrass plants as well. This species is likely an excellent colonizer of senescent tissue and frequently clouds etiology assays because it also grows rapidly compared with more aggressive pathogens.

Most *Pythium* spp. that infect grasses are homothallic and generally produce oospores or sporangia (or both) on artificial media. All species produce coenocytic mycelia. They are distinguished by characteristics of their antheridia, oogonia, and sporangia (Figs. 95–97). Oospores range from 15 to 30 μm in diameter, depending on the species.

Disease Cycle

Little is known regarding the disease cycle of Pythium root rot in turfgrasses. *Pythium* spp. are ubiquitous in the environment and capable of long-distance transport in water, soil, or infected plant tissue. Thick-walled oospores can survive unfavorable conditions in the soil for months or years. Hyphae and sporangia may also survive in the soil or infected plant tissue for shorter periods. When temperature and moisture are conducive to disease development, oospores germinate and infect the host plant directly or give rise to sporangia. Sporangia may also germinate and infect the plant directly or release zoospores, which are small, flagellated spores that enable *Pythium* spp. to spread rapidly under wet conditions. Upon coming in contact with a suitable host, zoospores encyst and then germinate to form germ tubes that initiate host infection.

Epidemiology

There are many *Pythium* spp. that have the ability to cause root rot of turfgrasses. These species grow and infect turfgrass roots at different temperatures, which can range from 10 to 35°C (50 to 95°F). As a result, Pythium root rot can develop at any time during the growing season if the soil remains saturated for prolonged periods. Poor surface or subsurface drainage, overirrigation, heavy rainfalls, and excessive accumulation of thatch and organic matter are the most common factors that lead to a Pythium root rot outbreak. Symptoms are most severe when heat and/or drought stress follows a prolonged wet period.

Management

Turfgrasses should not be established in poorly drained areas that remain saturated for extended periods. Golf course putting greens must be aerified and topdressed regularly to control the accumulation of thatch and reduce organic matter. In general, 15–20% of the putting green surface area should be impacted by hollow-tine aerification annually, and 5,000 lb (2,272 kg) of topdressing should be applied per 1,000 ft^2 (93 m^2) each year. Pruning or removal of trees surrounding putting greens to increase sunlight penetration and air movement reduces Pythium root rot activity. Installation of high-powered fans also helps to alleviate the problem where air movement is restricted. For golf course putting greens with poor internal drainage, reconstruction is the only practical long-term solution for Pythium root rot.

Where this disease has been a persistent problem, labeled fungicides should be applied every 14–21 days during the growing season or when rainfall occurs 2–3 days in a row. For curative applications, ethazole should be applied first, followed by an application of cyazofamid, mefenoxam, or propamocarb 2–3 days later. To minimize the potential for foliar burn, ethazole must be watered-in immediately after application with at least 3 mm of water. Other fungicides should also be watered-in to drive the active ingredient into the root zone where the *Pythium* spp. are most active.

Selected References

Abad, Z. G., Shew, H. D., and Lucas, L. T. 1994. Characterization and pathogenicity of *Pythium* species isolated from turfgrasses with symptoms of root and crown rot in North Carolina. Phytopathology 84:913-921.

Couch, H. B. 1995. Diseases of Turfgrasses. 3rd ed. Krieger Publishing, Malabar, FL.

Hampy, H. D., VanRyzin, B. J., Butler, E. L., and Kerns, J. P. 2021. Etiology and management of Pythium root rot. Int. Turfgrass Soc. Res. J. doi:10.1002/its2.24

Hsiang, T., Wu, C., Yang, L., and Liu, L. 1995. Pythium root rot associated with cool-season dieback of turfgrass in Ontario and Quebec. Can. Plant Dis. Surv. 75:191-195.

Nelson, E. B., and Craft, C. M. 1991. Identification and comparative pathogenicity of *Pythium* spp. from roots and crowns of turfgrasses exhibiting symptoms of root rot. Phytopathology 81:1529-1536.

Schmitthenner, A. F. 1980. *Pythium* species: Isolation, biology and identification. Pages 33-36 in: Advances in Turfgrass Pathology. P. O. Larsen and B. G. Joyner, eds. Harcourt Brace Jovanovich, Duluth, MN.

(Prepared by J. P. Kerns and E. L. Butler)

Pythium Root Dysfunction

Pythium root dysfunction was first described in 1985 when distinct patches of dead *Agrostis stolonifera* were seen in golf course putting greens. Symptoms were most severe on the apexes of slopes and did not conform to the patterns typically associated with Pythium root rot. The roots in affected areas were not necrotic, but they lacked root hairs, leading to the name "Pythium-induced root dysfunction." The disease became fairly common in the late 1990s and early 2000s, corresponding with a boom in golf course construction. This disease is frequently confused with Pythium root rot and the names are used interchangeably, but they are separate diseases, each with a distinct etiology, epidemiology, and management strategy. Pythium root dysfunction has been documented only in *A. stolonifera* putting greens and is associated primarily with those that are less than 10 years old and constructed with root zones that are high in sand content (85% or greater).

Symptoms and Signs

Symptoms of Pythium root dysfunction typically appear during the summer months when creeping bentgrass is subjected to heat and drought stress, but patches may also develop from fall to spring during periods of warm and dry weather. Initially, symptoms appear as circular areas of wilt, chlorosis, or drought stress 4–16 cm in diameter (Fig. 210). These areas progress to larger, circular to irregular patches of yellow to orange foliar decline ranging up to 50 cm in diameter (Fig. 211). If affected areas are left untreated, large areas of turf can eventually be killed (Fig. 212).

Symptoms typically appear first on mounds and along slopes, because these areas are most prone to drought stress. Pythium root dysfunction symptoms are generally most severe on putting greens with good air circulation and sunlight exposure.

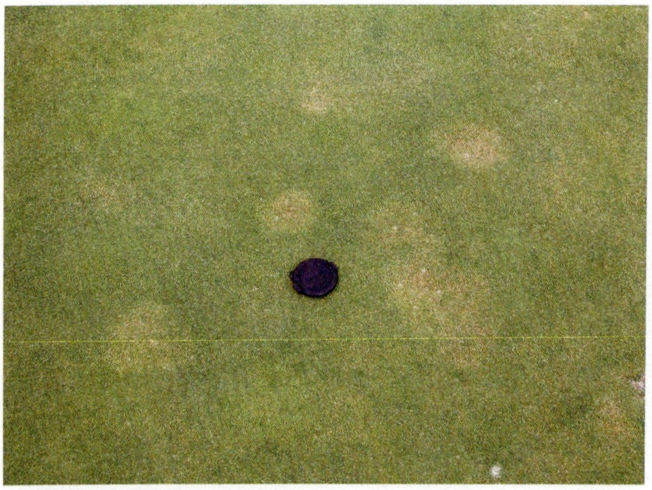

Fig. 210. Initial symptoms of Pythium root dysfunction on an *Agrostis stolonifera* putting green. (Courtesy L. P. Tredway—© APS)

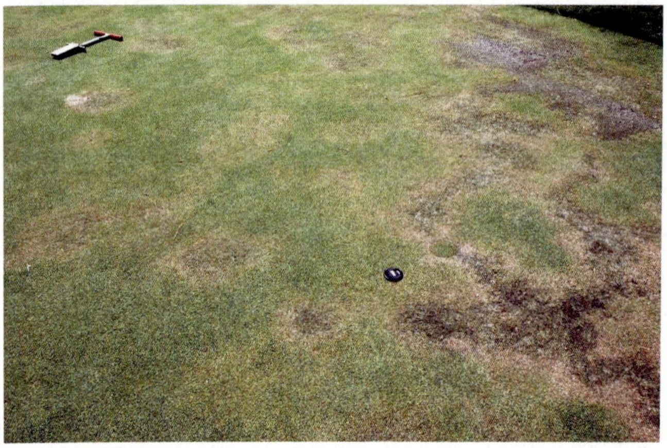

Fig. 211. Advanced symptoms of Pythium root dysfunction on an *Agrostis stolonifera* putting green. (Courtesy L. P. Tredway—© APS)

Fig. 212. Pythium root dysfunction left untreated on *Agrostis stolonifera* progressing to large dead patches. (Courtesy L. P. Tredway—© APS)

Many times, golf course superintendents report symptoms characteristic of Pythium root dysfunction after a tournament or other event, especially if the greens were mowed lower than normal or allowed to dry out.

Roots infected by the Pythium root dysfunction pathogens have bulbous root tips, lack root hairs, and are light tan (Fig. 213). In the field, sand does not cling to roots affected by *Pythium* spp. as it normally does to healthy roots, presumably because of the lack of root hairs. Root depth and density may not be impacted early in the season, but infected roots die rapidly upon heat exposure and reduced rooting is often observed in mid- to late summer (Fig. 214).

Pythium root dysfunction can be difficult to diagnose at the peak of symptom expression in summer because the pathogens are not highly active at this time. Furthermore, infected roots are easily detached from symptomatic plants during sample collection or processing. Diagnosticians should handle suspected samples carefully to avoid detaching the dead and dying roots.

During periods of pathogen activity, oospores and coenocytic hyphae are readily observed in cortical cells of creeping bentgrass roots (Fig. 215). Frequently, bisporous oospores are observed along the vascular cylinders of affected roots (Fig. 216). The density of oospores present in the root tissue does not seem to correlate with the severity of disease. The number of oospores can be low, but the potential for root disease still exists. Pythium root dysfunction can be diagnosed by observation of

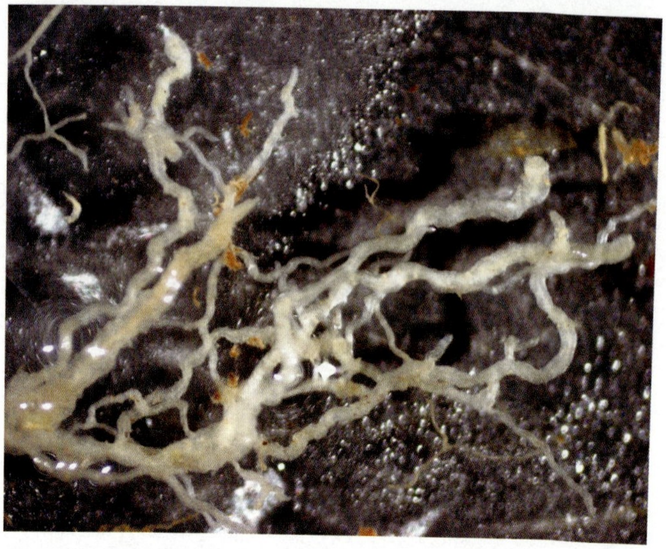

Fig. 213. Lack of root hairs, bulbous root tips, and beige coloration, characteristic symptoms of Pythium root dysfunction. (Courtesy L. P. Tredway—© APS)

Fig. 214. *Agrostis stolonifera* with reduced root depth and density caused by Pythium root dysfunction (left) compared with roots from an unaffected area (right). (Courtesy L. P. Tredway—© APS)

characteristic plant and stand symptoms combined with microscopic confirmation of the pathogen's presence.

Causal Agents

Three *Pythium* spp. have demonstrated the ability to incite Pythium root dysfunction: *P. aristosporum, P. arrhenomanes,* and *P. volutum. P. aristosporum* and *P. arrhenomanes* are frequently isolated from creeping bentgrass turf in the Midwestern and Mid-Atlantic areas of the United States. *P. volutum* has been associated primarily with creeping bentgrass in the southeastern United States. Many species of *Pythium* are distributed worldwide, but these organisms are normally found in the Northern Hemisphere.

Most *Pythium* spp. that infect grasses are homothallic and generally produce oospores or sporangia (or both) on artificial media. All species produce coenocytic mycelia. They are distinguished by characteristics of their antheridia, oogonia, and sporangia (Figs. 95–97).

P. aristosporum, P. arrhenomanes, and *P. volutum* produce filamentous-inflated sporangia and moderately thick hyphae up to 6.5 μm wide. Oogonia and oospores are not produced in abundance in root tissues but can be induced in grass-leaf water cultures. Oogonia and oospores are 21–30 μm in diameter, large compared with those of many *Pythium* spp.

Disease Cycle

Pythium root dysfunction symptoms are typically observed within 1 year of planting newly constructed or renovated creeping bentgrass greens but have been observed as soon as 3 months. The source of *Pythium* spp. inoculum in newly constructed or fumigated greens is unknown, but *Pythium* spp. are known to spread readily in water, soil media, or plant tissue.

The causal agents of Pythium root dysfunction are most active in fall to spring when creeping bentgrass roots are actively growing. Hyphal growth can be observed in root tips, root hairs, and cortical tissues, and these infections appear to hamper the absorption of water and nutrients from the soil. Aboveground symptoms are often absent at this time but may include growth suppression or increased susceptibility to drought or nutrient deficiency.

The progression of symptoms in distinct circular patches suggests that hyphal growth from root to root may be the primary means of dispersal for these pathogens. Sporangia and zoospores are rarely observed in creeping bentgrass roots exhibiting symptoms of Pythium root dysfunction, although they can be induced in grass-leaf water cultures.

When subjected to heat stress, infected roots die back rapidly, leading to decline of the turf. The pathogens are not highly active at this time and likely survive in the soil or infected roots as oospores or hyphae.

The symptoms typically decline in severity over 5–10 years after establishment and disappear entirely in many cases. The cause of this decline is unknown but is likely related to changes in the soil microbial population and/or accumulation of organic matter in the soil profile over time.

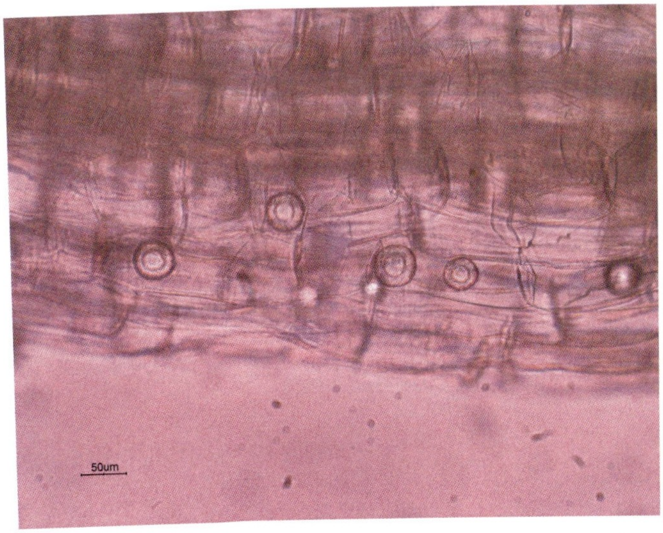

Fig. 215. Oospores of *Pythium volutum* in cortical cells of an *Agrostis stolonifera* root. (Courtesy L. P. Tredway—© APS)

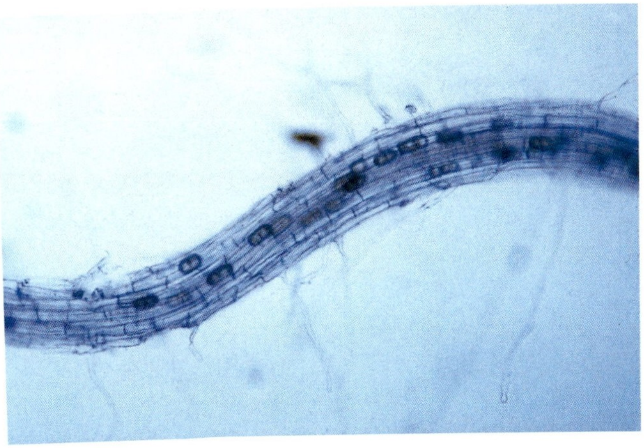

Fig. 216. Bisporous oospores of a *Pythium* sp. embedded in the cortical cells of an *Agrostis stolonifera* root and lining the vascular cylinder. (Courtesy P. H. Dernoeden—© APS)

Epidemiology

Symptoms are most common during hot weather in summer but may also appear during warm, dry conditions in fall, winter, or spring. Infection of creeping bentgrass roots results in the loss of root hairs, thus reducing water and nutrient absorption from the soil. Infected roots die back rapidly when soil temperatures exceed 29°C (84°F). The severity of Pythium root dysfunction symptoms is enhanced by low fertility, drought stress, and low soil oxygen levels.

Most root infection appears to occur in the fall, winter, and spring when soil temperatures are most favorable for the growth of the pathogens. *P. aristosporum, P. arrhenomanes,* and *P. volutum* grow optimally at 25°C (77°F) but can infect turfgrass roots at temperatures as low as 13°C (55°F). In greenhouse inoculations, *P. arrhenomanes* was more aggressive in creeping bentgrass roots at 25°C (77°F) than at 35°C (95°F). Wet soil conditions caused by rainfall or overirrigation are conducive to infection during times of active pathogen growth.

Management

Creeping bentgrass greens should be constructed with at least 15% organic matter by volume to provide adequate water- and nutrient-holding capacity. After establishment, creeping bentgrass should be fertilized with 4–6 lb of nitrogen per 1,000 ft^2 (195–292 kg/ha) annually. Regular soil and tissue testing ensures that all other nutrients are present in adequate amounts. Mowing heights above 3 mm during the summer increase photosynthesis and alleviate stress. Severe drought stress can be avoided by irrigating before symptoms of wilt become widespread, and areas that dry out more quickly can be supplemented by syringing and hand-watering. Soil surfactants increase the uniformity of soil moisture across putting greens and assist with water management. Frequent cultivation and topdressing are needed to maximize oxygen levels in the soil. In general, 15–20% of the putting green surface area should be impacted by hollow-tine aerification annually, and 5,000 lb (2,272 kg) of topdressing should be applied per 1,000 ft^2 (93 m^2) each year. Biweekly solid-tine aeration during the summer increases root survival by creating temporary channels for air and water movement through the soil profile.

Standard fungicides, such as mefenoxam, propamocarb, and fosetyl-Al, have provided poor to moderate control of Pythium root dysfunction when applied alone. The most effective fungicides have been those belonging to the quinone outside inhibitor (azoxystrobin, fluoxastrobin, and pyraclostrobin) and quinone inside inhibitor (cyazofamid) chemical classes. Applications followed immediately by 3 mm of irrigation have been slightly more effective than standard foliar applications. The same result may also be obtained by applying larger volumes of water (4–6 gallons/1,000 ft^2 or 1,600–2,400 L/ha) to drive the fungicide deeper into the canopy.

Selected References

Feng, Y., and Dernoeden, P. H. 1999. *Pythium* species associated with root dysfunction of creeping bentgrass in Maryland. Plant Dis. 83:516-520.

Hodges, C. F., and Campbell, D. A. 1994. Infection of adventitious roots of *Agrostis palustris* by *Pythium* species at different temperature regimes. Can. J. Bot. 72:378-383.

Hodges, C. F., and Coleman, L. W. 1985. *Pythium*-induced root dysfunction of secondary roots of *Agrostis palustris*. Plant Dis. 69:336-340.

Kerns, J. P., Shew, H. D., Benson, D. M., and Tredway, L. P. 2009. Impact of irrigation frequency, organic matter content and creeping bentgrass cultivar on the development of Pythium root dysfunction. Int. Turfgrass Soc. Res. J. 11:227-239.

Kerns, J. P., and Tredway, L. P. 2007. First report of Pythium root dysfunction of creeping bentgrass caused by *Pythium volutum* in North Carolina. Plant Dis. 91:632.

Kerns, J. P., and Tredway, L. P. 2008. Pathogenicity of *Pythium* species associated with Pythium root dysfunction of creeping bentgrass and their impact on root growth and survival. Plant Dis. 92:862-869.

Kerns, J. P., and Tredway, L. P. 2008. Influence of temperature on pathogenicity of *Pythium volutum* toward creeping bentgrass. Plant Dis. 92:1669-1673.

Kerns, J. P., and Tredway, L. P. 2010. Pythium root dysfunction of creeping bentgrass. Plant Health Prog. doi:10.1094/PHP-2010-0125-01-DG

(Prepared by L. P. Tredway, J. P. Kerns, and E. L. Butler)

Spring Dead Spot

Spring dead spot is the most important disease of *Cynodon dactylon* and *Cynodon* hybrids in temperate climates where these grasses are exposed to cold to freezing conditions during winter dormancy. The disease also occurs in *Buchloë dactyloides* and is observed with increasing frequency in *Zoysia* spp. The etiology of spring dead spot is unique among turfgrass diseases for two reasons. First, the disease is caused by at least three pathogen species that vary in their geographical distribution and response to certain management practices. Second, the spring dead spot pathogens do not appear to kill the turf directly. Instead, necrosis of roots, rhizomes, and stolons induced by these pathogens increases turf susceptibility to cold and freezing conditions. For this reason, effective spring dead spot management practices are generally consistent with those that improve turf cold tolerance.

Symptoms and Signs

The most widely recognized symptoms appear in spring when warm-season grasses resume growth after winter dormancy. Circular patches of turf appear to remain dormant (Figs. 217 and 218), but they are completely dead and eventually collapse to leave depressions in the turf stand (Fig. 219). The roots, rhizomes, and stolons of infected plants appear black and severely rotted (Fig. 220). *Ophiosphaerella* spp. produce septate, darkly pigmented hyphae and mats of mycelia on infected tissues (Fig. 179). Ascocarps (pseudothecia) of the causal fungi are occasionally found on dead tissues as well.

Spring dead spot infection centers are perennial, recurring in the same locations and increasing in size up to 1 m in diameter over time. *O. herpotricha* infection centers expand more rapidly than those caused by *O. korrae* or *O. narmari,* but the turf within them may be less severely affected (Fig. 221). The centers of spring dead spot patches are sometimes recolonized by turf or weed species to form a ring symptom. Rings that

Fig. 217. Patches of *Cynodon* sp. turf affected by spring dead spot that remain dormant as the surrounding turf resumes growth. (Courtesy L. P. Tredway—© APS)

Fig. 218. Spring dead spot symptoms in a *Cynodon* sp. putting green. (Courtesy L. P. Tredway—© APS)

Fig. 219. Severe symptoms of spring dead spot persisting into summer in a *Cynodon* sp. golf course fairway. (Courtesy L. P. Tredway—© APS)

coalesce can appear as serpentine arcs. In severe cases, patches may coalesce to affect large, irregularly shaped areas that resemble winter-kill or winter desiccation. Recovery from spring dead spot injury is typically very slow, taking the entire growing season in severe cases (Fig. 222). Metabolites with toxic effects on Bermudagrass have been extracted from cultures of *O. herpotricha* and spring dead spot infection centers and may inhibit turf recovery from the disease.

Spring dead spot symptoms are sometimes visible in the fall as faint chlorotic patches, particularly following periods of cool, wet weather. During winter dormancy, affected patches may appear thin or slightly sunken compared with adjacent healthy turf.

Causal Agents

O. korrae, O. herpotricha, and *O. narmari* are documented as the causal agents of spring dead spot (Table 5) and differ in their geographical distributions. *O. korrae* is most common in California and the southeastern United States, whereas *O. herpotricha* is prevalent throughout the Midwest and the northern transition zone of the United States (Maryland, Tennessee, and Virginia) and in Argentina. *O. narmari* is the predominant cause in Australia and New Zealand but is found only occasionally in North America. Although most spring dead spot populations are dominated by a single species, multiple species are sometimes isolated from a location or even a particular infection center.

Ophiosphaerella spp. colonize the surfaces of roots, rhizomes, and stolons with dark brown, septate hyphae approximately 5 µm wide (Fig. 223). Threads of runner hyphae may fuse to form large, dark, oblong infection mats, 60–400 µm in diameter. Ectotrophic hyphae often build up at the junctions between primary and secondary roots. Pseudothecia, thick walled and black, occasionally develop on stolons (Fig. 184) and roots or on sheaths near stem bases. Ascospores of *O. narmari* are about 35–105 µm long, whereas those of *O. korrae* (Fig. 185) are 105–180 µm long and those of *O. herpotricha* are 140–180 µm long.

Ophiosphaerella spp. exhibit subtle differences in colony morphology, but this is not a reliable identification method because of the substantial variation within species. Use of species-specific PCR probes or sequencing of internal transcribed spacer rDNA or similar genetic loci have proved to be accurate and efficient methods.

Disease Cycle and Epidemiology

Ophiosphaerella spp. persist in infected plant tissue and are likely disseminated by sod, sprigs, or plant debris. Spring dead spot symptoms may begin to appear as soon as 2 years after establishment of a susceptible turfgrass. Individual infection centers recur in the same location, expanding in size with each successive year. The overall severity of the disease varies from year to year depending on the suitability of conditions for infection and symptom expression. For unknown reasons, spring dead spot may dissipate over time in some locations while continuing to persist and increase in severity in others.

The optimal temperature range for in vitro growth of *O. korrae* and *O. herpotricha* is 20–25°C (68–77°F), but in the field most infection is thought to occur when soil temperatures are below 21°C (70°F). At these temperatures, root growth of

Fig. 220. Necrosis of Bermudagrass roots and stolons caused by *Ophiosphaerella korrae*. (Courtesy L. P. Tredway—© APS)

Fig. 221. Artificially inoculated plot with smaller but more severe patches caused by *Ophiosphaerella korrae* (left) compared with larger but less severely affected patches caused by *O. herpotricha* (right). (Courtesy L. P. Tredway—© APS)

Fig. 222. Slow recovery from spring dead spot damage by growth of stolons from the outer perimeter. Patches are often colonized by weed species. (Courtesy L. P. Tredway—© APS)

Fig. 223. Dark mycelial network and black pseudothecia of *Ophiosphaerella korrae* on a *Cynodon dactylon* stolon. (Courtesy R. W. Smiley—© APS)

warm-season turf species is slowed and the fungal pathogens have a competitive advantage. *Ophiosphaerella* spp. initially colonize the surfaces of roots, rhizomes, and stolons with ectotrophic hyphae as previously described. Penetration of the host occurs directly via thinner, lightly pigmented hyphae that grow intracellularly to infest the cortical tissue, first inducing dark, elliptical lesions followed by widespread necrosis. Pseudothecia are occasionally produced on necrotic plant tissue, but the role of ascospores in dissemination of *Ophiosphaerella* spp. is unknown.

Extensive necrosis of roots, rhizomes, and stolons by *Ophiosphaerella* spp. typically does not induce symptoms on the turfgrass foliage. Characteristic spring dead spot symptoms are induced by cold or freezing conditions, to which infected plants are less tolerant than healthy plants. Meristematic tissue of *Ophiosphaerella*-infected plants has shown less capacity to supercool compared with that of healthy plants, perhaps because of reduced carbohydrate reserves. Exposure to temperatures of –2°C (28°F) or lower during winter dormancy results in rapid mortality of infected Bermudagrass plants. Other forms of winter injury, such as desiccation, may also contribute to expression of spring dead spot symptoms.

Management

Spring dead spot can be effectively managed with an integrated approach including cultural practices and preventive fungicide applications. Bermudagrass cultivars vary in their resistance to *Ophiosphaerella* spp., and cultivars with improved winter hardiness generally exhibit some tolerance to spring dead spot. Core aeration and other practices that alleviate compaction and improve drainage and root growth dramatically reduce severity of the disease. Fertilizer and irrigation applied judiciously maximize root growth and improve winter hardiness of the turfgrass. Excessive application of nitrogen or potassium in the fall has been shown to enhance spring dead spot severity.

Ophiosphaerella spp. respond differently to soil pH and nitrogen sources. *O. herpotricha* can be effectively suppressed through application of ammonium sulfate or elemental sulfur to acidify the soil. These practices have not provided consistent suppression of *O. korrae.* A long-term field trial demonstrated effective suppression of *O. korrae* with calcium nitrate, but the underlying mechanism remains unknown.

Fungicides for spring dead spot prevention should be applied preventively in the fall. Demethylation inhibitor and succinate dehydrogenase inhibitor chemistries generally provide good to excellent control. Two applications, 28 days apart, when soil temperatures are conducive to pathogen infection are most often recommended. Fungicides should be applied in a large volume of water or drenched into the soil for best results. Spring applications targeting spring dead spot have not provided a consistent benefit, but they may help to facilitate recovery if disease symptoms are present. *Ophiosphaerella* spp. vary in their sensitivity to fungicides, but soil type and other environmental conditions may also influence fungicide efficacy. Products should be selected on the basis of local efficacy trials when possible.

Cultural practices can help to speed recovery from spring dead spot symptoms during spring and summer. Cultivation to disrupt the matt of dead turf can facilitate spread of the desirable turf into the affected patch. Light and frequent fertilization and irrigation are recommended, but excessive amounts, which can exacerbate the disease in the long term, should be avoided. Preemergent herbicides that inhibit root development should be avoided when spring dead spot symptoms are evident, because they can slow recovery from spring dead spot injury.

Selected References

Baird, J. H., Martin, D. L., Taliaferro, C. M., Payton, M. E., and Tisserat, N. A. 1998. Bermudagrass resistance to spring dead spot caused by *Ophiosphaerella herpotricha*. Plant Dis. 82:771-774.

Caasi, O. C., Walker, N. R., Marek, S. M., Enis, J. N., and Mitchell, T. K. 2010. Infection and colonization of turf-type bermudagrass by *Ophiosphaerella herpotricha* expressing green or red fluorescent proteins. Phytopathology 100:415-423.

Dernoeden, P. H., Crahay, J. N., and Davis, D. B. 1991. Spring dead spot and bermudagrass quality as influenced by nitrogen source and potassium. Crop Sci. 31:1674-1680.

Endo, R. M., Ohr, H. D., and Krausman, E. M. 1985. *Leptosphaeria korrae,* a cause of spring dead spot disease of bermudagrass in California. Plant Dis. 69:235-237.

Iriarte, F. B., Fry, J. D., Martin, D. L., Todd, T. C., and Tisserat, N. A. 2005. Effect of cold acclimation and freezing on spring dead spot severity in bermudagrass. HortScience 40:421-423.

Martin, D. L., Bell, G. E., Baird, J. H., Taliaferro, C. M., Tisserat, N. A., Kuzmic, R. M., Dobson, D. D., and Anderson, J. A. 2001. Spring dead spot resistance and quality of seeded bermudagrasses under different mowing heights. Crop Sci. 41:451-456.

McCarty, L. B., DiPaola, J. M., and Lucas, L. T. 1991. Regrowth of bermudagrass infected with spring dead spot following low temperature exposure. Crop Sci. 31:182-184.

McCarty, L. B., Lucas, L. T., and DiPaola, J. M. 1992. Spring dead spot occurrence in bermudagrass following fungicide and nutrient applications. HortScience 27:1092-1093.

Nus, J. L., and Shashikumar, K. 1993. Fungi associated with spring dead spot reduce freezing resistance in bermudagrass. HortScience 28:306-307.

Smith, A. M. 1971. Spring dead spot of couch grass turf in New South Wales. J. Sports Turf Res. Inst. 47:54-59.

Tisserat, N. A., Pair, J. C., and Nus, A. 1989. *Ophiosphaerella herpotricha,* a cause of spring dead spot of bermudagrass in Kansas. Plant Dis. 73:933-937.

Tisserat, N., Wetzel, H., III, Fry, J., and Martin, D. L. 1999. Spring dead spot of buffalograss caused by *Ophiosphaerella herpotricha* in Kansas and Oklahoma. Plant Dis. 83:199.

Tredway, L. P., and Butler, E. L. 2007. First report of spring dead spot of zoysiagrass caused by *Ophiosphaerella korrae* in the United States. Plant Dis. 91:1684.

Tredway, L. P., Soika, M. D., Butler, E. L., and Kerns, J. P. 2021. Impact of nitrogen source, fall fertilizers, and preventive fungicides on spring dead spot caused by *Ophiosphaerella korrae* and *O. herpotricha*. Crop Sci. 61:3187-3196.

Tredway, L. P., Tomaso-Peterson, M., Perry, H., and Walker, N. R. 2009. Spring dead spot of bermudagrass: A challenge for researchers and turfgrass managers. Plant Health Prog. doi:10.1094/PHP-2009-0710-01-RV

Venkatasubbaiah, P., Tisserat, N. A., and Chilton, W. S. 1994. Metabolites of *Ophiosphaerella herpotricha,* a cause of spring dead spot of bermudagrass. Mycopathologia 128:155-159.

Walker, J., and Smith, A. M. 1972. *Leptosphaeria narmari* and *L. korrae* spp. nov., two long-spored pathogens of grasses in Australia. Trans. Brit. Mycol. Soc. 58:459-466.

Walker, N. R., Mitchell, T. K., Morton, A. N., and Marek, S. M. 2006. Influence of temperature and time of year on colonization of bermudagrass roots by *Ophiosphaerella herpotricha*. Plant Dis. 90:1326-1330.

Wetzel, H. C., Hulbert, S. H., and Tisserat, N. A. 1999. Molecular evidence for the presence of *Ophiosphaerella narmari* n. comb., a cause of spring dead spot of Bermuda grass, in North America. Mycol. Res. 103:981-989.

(Prepared by L. P. Tredway)

Summer Patch

Summer patch was first described as a disease of cool-season turfgrasses in 1984. Before that time, it was an unidentified component of Fusarium blight, which was ultimately proved to be a disease complex consisting of summer patch, necrotic ring spot, and perhaps other diseases. Summer patch is a common root disease of *Poa* and *Festuca* spp. grown in temperate and subtropical climates. It has also been recently recognized as a common disease in *Agrostis* spp. putting greens in transition zones or subtropical climates.

Symptoms and Signs

Summer patch is most common in turfgrass stands that are 3 years old or older. In *P. pratensis* and *P. annua,* summer patch infection centers are initially 3–8 cm in diameter (Fig. 224), expanding up to 30 cm or larger over multiple seasons. Mild symptoms may appear in late spring to early summer as patches or rings of slow-growing, thinned, or wilted turf. The symptoms intensify as heat and other stresses accumulate, with the foliage rapidly turning orange to reddish brown before collapsing and turning tan or straw colored (Fig. 225). The outer perimeters of infection centers often remain orange to reddish brown, indicative of active infection as the pathogen spreads outward in a radial fashion (Fig. 226). The centers of patches may become recolonized with *P. pratensis* or other turf species to form a "frog-eye" symptom (Fig. 227). In severe cases, patches may coalesce to produce large, irregularly shaped areas of turf that resemble those affected by heat stress, insect damage, anthracnose, and other diseases.

Examination of infected roots, rhizomes, or stolons reveals a sparse network of dark brown to black ectotrophic hyphae 2–5 µm wide (Fig. 179). Hyphopodia are darkly pigmented and globose, 6–12 µm in diameter, and formed singly or in aggregates referred to as "growth cessation structures" (Fig. 228). In the

Fig. 224. Initial symptoms of summer patch of *Poa pratensis,* caused by *Magnaporthiopsis poae,* forming small spots that resemble those of dollar spot. (Courtesy R. W. Smiley—© APS)

Fig. 225. Patch and ringlike symptoms of summer patch of *Poa pratensis,* caused by *Magnaporthiopsis poae.* (Courtesy L. P. Tredway—© APS)

Fig. 226. Summer patch of *Poa pratensis,* caused by *Magnaporthiopsis poae.* The orange to reddish brown perimeters of patches and rings are indicative of active infection. (Courtesy L. P. Tredway—© APS)

Fig. 227. Rings of wilted and declining *Poa pratensis,* symptomatic of summer patch caused by *Magnaporthiopsis poae.* (Courtesy B. B. Clarke—© APS)

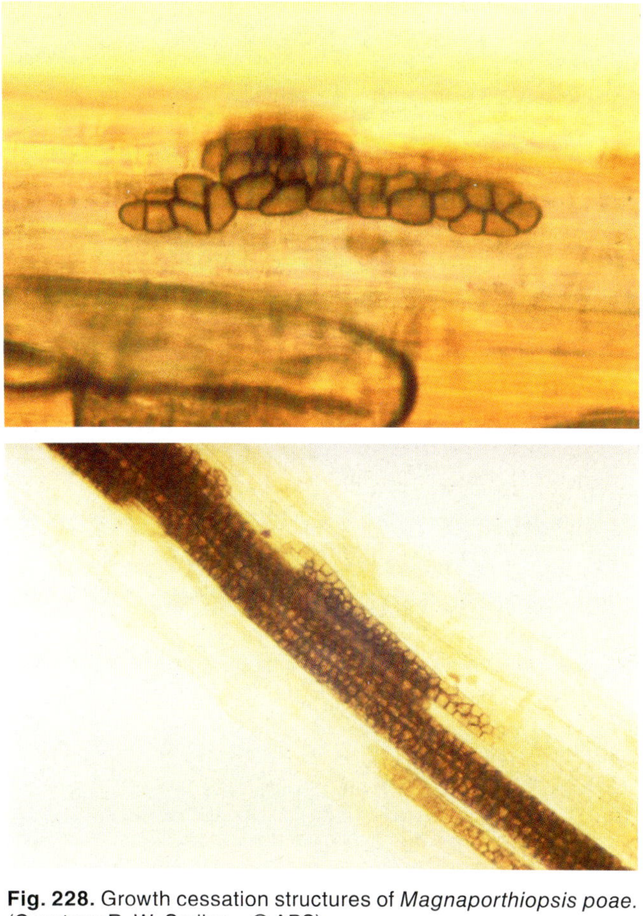

Fig. 228. Growth cessation structures of *Magnaporthiopsis poae.* (Courtesy R. W. Smiley—© APS)

latter stages of infection, vascular discoloration and cortical rot are extensive, with roots, rhizomes, and crowns turning dark brown to black (Fig. 229). Perithecia have not been observed under field conditions. No leaf lesions are associated with summer patch, but white bands sometimes develop on the foliage of infected plants as a result of heat stress.

In temperate climates, *Agrostis* spp. exhibit tolerance to summer patch, often colonizing infection centers where *Poa* spp. have been killed or severely damaged (Fig. 230). In subtropical or tropical climates where *Agrostis* spp. are subjected to prolonged heat stress, summer patch is now recognized as a widespread problem that may have been previously misdiagnosed as take-all patch or other diseases. Circular patches, typically 15–30 cm in diameter, initially show signs of wilt or chlorosis followed by orange dieback (Fig. 231). Ectotrophic hyphae and necrosis of roots and stolons are observed as described above.

In fine-leaf species of *Festuca* such as *F. rubra* and *F. brevipila,* summer patch infection centers are typically 8–15 cm in diameter. Infected turf initially turns chlorotic to bronze and then becomes light tan before collapsing to leave sunken depressions in the turf stand (Fig. 232).

Causal Agents

Magnaporthiopsis poae is the most widely distributed incitant of summer patch and has been associated with *Poa, Agrostis,* and *Festuca* spp. Two additional species, *M. meyeri-festucae* and *M. agrostidis* (Table 5), have been characterized as pathogens of *Festuca* spp. in the U.S. state of New Jersey and of *Agrostis* spp. in New South Wales, Australia, respectively.

Considerable variation in colony morphology among and within *Magnaporthiopsis* spp. has been reported. Mycelial growth may be appressed or aerial. Colonies are white at first but later darken to beige, golden brown, olive green, or occasionally black. In culture, hyphae often curl back toward the center of the colony. Conidiophores (phialides) and conidia (phialospores) can be observed in culture and vary among species. *M. poae* produces hyaline, straight or slightly curved conidiophores, 6–15 × 2–5 µm. Conidia are straight to slightly curved, 3–8 × 1–3 µm, and produced in aggregates atop conidiogenous cells. Conidia of *M. meyeri-festucae* are similar to those of *M. poae,* but conidiophores are pigmented brown and branching, with distinct conidiogenous cells that are bottle shaped and 7.5–14.5 × 2.5–5 µm. *M. agrostidis* produces complex conidiophores and conidiogenous cells similar to those of *M. meyeri-festucae,* but conidia are distinctly curved, 4–6 × 1 µm, and tapered toward the basal ends.

Perithecia of *M. poae* are black and spherical (252–556 µm in diameter), with long (357–756 µm), cylindrical necks (Fig. 184). They have been observed only in laboratory cultures contain-

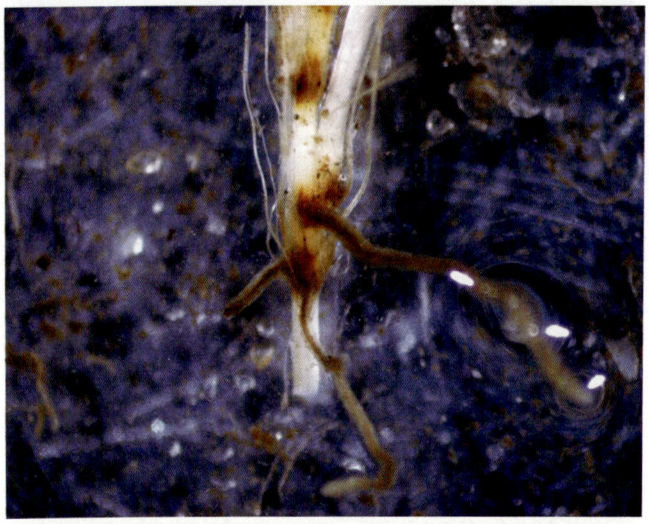

Fig. 229. Necrosis of *Agrostis stolonifera* roots and crown caused by *Magnaporthiopsis poae.* (Courtesy L. P. Tredway—© APS)

Fig. 230. Summer patch of *Poa annua* on a golf green, with *Agrostis stolonifera* surviving in patches of dying *P. annua.* (Courtesy P. H. Dernoeden—© APS)

Fig. 231. Summer patch, caused by *Magnaporthiopsis poae,* in an *Agrostis stolonifera* putting green in the transition zone of the United States. (Courtesy L. P. Tredway—© APS)

Fig. 232. Severe summer patch of *Festuca rubra*. (Courtesy P. H. Dernoeden—© APS)

ing both mating types. Asci are unitunicate, clavate, straight or slightly curved, and 63 × 108 μm. Each ascus contains eight ascospores, which at maturity are fusoid and 23–42 × 4–6 μm (Fig. 185). The ascospores have three septa; the two intermediate cells are dark brown, and the two terminal cells are hyaline. The sexual states of *M. meyeri-festucae* and *M. agrostidis* have not been observed.

Disease Cycle

Magnaporthiopsis spp. are believed to survive the winter as mycelium in previously colonized plant debris and in perennial host tissue. Root infection commences in the spring when soil temperatures reach 18–21°C (64–70°F). The fungus moves from plant to plant by growth of ectotrophic hyphae along the surfaces of roots, rhizomes, and stolons. Symptoms develop during hot, rainy weather or when high temperatures follow periods of heavy rainfall. Patches may continue to expand through the summer and early autumn. As in many diseases caused by ectotrophic root-infecting fungi, infection centers are perennial and may increase by 8–20 cm in diameter per year. *M. poae* may be spread by aerification and dethatching equipment or by the transport of infested sod or other plant material. The role of ascospores or conidia in dissemination is unknown because they are rarely observed in nature.

Epidemiology

In culture, *Magnaporthiopsis* spp. grow at temperatures of 10–37°C (50–99°F), with optimal growth at 25–30°C (77–86°F). In the field, colonization and infection of roots begins in the spring when soil temperatures reach 18–21°C (64–70°F) and continue throughout the summer in most climates. In growth chamber studies, root infection and necrosis have been observed at 20–37°C (68–99°F), with optimal disease development occurring at 28–30°C (82–86°F). Wet soil conditions caused by poor drainage, frequent rainfall, or excessive irrigation are also conducive to infection by *Magnaporthiopsis* spp.

Summer patch symptoms are rarely observed during the spring infection period, except during spells of abnormally hot and/or dry weather. Aboveground symptom expression typically begins in early to midsummer, induced by physiological stresses such as heat. Daytime high temperatures of 28–35°C (82–95°F) with nighttime temperatures above 20°C (68°F) induce symptom expression in *P. pratensis*. In *Agrostis* spp., daytime temperatures of 30–40°C (86–104°F) induce aboveground symptoms; the most severe symptoms are produced at 35°C (95°F).

Other stresses, such as soil compaction, drought, low fertility, or mechanical wear, may also encourage expression of summer patch symptoms.

Summer patch is most severe on sunny, exposed, poorly drained, and compacted sites during hot, wet years. Heat stress plays an important role in disease development. Drought stress usually is not a predisposing factor; however, once severe root rot has occurred, drought stress may enhance the severity of foliar symptoms. *M. poae* can spread along roots, crowns, and stem tissue at a rate of up to 2.5 cm per week under ideal conditions for growth of the fungus. Symptom expression increases with the use of quick-release fertilizers or nitrate sources of nitrogen. The disease is often more severe in turfgrass maintained at a low mowing height and with frequent light irrigation. Summer patch is greatly enhanced in turfgrasses grown in soil with high pH compared with those grown in low pH soils. The application of lime, particularly fine grades, favors disease development, but this effect can be countered with the use of acidifying fertilizers such as ammonium sulfate. The disease is most prevalent in turf that is at least 2 years old.

Management

Effective summer patch management can be achieved with an integrated approach to suppress pathogen growth and minimize physiological stress on the turfgrass. Where possible, soil pH should be maintained at approximately 6.0 through use of acidifying fertilizers such as ammonium sulfate or sulfur-coated urea. Lime should be applied only as indicated by soil test results, and only coarse grades of lime should be used to avoid a rapid increase in soil pH. Cultural practices that alleviate stress and promote root development reduce summer patch severity. Turf should not be mowed below recommended heights, particularly during periods of heat stress. Other practices that may aid in disease control are aerifying (in the absence of foliar symptoms), improving drainage, reducing compaction, and syringing to reduce heat stress.

Certain fungicides are effective for summer patch control but should be applied preventively for best results. Applications beginning in the spring when soil temperature is 18°C (64°F) and repeated at 14- to 28-day intervals provide effective protection. Curative applications, after infection or symptom expression has already occurred, may prevent further disease progress, but turf recovery is often limited by high temperatures. For best results, fungicides should be applied in at least 4 gallons of water per 1,000 ft^2 (1,600 L/ha) or lightly irrigated into the thatch immediately after application.

Selected References

Davis, D. B., and Dernoeden, P. H. 1991. Summer patch and Kentucky bluegrass quality as influenced by cultural practices. Agron. J. 83:670-677.

Hill, W. J., Heckman, J. R., Clarke, B. B., and Murphy, J. A. 2003. Summer patch disease severity on Kentucky bluegrass in response to fertilizer source. J. Plant Nutr. 26:1499-1512.

Hutchens, W. J., Gannon, T. W., Shew, H. D., and Kerns, J. P. 2019. Effect of post-application irrigation on fungicide movement and efficacy against *Magnaporthiopsis poae*. Crop Prot. 122:106-111.

Kackley, K. E., Grybauskas, A. P., Dernoeden, P. H., and Hill, R. L. 1990. Role of drought stress in the development of summer patch in field-inoculated Kentucky bluegrass. Phytopathology 80:655-658.

Landschoot, P. J. 1993. Taxonomy and biology of ectotrophic root-infecting fungi associated with patch diseases of turfgrasses. Pages 41-71 in: Turfgrass Patch Diseases Caused by Ectotrophic Root-Infecting Fungi. B. B. Clarke and A. B. Gould, eds. American Phytopathological Society, St. Paul, MN.

Landschoot, P. J., and Jackson, N. 1989. *Magnaporthe poae* sp. nov., a hyphopodiate fungus with a *Phialophora* anamorph from grass roots in the United States. Mycol. Res. 93:59-62.

Luo, J., Vines, P. L., Grimshaw, A., Hoffman, L., Walsh, E., Bonos, S. A., Clarke, B. B., Murphy, J. A., Meyer, W. A., and Zhang, N. 2018. *Magnaporthiopsis meyeri-festucae,* sp. nov. associated with a summer patch-like disease of fine fescue turfgrasses. Mycologia 109:780-789.

Smiley, R. W. 1987. The etiologic dilemma concerning patch diseases of bluegrass turfs. Plant Dis. 71:774-781.

Smith, J. D., Jackson, N., and Woolhouse, A. R. 1989. Fungal Diseases of Amenity Turf Grasses. E. & F. N. Spon, London.

Thompson, D. C., Clarke, B. B., and Heckman, J. R. 1995. Nitrogen form and rate of nitrogen and chloride application for the control of summer patch in Kentucky bluegrass. Plant Dis. 79:51-56.

Tredway, L. P. 2006. Genetic relationships among *Magnaporthe poae* isolates from turfgrass hosts and relative susceptibility of 'Penncross' and 'Penn A-4' creeping bentgrass. Plant Dis. 90:1531-1538.

Wong, P. T. W., Khemmuk, W., Geering, A. D. W., and Shivas, R. G. 2015. *Magnaporthiopsis agrostidis*. Persoonia 35:322-323.

(Prepared by L. P. Tredway)

Take-All Patch

Take-all patch is a severe disease of *Agrostis* spp. in temperate climates throughout the world. The pathogen, *Gaeumannomyces avenae,* has been documented to also infect *Festuca* and

Poa spp., but these species are only occasionally damaged and normally survive in areas where *Agrostis* spp. have been killed.

Symptoms and Signs

Symptoms of take-all patch usually appear in late spring or early summer. The initial symptoms are small, circular, light brown or reddish brown patches (Fig. 233), which may resemble those of Microdochium patch. Take-all patch often persists into the summer, however, whereas the symptoms of Microdochium patch usually subside in late spring. Patches typically recur in successive years and expand up to 15 cm per year to reach diameters of 1 m or more. The peripheries of patches or sometimes entire patches may turn reddish brown or bronze when the disease is actively developing (Fig. 234). The centers of patches are often recolonized by *Agrostis, Poa,* or *Festuca* spp. to produce a ringlike symptom (Figs. 235–237). In mixed stands, *Agrostis* spp. may appear yellow brown to brown, leaving a thinned stand dominated by *Poa* or *Festuca* spp.

Symptoms are most striking after plants have been stressed by hot, dry weather in late summer. Mild infections that are not apparent in late spring often develop symptoms under such conditions. Recently stressed grass appears bronze to reddish brown and then fades to a dull brown. During the winter, affected patches may appear gray.

In the initial stages of infection, root surfaces are colonized by dark strands of ectotrophic hyphae (Fig. 179) oriented parallel to the root axis. The vascular cylinder is typically dark brown, and the cortex remains transparent. As the disease progresses, the stolons, rhizomes, roots, and bases of shoots turn dark brown to black; the symptoms are similar to those of necrotic ring spot (Fig. 190). Perithecia (fruiting structures) of the pathogen (similar to those shown in Figure 184) may develop during late spring or autumn in leaf sheaths around crowns and stem bases.

Fig. 233. Take-all patch of *Agrostis stolonifera* on a golf course fairway. (Courtesy B. B. Clarke—© APS)

Fig. 234. Take-all patch of *Agrostis stolonifera* on a golf green. (Courtesy P. H. Dernoeden—© APS)

Fig. 235. Take-all patch of *Agrostis stolonifera* on a golf course fairway with *Festuca rubra* recolonizing the centers of patches. (Courtesy B. B. Clarke—© APS)

Fig. 236. Take-all patch of an *Agrostis* sp., caused by *Gaeumannomyces avenae,* on a bowling green. (Courtesy S. G. Fushtey—© APS)

Fig. 237. Ringlike symptom of take-all patch produced as the center of the patch is recolonized by *Agrostis* and *Poa* spp. (Courtesy P. L. Koch—© APS)

Causal Agent

Take-all patch is caused by *G. avenae* (Table 5). It forms dark brown to black ectotrophic runner hyphae on roots, crowns, stolons, and rhizomes of susceptible plants. Infection hyphae originate from simple hyphopodia (Fig. 238) or mycelial mats and penetrate host tissue. Dark brown or black, flask-shaped, beaked, ostiolate perithecia with curved necks are formed under and protrude through the lower leaf sheath. Asci are unitunicate, eight spored, and 90–160 × 12–16 µm when mature. Ascospores (similar to those shown in Figure 185) are hyaline to faintly yellow, multiseptate, and 85–140 µm long.

Disease Cycle

G. avenae survives as mycelium in previously infected plant tissue. Infection of host roots, stolons, rhizomes, and crowns occurs during the autumn and spring. Initially, ectotrophic runner hyphae colonize the surfaces of host roots. Runner hyphae give rise to simple hyphopodia and mycelial mats, which initiate infection of the root cortex. Extensive colonization of host tissues may occur under favorable conditions (cool, moist weather). Aboveground symptoms typically appear in late spring or early summer when heat or drought stress induces a decline of infected plants. In severe cases, symptoms may be observed in autumn as well as spring. The pathogen is presumably spread by transport of infested soil or plant material. Perithecia may be produced in the autumn and possibly throughout mild winters. Infection of wheat roots by ascospores has been demonstrated under controlled conditions, but their importance in the dissemination of the pathogen in *Agrostis* spp. is unknown.

Epidemiology

Take-all patch is most damaging in *Agrostis* spp. planted in fumigated soil, on recently cleared forested sites, or in substrates with a high sand content. Other soil conditions that favor the disease include light texture, low organic matter content, low or unbalanced fertility, manganese deficiency, and pH above 6.5. The incidence of take-all patch is more strongly associated with rhizosphere pH than bulk soil pH. Liming favors the disease, although the response may not appear for several years after lime is applied. A more rapid increase in disease severity follows the application of lime with a fine particle size. Take-all patch is most severe when cool, wet weather in the autumn or spring is followed by hot, dry conditions. If left untreated, the disease may decline in severity after several years, presumably because of a buildup of antagonistic or competing microorganisms. In some instances, however, take-all patch can intensify over a period of 10 years or longer.

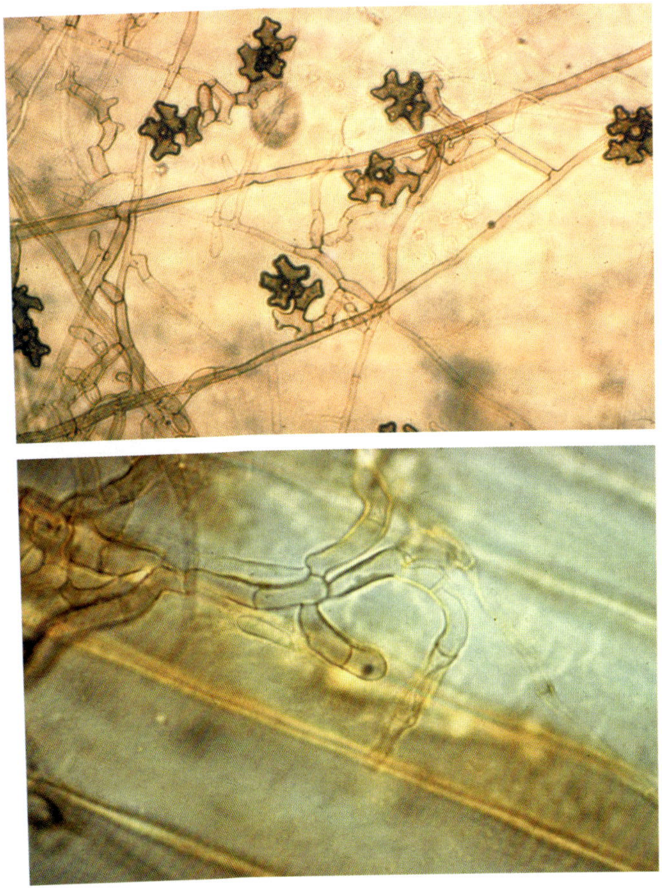

Fig. 238. Lobed hyphopodia on hyphae of *Gaeumannomyces graminis* (top) and an unlobed (simple) hyphopodium of *G. avenae* (bottom). (Courtesy P. J. Landschoot—© APS)

Management

An integrated management approach is most effective for the control of take-all patch. All *Agrostis* spp. and cultivars are susceptible, but cultivars of *A. stolonifera* with moderate levels of resistance to the disease are available. Where practical, planting a mixture of *Agrostis* spp. and *F. rubra* can effectively reduce disease severity.

To further reduce the severity of take-all patch, soil pH should be maintained below 6.5, preferably 5.5–6.0. The application of acidifying fertilizers, such as ammonium sulfate, in cool weather is recommended for reducing soil pH gradually over time. When liming is necessary, only the coarsest grades of lime (20–30 mesh) should be applied to avoid rapid changes in pH. The use of alkaline topdressing materials and nitrate forms of nitrogen should be avoided, and the pH of water used for irrigation should be monitored.

Manganese deficiency greatly enhances the development of take-all patch. On deficient sites, supplemental applications of manganese totaling 2 lb/acre (2.2 kg/ha) in the autumn or spring can significantly reduce disease severity. Phosphorus and potassium should be applied to ensure that these elements are present in adequate amounts as determined by soil tests or tissue analyses.

Factors that restrict root growth may increase the severity of take-all patch. Excessive irrigation and nitrogen applications should be avoided, adequate surface and subsurface drainage should be provided, and turf should be aerified when symptoms are absent to alleviate soil compaction. After symptoms of take-all patch appear, adequate irrigation and the use of fertilizer in frequent, light applications help to compensate for damage to the root system.

Replacement of diseased patches by plugging or resodding may be practical on golf and bowling greens where small areas are affected.

Fungicides are available for the control of take-all patch, but they must be applied as a preventive measure for best results. Applications in autumn (prior to dormancy) and early spring are most effective. Increasing spray water volume or posttreatment irrigation can improve fungicide efficacy. Local extension personnel can provide a list of currently available products.

Selected References

Chastagner, G. A. 1997. Management of necrotic ring spot and take-all patch with DMI fungicides in the Pacific Northwest of the USA. Int. Turfgrass Soc. Res. J. 8:833-846.

Dernoeden, P. H. 1987. Management of take-all patch of creeping bentgrass with nitrogen, sulfur, and phenyl mercury acetate. Plant Dis. 71:226-229.

Dernoeden, P. H. 1993. Integrating strategies for the management of patch diseases caused by root invading ectotrophic fungi. Pages 123-161 in: Turfgrass Patch Diseases Caused by Ectotrophic Root-Infecting Fungi. B. B. Clarke and A. B. Gould, eds. American Phytopathological Society, St. Paul, MN.

Heckman, J. R., Clarke, B. B., and Murphy, J. A. 2003. Optimizing manganese fertilization for the suppression of take-all patch disease on creeping bentgrass. Crop Sci. 43:1395-1398.

Landschoot, P. J., Gould, A. B., and Clarke, B. B. 1993. Ecology and epidemiology of ectotrophic root-infecting fungi associated with patch diseases of turfgrasses. Pages 73-105 in: Turfgrass Patch Diseases Caused by Ectotrophic Root-Infecting Fungi. B. B. Clarke and A. B. Gould, eds. American Phytopathological Society, St. Paul, MN.

Latin, R. 2005. Managing take-all patch of creeping bentgrass on sand-based greens. Golf Course Manage. 73:101-105.

Nilsson, H. E., and Smith, J. D. 1981. Take-all of grasses. Pages 433-448 in: Biology and Control of Take-All. M. J. C. Asher and P. J. Shipton, eds. Academic Press, New York.

Walker, J. 1980. *Gaeumannomyces, Linocarpon, Ophiobolus,* and several other genera of scolecospored ascomycetes and *Phialophora* conidial states, with a note on hyphopodia. Mycotaxon 11:1-129.

(Prepared by R. W. Smiley, P. H. Dernoeden, and B. B. Clarke; revised by L. P. Tredway)

Take-All Root Rot and Bermudagrass Decline

Bermudagrass decline was the first described root decline or rot of warm-season turfgrass. Root rots caused by the same pathogen, *Gaeumannomyces graminis* var. *graminis,* were subsequently reported on *Eremochloa ophiuroides, Paspalum vaginatum, Pennisetum clandestinum, Stenotaphrum secundatum,* and *Zoysia japonica.* The root rot disease of these grasses has been named "take-all root rot." A comprehensive taxonomic and molecular study resolved the genus *Gaeumannomyces,* resulting in the identification of cryptic, novel species associated with take-all root rot (Table 5). Additional research has identified species of *Magnaporthiopsis* and *Candidacolonium* also associated with Bermudagrass decline. Take-all root rot and Bermudagrass decline are grouped in this section because the root and shoot symptomology is similar.

Symptoms

Bermudagrass decline of *Cynodon* spp. hybrids and take-all root rot share similar symptomology. The diseases first appear in irregularly shaped, chlorotic patches up to 1 m in diameter (Figs. 239 and 240). Leaf spots and other foliar symptoms are either absent or not directly related to the occurrence of this disease. The lower leaves are the first to become chlorotic, and they die as the chlorosis progresses into the upper canopy. At the time chlorosis becomes evident in the foliage, roots and associated stolons and rhizomes turn off-white to dark brown and may exhibit isolated black lesions. Root length is often shortened, and the entire root system may eventually turn black and become completely rotted (Fig. 241). Less severely infected roots, rhizomes, and stolons may bear dark strands of hyphae oriented primarily parallel to the root axes.

As the disease progresses, affected patches may become devoid of living plants and coalesce to form large areas of dead or severely thinned turf. Unlike other diseases caused by ectotrophic root-infecting fungi, which are characterized by distinct, circular patches of necrotic tissue, Bermudagrass decline and take-all root rot typically kill turf in an irregular pattern, sometimes exceeding 5 m in diameter. New tillers of the affected turfgrass species may recolonize dead areas, but they soon become infected and begin to decline. Extensive dieback of the roots and stolons occurs, and roots may no longer be able to hold stolons in place. Affected stolons are easily lifted from the soil surface. Take-all root rot and large patch (see Large Patch) are often active simultaneously during autumn months. These symptoms can easily be confused with those of diseases caused by *Pythium* spp. and damage caused by plant-parasitic nematodes.

Symptoms of Bermudagrass decline and take-all root rot in the southeastern United States occur during warm to hot, humid periods of summer through early autumn, which typically are periods of high rainfall. In other regions, high rainfall also appears to be a key factor in disease development. Such periods are normally associated with low-intensity light, which is not conducive to the growth of *Cynodon* spp. hybrids.

In the United States, Bermudagrass decline symptoms have been observed primarily in *Cynodon* spp. maintained at mowing heights of 0.125 in. (3.2 mm) or less. This disease is com-

Fig. 239. Bermudagrass decline of *Cynodon dactylon* × *C. transvaalensis.* (Courtesy M. Dieckhoff—© APS)

Fig. 240. Bermudagrass decline of *Cynodon dactylon* × *C. transvaalensis* and take-all root rot of *Stenotaphrum secundatum.* (Courtesy M. Tomaso-Peterson—© APS)

Fig. 241. Root rot caused by Bermudagrass decline and take-all root rot. (Courtesy M. Tomaso-Peterson—© APS)

mon in *Cynodon* sp. ultradwarf greens and may be referred to as take-all root rot. Symptoms in ultradwarf greens often appear as irregular chlorotic patches or general thinning of turf density along the perimeters (Figs. 242 and 243). Foliar necrosis in the turf canopy, resulting in plant death, thinning turf, and bare patches can occur under high disease pressure. The ectotrophic fungi associated with rotted roots can also be readily isolated from asymptomatic plants of the same species that are maintained at a higher mowing height, such as the turf on aprons and fairways around affected greens and on sports fields.

Causal Agents

Historically, *G. graminis* var. *graminis* was reported as the principal causal agent of Bermudagrass decline and take-all root rot. Phylogenetic analyses and taxonomic studies were used to resolve the genus *Gaeumannomyces,* resulting in eight cryptic species previously identified as *G. graminis* var. *graminis*. Research also identified additional ectotrophic root-infecting fungi in the genera *Magnaporthiopsis* and *Candidacolonium* associated with root rot of *Cynodon* spp. ultradwarf grasses (Table 5). As a group, *Gaeummanomyces* spp. and other ectotrophic root-infecting fungi are characterized by the formation of dark brown to black runner hyphae and growth cessation structures on the surfaces of roots, stolons, and rhizomes of susceptible plants (Figs. 244 and 245). Infection hyphae originate from both simple and strongly lobed hyphopodia (appressorium-like hyphal swellings) (Fig. 246) and penetrate host tissue. *Gaeumannomyces* spp. form dark brown to black, flask-shaped, beaked, ostiolate perithecia with curved necks that protrude through the lower leaf sheaths. Perithecia are rarely observed on affected plants unless they have been subjected to lengthy periods of incubation in a laboratory. Rapid identification of ectotrophic root-infecting fungi can be accomplished with PCR and real-time PCR multiplex assays. A complex of these fungi may be identified in a single root sample with some characterized as nonpathogenic or secondary pathogens.

Fig. 242. Bermudagrass decline in a *Cynodon dactylon* × *C. transvaalensis* putting green. (Courtesy M. Tomaso-Peterson—© APS)

Fig. 243. Bermudagrass decline in a *Cynodon dactylon* × *C. transvaalensis* putting green. (Courtesy M. Tomaso-Peterson—© APS)

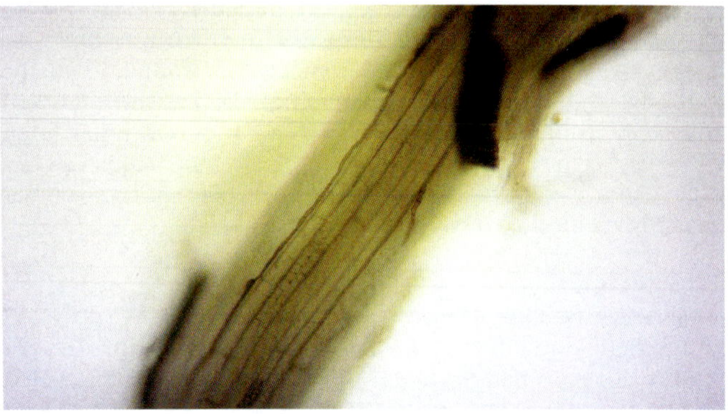

Fig. 244. Runner hyphae of *Gaeumannomyces graminis*. (Courtesy M. Tomaso-Peterson—© APS)

Fig. 245. Growth cessation structures of *Gaeumannomyces graminis*. (Courtesy M. Tomaso-Peterson—© APS)

Fig. 246. Lobed hyphopodia of *Gaeumannomyces graminis* on a *Cynodon dactylon* × *C. transvaalensis* stolon. (Courtesy M. Tomaso-Peterson—© APS)

Disease Cycle

Little is known about the life history of *G. graminis* and many of the other ectotrophic fungi that colonize warm-season turfgrasses. However, *G. graminis* is well known as a universal occupant of grasslands. The pathogenicity and survival of this fungus may be similar to that of *G. avenae* (see Take-All Patch), except that these warm-season pathogens have a greater tolerance for temperatures above 25°C (77°F). These ectotrophic fungi survive adverse environmental conditions through mycelial colonization of roots, rhizomes, and stolons. The fungi spread from plant to plant by growing along the surfaces of roots or stolons. They may also be disseminated by the movement of infected roots, rhizomes, and stolons (by aerification and vertical mowing equipment, for example) and the transport of infested vegetative material, such as turfgrass sod and sprigs. Sod severely weakened by root rot may be difficult to harvest. There does not appear to be a relationship between the incidence of Bermudagrass decline or take-all root rot and the soil type or the cultivar or age of the host.

Management

All cultivars and hybrids of warm-season grasses appear to be susceptible to infection by this group of ectotrophic fungi.

Bermudagrass decline and take-all root rot can be controlled by practices that relieve plant stress. Perhaps the most effective means of reducing symptom expression is to raise the mowing height of affected grasses during periods of high rainfall. Core aeration should be performed frequently to relieve or prevent soil compaction, and cores should be removed rather than spread over the turf surface as a topdressing. Solid tines should be used to reduce root injury; and after aerification, the areas should be topdressed frequently with a pasteurized, pathogen-free sand or soil mixture.

Adequate plant nutrition should be maintained, with particular attention to nitrogen, potassium, and micronutrients. Nitrogen should be applied in a 1:1 or 1:2 ratio with potassium. Slow-release sources of nitrogen or acidifying fertilizers, such as ammonium sulfate and manganese sulfate, should be used. Fertilizers containing nitrate sources of nitrogen should be avoided. Micronutrients (manganese, copper, iron, and zinc) should also be applied if they are present at low levels in plant tissue. Once visual symptoms of root decline are apparent, nutrients should be applied to the foliage because root absorption may be compromised by extensive decay. Hydrated lime should not be used to control algae because it raises the soil pH and may enhance symptom expression. If it is necessary to apply magnesium, magnesium sulfate should be used instead of dolomitic lime.

Acropetal penetrant fungicides (e.g., benzimidazoles, demethylation inhibitors, and strobilurins) may help reduce the incidence of root decline in *Stenotaphrum* spp. when used preventively. Demethylation inhibitor fungicides that are safe on *Cynodon* spp. have been effective when applied in late summer and irrigated in immediately after application. In vitro fungicide assays have shown that these ectotrophic fungi lack sensitivity to succinate dehydrogenase inhibitor fungicides. Fungicide labels should be consulted for selection and use in managing these root rot diseases.

Selected References

Bronzato-Badial, A., King, J., and Tomaso-Peterson, M. 2020. Monitoring ectotrophic root-infecting fungi associated with Bermudagrass putting green roots using quantitative multiplex assays. Plant Health Prog. 21:144-151.

Butler, L., Stephens, C., Kerns, J., and Gannon, T. 2019. Take-all root rot: A complex disease. North Carolina State Extension. https://turfpathology.ces.ncsu.edu/2019/take-all-root-rot-a-complex-disease

Câmara, M. P. S., O'Neill, N. R., Dernoeden, P. H., and Palm, M. E. 2000. *Ophiosphaerella agrostis* sp. nov. and its relationship to other species of *Ophiosphaerella*. Mycologia 92:317 325.

Elliott, M. L. 1991. Determination of an etiological agent of bermudagrass decline. Phytopathology 81:1380-1384.

Elliott, M. L. 1993. Bermudagrass decline: Transmission of the causal agent *Gaeumannomyces graminis* var. *graminis* by vegetative planting material. Int. Turfgrass Soc. Res. J. 7:329-334.

Elliott, M. L. 1995. Disease response of bermudagrasses to *Gaeumannomyces graminis* var. *graminis*. Plant Dis. 79:699-702.

Elliott, M. L. 1995. Effect of systemic fungicides on a bermudagrass putting green infested with *Gaeumannomyces graminis* var. *graminis*. Plant Dis. 79:945-949.

Elliott, M. L., Hagan, A. K., and Mullen, J. M. 1993. Association of *Gaeumannomyces graminis* var. *graminis* with a St. Augustinegrass root rot disease. Plant Dis. 77:206-209.

Elmore, W. C., Gooch, M. D., and Stiles, C. M. 2002. First report of *Gaeumannomyces graminis* var. *graminis* on seashore paspalum in the United States. Plant Dis. 86:1405.

Freeman, T. E., and Augustin, B. J. 1984. Bermudagrass decline. Plant Pathol. Fact Sheet PP-31. Florida Extension Service, Institute of Food and Agricultural Sciences, University of Florida.

Hernández-Restrepo, M., Groenwald, J. Z., Elliott, M. L., Canning, G., McMillan, V. E., and Crous, P. W. 2016. Take-all or nothing. Stud. Mycol. 86:19-48.

Landschoot, P. J. 1993. Taxonomy and biology of ectotrophic root-infecting fungi associated with patch diseases of turfgrasses. Pages 41-71 in: Turfgrass Patch Diseases Caused by Ectotrophic Root-Infecting Fungi. B. B. Clarke and A. B. Gould, eds. American Phytopathological Society, St. Paul, MN.

Landschoot, P. J., and Jackson, N. 1989. *Gaeumannomyces incrustans* sp. nov., a root-infecting hyphopodiate fungus from grass roots in the United States. Mycol. Res. 93:55-58.

Landschoot, P. J., and Jackson, N. 1989. *Magnaporthe poae* sp. nov., a hyphopodiate fungus with a *Phialophora* anamorph from grass roots in the United States. Mycol. Res. 93:59-62.

Luo, J., Vines, P. L., Grimshaw, A., Hoffman, L., Walsh, E., Bonos, S. A., Clarke, B. B., Murphy, J. A., Meyer, W. A., and Zhang, N. 2017. *Magnaporthiopsis meyeri-festucae*, sp. nov., associated with a summer patch-like disease of fine fescue turfgrasses. Mycologia 109:780-789.

Martin, B. 2017. Take-all root rot on the increase in the Carolinas. Golf Course Manage. 9:72-77.

Stephens, C. M., and Kerns, J. P. 2020. First report of *Gaeumannomyces graminicola* causing bermudagrass decline of ultradwarf bermudagrass putting greens in North Carolina. Plant Dis. 104:1552.

Tucker, M. A. 2020. Identification and distribution of ectotrophic root-infecting fungi within an ultradwarf bermudagrass putting green and fungicide efficacy associated with these fungi. M.S. thesis. Mississippi State University.

Vines, P. L., Hoffmann, F. G., Meyer, F., Allen, T. W., Luo, J., Zhang, N., and Tomaso-Peterson, M. 2020. *Magnaporthiopsis cynodontis*, a novel turfgrass pathogen with widespread distribution in the United States. Mycologia 112:52-63.

Vines, P. L., Hoffmann, F. G., Meyer, F., Allen, T. W., and Tomaso-Peterson, M. 2021. *Gaeumannomyces nanograminis*, sp. nov., a hyphopodiate fungus identified from diseased roots of ultradwarf bermudagrass in the United States. Mycologia 113:938-948.

Vines, P. L., and Tomaso-Peterson, M. 2021. *Candidacolonium cynodontis* P. L. Vines & M. Tomaso-Peterson, gen. et sp. nov. I. Index Fungorum 488:1.

Wilkinson, H. T., and Kane, R. T. 1993. *Gaeumannomyces graminis* var. *graminis* infecting zoysiagrass in Illinois. Plant Dis. 77:100.

Wong, P. T. 2002. *Gaeumannomyces wongoonoo* sp. nov., the cause of a patch disease of buffalo grass (St. Augustine grass). Mycol. Res. 106:857-862.

(Prepared by R. H. Smiley, P. H. Dernoeden, and B. B. Clarke; revised by M. Tomaso-Peterson)

Other, Less Commonly Observed Diseases

Cephalosporium Stripe

Cephalosporium stripe, caused by *Cephalosporium gramineum,* is a disease of many grasses and cereals. There are no descriptions of this disease in turfgrasses. However, *C. gramineum* is commonly found colonizing senescing leaf tissues of numerous turfgrass species.

Symptoms and Signs

In wheat, infected leaves contain one or more thin, yellowish brown streaks that extend from the leaf sheath to the tip or cut end of the leaf. As an infected leaf ages, it senesces early, becoming chlorotic and then brown. Areas with many infected tillers appear unthrifty, grow slowly, and do not respond normally to applications of water and fertilizers.

Causal Agent

Hymenula cerealis (syn. *C. gramineum*) causes a vascular wilt of many grasses and cereals. It grows very slowly but sporulates profusely, producing small (about 2 × 4 μm), ovoid or crescent-shaped, unicellular conidia.

Disease Cycle

In turfgrasses, *C. gramineum* generally behaves as a saprophyte on chlorotic, senescing leaf tissue. Little is known about its survival and activity in turfgrasses. It is common on leaves of *Agrostis* and *Poa* spp. that have been damaged by mechanical or environmental stress. In cereals and other grasses, *C. gramineum* survives unfavorable periods as soilborne conidia or as mycelium in plant tissues. In cereals, the fungus is most active during the winter and early spring. It enters roots through wounds and must infect the vascular tissue for disease to develop. The tiny conidia in the xylem vessels are carried upward by the transpirational fluid and become lodged at the small passages through stem nodes and in the leaves. The fungus produces metabolic products that occlude the xylem vessels and cause thin, yellowish brown streaks over the entire length of the leaf.

Epidemiology and Management

Attempts to demonstrate the susceptibility of cool-season turfgrass species to isolates of *C. gramineum* from wheat have been unsuccessful, and there is no documentation that this fungus can behave as a primary pathogen in turfgrasses. In turfgrasses, it is primarily associated with environmentally stressed or mechanically injured tissues. Fungicide applications to control *C. gramineum* in turfgrass are generally not warranted.

Selected References

Bockus, W. W., Bowden, R. L., Hunger, R. M., Morrill, W. L., Murray, T. D., and Smiley, R. W., eds. 2010. Compendium of Wheat Diseases. 3rd ed. American Phytopathological Society, St. Paul, MN.

Gams, W. 1971. *Cephalosporium*-artige Schimmelpilze (Hyphomycetes). Gustav Fischer Verlag, Stuttgart, Germany.

(Prepared by R. W. Smiley, P. H. Dernoeden, and B. B. Clarke)

Cercosporoid and Cercosporoid-like Fungal Diseases

Cercosporoid fungi reside in the family Mycosphaerellaceae and cause leaf spot on members of Poaceae worldwide. Foliar symptoms may be confused with those of gray leaf spot. Species in four genera, *Catenulocercospora, Cercospora, Graminopassalora,* and *Passalora,* infect *Agrostis, Buchloë, Cynodon, Festuca, Poa,* and *Stenotaphrum* spp.

Porocercospora is a new genus, erected to accommodate *P. seminalis,* which produces pigmented conidia and infects *Buchloë* and *Cynodon* spp. *Passalora agrostidicola* also produces pigmented conidia and causes leaf spots on *Agrostis* spp.

Symptoms and Signs

Lesions (Fig. 247) occur on the leaf blades and sheaths. Newly initiated spots are uniformly brown to purple, elliptical to oblong, and most prevalent along the margins of leaves. As the spots age, they become elongated parallel to the leaf axis and grow as large as 1 × 4 mm. The centers of older spots become tan to gray. In warm, moist weather, sporulation of the pathogen may be so abundant that the spots appear white to gray. Severely affected leaves become chlorotic and may die, causing the turf to become thinned.

Brown stripe, caused by *Graminopassalora graminis,* may occur on nearly all grass species. The causal agent appears to be widespread throughout the world, but it is considered to be of minor importance in turfgrasses. Leaf lesions are small, chocolate-colored spots with gray centers. The lesions may elongate as they enlarge and may develop into stripes between the veins of leaves and leaf sheaths. When the disease is severe, a tip blight of leaves may occur. A disease of *Cynodon* and *Zoysia* spp. caused by *Bipolaris heveae* has been incorrectly referred to as brown stripe in the literature but is included under *Bipolaris* diseases in this volume.

False smut (leaf and glume spot) of *Buchloë dactyloides* is caused by *P. seminalis,* which infects the unfertilized ovary, reducing seed yield (Fig. 248). Dissection of an infected ovary reveals black, spongy mycelium filling the ovarian cavity.

Fig. 247. Cercospora leaf spot of *Stenotaphrum secundatum.* (Courtesy T. E. Freeman)

Causal Agents

Cercosporoid and cercosporoid-like fungi produce pigmented, erect, unbranched, multiseptate conidiophores on infected tissue (Fig. 249). Considerable variation occurs in the lengths of conidiophores (20–800 µm). Conidia (10–300 µm long) are solitary or in chains, straight to slightly curved, cylindrical to club shaped, zero to multiseptate, and hyaline or pigmented.

Fig. 248. Mycelium of *Porocercospora seminalis* filling the ovaries of a buffalograss inflorescence. (Courtesy K. L. Amundsen—© APS)

Catenulocercospora fusimaculans causes leaf spots on *Agrostis, Buchloë,* and *Stenotaphrum* spp. The conidia are hyaline, one to seven septate, obclavate (having thick bases and tapering toward the tips), curved, and 1.5–3 × 20–100 µm.

Cercospora agrostidis causes leaf spots on *Agrostis* spp. The conidia are hyaline, one to seven septate, straight or slightly curved, and 1.5–3.0 × 10–60 µm.

Cercospora cynodontis causes leaf spots on *Cynodon* spp. The conidia are hyaline, zero to five septate, and 2.5–4.0 × 17.5–65 µm.

Cercospora festucae causes leaf spots on *Festuca* spp. The conidia are hyaline, indistinctly multiseptate, obclavate, curved or wavy, and 2–4 × 40–300 µm.

G. graminis produces stromata in infected leaf tissue, often in the substomatal cavity. Conidiophores are produced in dense tufts (fascicles) that erupt through the leaf surface or emerge through stomata. Individual conidiophores are olive brown and 30–150 µm long. Conidia are 4–12 × 16–56 µm and one to three septate; they have rounded bases with obvious hila and slightly tapering tips (Fig. 249).

Passalora agrostidicola causes circular to irregular leaf spots on *Agrostis* spp. The conidia are lightly pigmented, solitary or in chains, straight to slightly curved, three to four septate with slight constrictions at the septa, and 1.5–2.0 × 18–38 µm.

P. seminalis causes false smut on *Buchloë* and *Cynodon* spp. The conidia are obclavate to cylindrical, medium brown, zero to 12 distoseptate, and 7–8 × 60–90 µm.

To confirm the identity of these pathogens, analyses of three partial nuclear genes are most reliable: 28S nrRNA gene

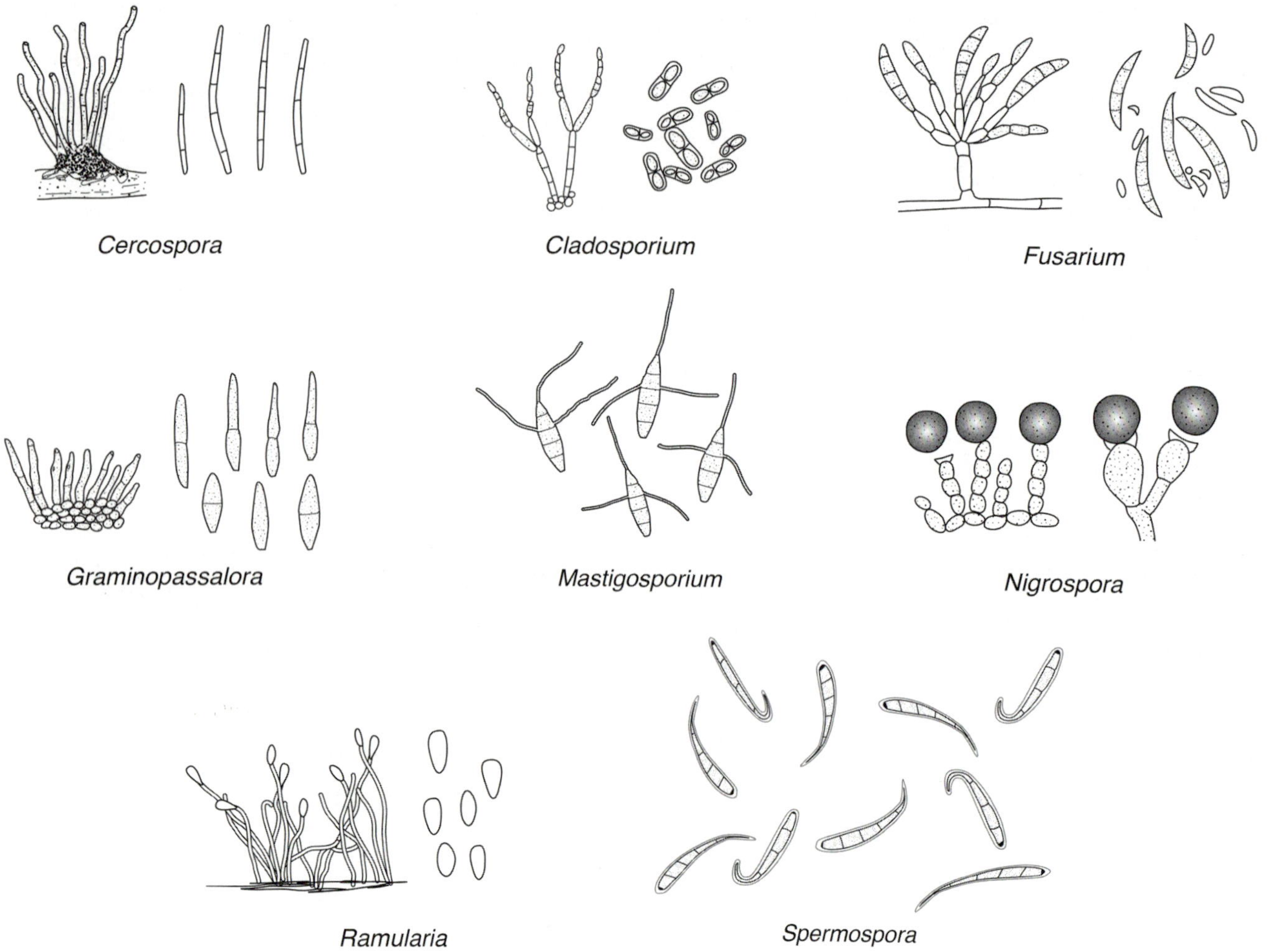

Fig. 249. Conidia and conidiophores of *Cercospora, Cladosporium, Fusarium, Graminopassalora, Mastigosporium, Nigrospora, Ramularia,* and *Spermospora* spp. (Courtesy L. P. Tredway—© APS. Drawing by S. McKinley.)

(LSU); internal transcribed spacer regions and intervening 5.8S nrRNA gene (ITS); and the nrDNA operon, RNA polymerase II second largest subunit (*RPB2*).

Disease Cycle

The pathogens survive adverse periods as conidia and as dormant mycelium in and among infected leaves and plant litter. Spores are produced and dispersed during periods of warm, humid weather. Spore germination and infection lead to the development of new leaf spots and false smut, from which new generations of spores may be produced. *G. graminis* survives through the winter as mycelial stromata in infected leaves and plant litter. Conidiophores are borne on stromata in the spring, and conidia are splashed, blown, or otherwise disseminated to healthy leaves.

Epidemiology and Management

A prolonged period of leaf wetness during warm weather is required for the development of leaf spot and false smut. *G. graminis* sporulates and causes disease mostly during cool, moist weather in spring and autumn. Control measures for brown stripe have not been reported. Watering practices that allow leaves to remain wet for an extended period should be avoided. Irrigation during the morning, when dew is already present, rather than during the late afternoon or evening, is preferred. Turf should be watered infrequently, but deeply, in warm weather. If turf is heavily shaded or if air movement is restricted, selective pruning of trees or shrubs and removal of barriers that impede air movement are recommended. Application of fertilizer should be increased slightly if the turf becomes significantly damaged. Several cultivars of *Stenotaphrum secundatum* exhibit resistance to Catenulocercospora leaf spot. Fungicides also provide effective control of the disease.

Selected References

Amaradasa, B. S., Madrid, H., Groenewald, J. Z., Crous, P. W., and Amundsen, K. 2014. *Porocercospora seminalis* gen. et comb. nov., the causal organism of buffalograss false smut. Mycologia 106:77-85.

Braun, U., Crous, P. W., and Nakashima, C. 2015. Cercosporoid fungi (*Mycosphaerellaceae*) 3. Species on monocots (*Poaceae*, true grasses). IMA Fungus 6:25-97.

Farr, D. F., and Rossman, A. Y. n.d. Fungal Databases. Systematic Mycology and Microbiology Laboratory, U.S. Department of Agriculture–Agricultural Research Service, Washington, DC. https://nt.ars-grin.gov/fungaldatabases/

Freeman, T. E. 1959. A leafspot of St. Augustine grass caused by *Cercospora fusimaculans*. Phytopathology 49:160-161.

Smith, J. D., Jackson, N., and Woolhouse, A. R. 1989. Fungal Diseases of Amenity Turf Grasses. E. & F. N. Spon, London.

Videira, S. I. R., Groenwald, J. Z., Nakashima, C., Braun, U., Barreto, R. W., de Wit, P. J. G. M., and Crous, P. W. 2017. Mycosphaerellaceae—Chaos or clarity? Stud. Mycol. 87:257-421.

(Prepared by R. W. Smiley, P. H. Dernoeden, and B. B. Clarke; revised by M. Tomaso-Peterson)

Cladosporium Eyespot

Cladosporium eyespot, also known as timothy eyespot, is a common disease of *Phleum* spp. in cool temperate climates, where this genus is an important turfgrass. The Cladosporium eyespot pathogen also causes disease in *Cynodon, Festuca,* and *Poa* spp.

Symptoms

Leaf spots (Fig. 250) are small, oval, and up to 3 mm long. The lesions are light brown to gray in the centers, each bordered by a narrow, bright or dark purplish brown margin. The margin may fade to brown or intensify to black. Tissue surrounding the lesions becomes chlorotic, and severely infected leaves may die. These symptoms are similar to those of Mastigosporium leaf spot.

Fig. 250. Cladosporium eyespot of *Phleum bertolonii.* (Courtesy P. Weibull—© APS)

Causal Agent

Cladosporium phlei produces long (up to 250 µm), erect, dark brown conidiophores, formed in groups (Fig. 249). Sporulation is sparse. Conidia are ovoid to cylindrical, echinulate (spiny), pale to dark brown, mostly one to three septate, and 5–12 × 14–36 µm (average 9 × 23 µm). Conidia are produced singly or in short chains.

Disease Cycle, Epidemiology, and Management

Little is known about the disease cycle and management of *C. phlei* in turfgrass. The fungus is most often found in shaded sites and in areas with frequent rainfall during periods of mild to warm weather. Cladosporium eyespot of *Phleum* spp. has been recorded during the winter, spring, and summer in cool temperate climates, which suggests that the pathogen is adaptable to a wide range of temperatures. Spores of *C. phlei* germinate at temperatures of 3–33°C (37–91°F); the optimal temperature is 24°C (75°F). Removal of shade and infrequent irrigation may be beneficial in controlling the disease.

Selected References

Farr, D. F., and Rossman, A. Y. n.d. Fungal Databases. Systematic Mycology and Microbiology Laboratory, U.S. Department of Agriculture–Agricultural Research Service, Washington, DC. https://nt.ars-grin.gov/fungaldatabases/

Mäkelä, K., and Sundheim, L. 1982. Cladosporium eyespot. Weibulls Gräs-Tips 23.

Petersen, V., and Hermansen, J. E. 1971. Leaf spot diseases on graminicolous species at a locality in Greenland. Friesia 10:25-29.

Smith, J. D., Jackson, N., and Woolhouse, A. R. 1989. Fungal Diseases of Amenity Turf Grasses. E. & F. N. Spon, London.

(Prepared by R. W. Smiley, P. H. Dernoeden, and B. B. Clarke; revised by M. Tomaso-Peterson)

Fusarium Diseases

Species of *Fusarium* cause leaf spot, foliar blight, and (less commonly) rotting of roots and living stems. These fungi are most active on new seedlings (see Damping-Off (Seedling Diseases)) and on older plants weakened by heat or cold stress (see Microdochium Patch). *Fusarium* spp. typically also colonize senescing plants damaged by root pathogens, such as those causing necrotic ring spot and summer patch.

Symptoms and Signs

In warm, wet weather, *Fusarium* spp. cause leaf spots that appear uniformly on plants over large areas (Fig. 251). The leaf spots (Fig. 252) are irregularly shaped and occur mostly on older leaves. Infected spots become water soaked and then turn tan, usually with brown to purplish brown margins separating infected tissue from healthy tissue. Initial infections often occur at the ends of leaves, resulting in tip blight.

In hot, dry weather, *Fusarium* spp. cause plants in small areas to turn light green and rapidly fade to tan and then become straw colored as they die. Affected areas generally appear in irregular patterns. A dark brown dry rot of roots, crowns, rhizomes, and stolons may develop in affected areas (Fig. 253). When temperatures and moisture levels are high, white to pink mycelium and masses of spores may be produced on the lower stems and crowns of infected plants. Irregularly shaped leaf lesions with light brown margins may or may not develop on the blighted turf. Similar diseases caused by *Drechslera poae* and *Bipolaris sorokiniana* can usually be distinguished by the presence of oval leaf or sheath lesions with restricted, dark margins.

Large numbers of conidia may be produced on senescent tissue of plants mechanically injured or weakened by heat and drought stress in the summer. *Fusarium* spp. behave as saprophytes under these conditions.

Fig. 251. Fusarium leaf spot of *Poa pratensis* caused by *Fusarium acuminatum.* (Courtesy R. W. Smiley—© APS)

Fig. 252. Fusarium leaf spot of *Poa pratensis* caused by *Fusarium heterosporum.* (Courtesy R. W. Smiley—© APS)

Fig. 253. Fusarium crown and root rot of *Poa pratensis* caused by *Fusarium crookwellense.* (Courtesy R. W. Smiley—© APS)

Causal Agents

Fusarium spp. produce crescent-shaped macroconidia, each with a basal cell with a distinctive heel (Fig. 249). Some species also produce microconidia. In some species, the macroconidia and microconidia are similar in shape, but in other species they are very different. Several species produce thick-walled chlamydospores (resting spores).

Disease Cycle

Fusarium spp. survive the winter as mycelia in infected plants and infested debris or, for some species, as chlamydospores in soil or thatch. These fungi sporulate profusely, especially in thatch that is rewetted after being dry. The mycelium is capable of infecting all parts of the turfgrass plant.

Epidemiology

Fusarium crown and root rot, leaf spots, and cool-season diseases may occur in stressed turf of any age. High temperatures and drought favor the development of crown and root rot. The disease occurs primarily in areas that are fully exposed to sunlight during most of the day in midsummer. Other conditions that favor diseases caused by *Fusarium* spp. include excessive or unbalanced applications of nitrogen in the spring or summer, mowing at a height lower than recommended for the specific turfgrass species, and the presence of a thick thatch layer.

Leaf spots occur when very humid conditions persist for prolonged periods. Diseases caused by *Fusarium* spp. and *B. sorokiniana* are similar in time of occurrence and type of damage caused.

Management

Diseases caused by *Fusarium* spp. may be reduced by minimizing stress. Turf should not be mowed to a height lower than recommended. Thatch should be removed mechanically if it accumulates to depths greater than 1.5 cm. Fertilizers that provide balanced ingredients should be applied, but excessive stimulation of grass with large amounts of nitrogen should be avoided. Turf should be watered thoroughly and as infrequently as possible to provide adequate moisture without causing drought stress. *Fusarium* diseases are difficult to control with fungicides, but these diseases are relatively rare in turfgrass systems.

Selected References

Nelson, P. E., Toussoun, T. A., and Marasas, W. F. O. 1983. *Fusarium* Species: An Illustrated Manual for Identification. Pennsylvania State University Press, University Park, PA.

Smiley, R. W., and Craven, M. M. 1979. *Fusarium* species in soil and thatch of *Poa pratensis* treated with fungicides. Soil Biol. Biochem. 11:355-363.

Smiley, R. W., and Thompson, D. C. 1985. Soil and atmospheric moistures associated with Fusarium crown rot and leaf blight of *Poa pratensis.* Plant Dis. 69:294-297.

Smith, J. D., Jackson, N., and Woolhouse, A. R. 1989. Fungal Diseases of Amenity Turf Grasses. E. & F. N. Spon, London.

(Prepared by R. W. Smiley, P. H. Dernoeden, and B. B. Clarke)

Leptosphaerulina Leaf Blight

Leptosphaerulina leaf blight occurs on most turfgrasses, particularly *Agrostis, Festuca, Lolium,* and *Poa* spp. The pathogen is saprobic on senescent turfgrass tissues. Symptoms of the disease may be confused with those of Nigrospora leaf blight, Ascochyta leaf blight, anthracnose, and dollar spot.

Symptoms and Signs

Large areas of turf may become uniformly blighted, or symptoms may appear in small spots or patches and in localized pockets of more heavily infected grass. Individual leaves generally die back from the tip and may have yellow, brown, or reddish brown lesions or generalized blighting extending down to the leaf sheath. Entire leaves and sometimes leaf sheaths appear water soaked and soon turn bleached white or tan. The overall effect of the blighting closely resembles bleaching from high-temperature stress, frost, or dull mower injury. Tiny, brown pseudothecia (fruiting bodies) of the pathogen, visible with a hand lens, develop in the dead tissues within lesions.

Causal Agent

Leptosphaerulina australis produces pale brown pseudothecia, 40–170 µm in diameter (Fig. 254). Each ascus contains eight ascospores, which are muriform, measure about 25–41 × 10–15 µm, and have two to six transverse septa and zero to three longitudinal septa. No asexual state of this fungus has been reported.

Disease Cycle

L. australis is a common soil saprophyte and unspecialized pathogen that colonizes weakened or senescing tissues of many different plant hosts. The fungus presumably survives adverse periods as pseudothecia or as mycelia in plant debris. It often is found in association with other turfgrass pathogens, including *Pyricularia oryzae* and *Pythium* spp.

Epidemiology

Extended periods of warm, humid weather and stressful soil conditions favor Leptosphaerulina leaf blight. However, the disease may appear at any time between spring and autumn. *L. australis* is regarded as nonpathogenic in turfgrasses, but it aggressively blights senescing leaves or tissues stressed by other pathogens or physiological disorders. The disease may become severe in turf growing on wet or compacted soil; in turf stressed by drought, low mowing, or the activity of other pathogens; and in recently transplanted sod.

Management

Accepted turfgrass management practices should be followed to prevent severe stress. Adequate amounts of complete fertilizers should be applied, and compacted soils should be aerated. Turf should be irrigated thoroughly (deeply) and as infrequently as possible to avoid moisture stress. Turf should not be watered in the late afternoon or evening. Low mowing heights should be avoided, and grass should not be mowed when it is wet. Applying herbicides or installing new sod during or just before an extended period of hot, humid weather should be avoided. Fungicide assays have shown significant variation among isolates of *L. australis,* but control measures using fungicides have not been reported.

Selected References

Couch, H. B. 1995. Diseases of Turfgrasses. 3rd ed. Krieger Publishing, Malabar, FL.

Graham, J. H., and Luttrell, E. S. 1961. Species of *Leptosphaerulina* on forage plants. Phytopathology 51:680-693.

Mitkowski, N. A., and Browning, M. 2004. *Leptosphaerulina australis* associated with intensively managed stands of *Poa annua* and *Agrostis palustris.* Can. J. Plant Pathol. 26:193-198.

Smith, J. D., Jackson, N., and Woolhouse, A. R. 1989. Fungal Diseases of Amenity Turf Grasses. E. & F. N. Spon, London.

(Prepared by R. W. Smiley, P. H. Dernoeden, and B. B. Clarke; revised by M. Tomaso-Peterson)

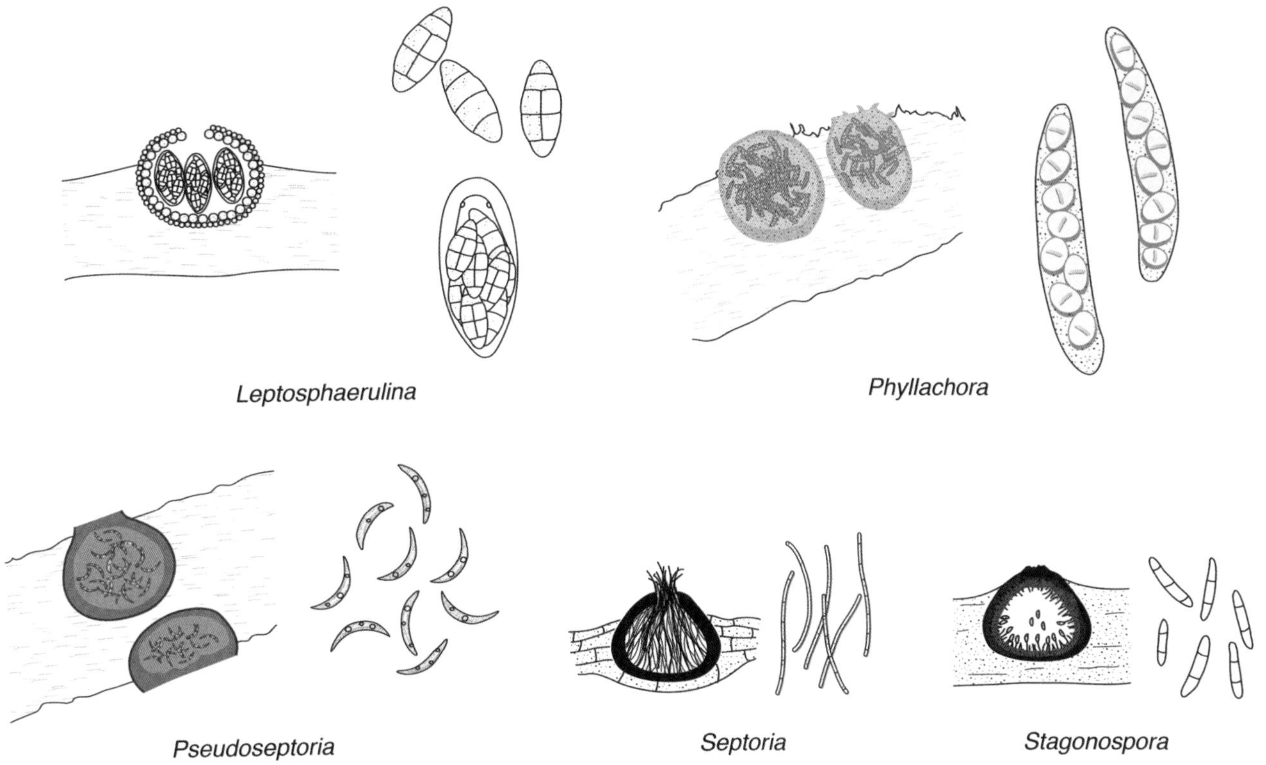

Fig. 254. Pycnidia and conidia of *Pseudoseptoria, Septoria,* and *Stagonospora* spp.; and ascocarps, asci, and ascospores of *Leptosphaerulina* and *Phyllachora* spp. (Courtesy L. P. Tredway—© APS. Drawing by S. McKinley.)

Mastigosporium Leaf Spot

Mastigosporium leaf spot, sometimes referred to as leaf fleck, eyespot, or purple leaf spot, occurs on several cool-season grass species in cool to cold humid environments in northern Europe, northern Asia, and northern North America. The disease is most common in unmowed or infrequently mowed grass stands.

Symptoms and Signs

Infection sites are initially small, pale, water-soaked spots on leaves. The spots soon turn dark. As the spots enlarge, they fade to sandy gray in the centers, with borders that turn wine red to violet. The spots become oval to nearly circular and 1–2 mm long. When the disease is severe, spots coalesce and the leaf dies. On *Phleum* spp., Mastigosporium leaf spot symptoms are nearly identical to those caused by Cladosporium eyespot.

Causal Agents

Mastigosporium rubricosum has a broad host range, occurring on *Agrostis, Festuca, Lolium, Phleum,* and *Poa* spp.; *M. kitzebergense, M. cylindricum,* and *M. album* have been observed only on *Phleum, Festuca,* and *Lolium* spp., respectively. These fungi produce short, hyaline conidiophores that are scattered or grouped on the leaf surface. Conidia produced by *M. rubricosum, M. kitzebergense,* and *M. cylindricum* are hyaline, slimy, mostly three septate, cylindrical to ellipsoidal, and 7–17 × 25–47 µm (Fig. 249). Conidia *of M. album* are longer, 8–16 × 33–85 µm, four to five septate, and each has one to three filiform appendages emanating from the distal end.

Disease Cycle

The pathogens survive the winter as mycelium in plant debris. A stromatic resting stage may also develop on some hosts. Conidia develop on conidiophores arising from the overwintering mycelium and are disseminated by water, wind, and possibly other means.

Epidemiology and Management

Mastigosporium spp. require cool, wet weather to infect their hosts. In *Phleum* spp., most damage occurs just after the snow melts in the spring and in the autumn. In *Agrostis* spp., the disease is most severe from midsummer to autumn in Nordic countries. In cool regions of North America, the disease becomes most severe in cold, rainy, or foggy weather during winter or late winter and, at higher elevations, during midsummer in shaded areas. Chlorothalonil provides effective control of the disease in grasses grown for seed production.

Selected References

Bollard, E. G. 1950. Studies in the genus *Mastigosporium*. I. General account of the species and their host ranges. Trans. Br. Mycol. Soc. 33:250-264.

Gunnerbeck, E. 1971. Studies on foliicolous deuteromycetes. I. The genus *Mastigosporium* in Sweden. Sven. Bot. Tidskr. 65:39-52.

Mäkelä, K. 1970. The genus *Mastigosporium* Riess in Finland. J. Finn. Mycol. Soc. 11:5-22.

Smith, J. D., Jackson, N., and Woolhouse, A. R. 1989. Fungal Diseases of Amenity Turf Grasses. E. & F. N. Spon, London.

(Prepared by R. W. Smiley, P. H. Dernoeden, and B. B. Clarke; revised by L. P. Tredway)

Microdochium Diseases

The two foliar diseases, sparse leaf patch of *Paspalum vaginatum* and leaf blight of *Poa pratensis* and *Agrostis stolonifera,* all managed as putting green and fairway turf, differ from Microdochium patch. *Microdochium paspali* (sparse leaf patch) and *M. poae* (leaf blight) infect foliar tissue during periods of moderate to high temperatures in the summer months, in contrast to *Monographella nivalis* (Microdochium patch), which actively infects both warm- and cool-season turfgrasses in the winter.

Symptoms and Signs

Sparse leaf patch. Leaf lesions on *P. vaginatum* are initially water soaked, fading to a light tan to straw color near the leaf tip and progressing to the base of the leaf and sheath. Crowns and roots are not affected. On closely mowed turf, sparse leaf patch first appears as small, dark brown to bronze irregular patches 2.5–10 cm in diameter (Fig. 255). Necrotic patches in the turf may coalesce as the disease progresses. Numerous, small patches in the turf may be confused with those of dollar spot; however, necrotic patches caused by sparse leaf patch are not bleached to a straw color (Fig. 256). Healthy green leaves may be interspersed among blighted turf. In the early morning or under conditions of high humidity, small tufts of white mycelium often appear at the margins of blighted patches and may be confused with signs of Pythium blight. Necrotic turf can appear in streaks caused by the spread of conidia and mycelium under moist conditions.

Leaf blight. Necrotic lesions first appear along the leaf margins and progress to cause leaf blight of *P. pratensis* and *A. stolonifera.* Infected leaves become chlorotic and die. Initial symptoms in the turf appear as small, water-soaked spots less than 2 cm in diameter that can be scattered throughout the

Fig. 255. Sparse leaf patch of *Paspalum vaginatum* on a putting green. (Courtesy P. F. Harmon—© APS)

Fig. 256. Numerous infection centers of sparse leaf patch of *Paspalum vaginatum* on a putting green. (Courtesy P. F. Harmon—© APS)

stand. Under favorable conditions, the disease spreads rapidly, resulting in large areas of irregular, blighted patches (Fig. 257). Frequently, with early morning dew or high humidity, white mycelium develops on infected leaves. Gelatinous, salmon-colored sporodochia may be observed on infected leaves in the later stages of disease development.

Causal Agents

M. paspali causes sparse leaf patch of *P. vaginatum*. Light yellow to salmon-colored sporodochia develop in culture after 7 days or following incubation of infected leaf tissue. Hyaline, slightly curved, falcate conidia (12.0 × 3.5 µm), zero to three septate and mostly two celled, are produced on irregularly branched, single-celled conidiophores (11.5 × 3.5 µm). Sessile conidia, in groups of three or more, may develop on mature hyphal branches. On potato dextrose agar, a *M. paspali* colony grows 12 mm in diameter per day at 25°C (77°F) and initially appears pale yellow, changing to pale orange to salmon colored with age. Aerial mycelium is sparse and cottony. Optimal growth temperatures for *M. paspali* in culture are 25–28°C (77–82°F).

M. poae causes leaf blight of *P. pratensis* and *A. stolonifera*. Salmon-colored colonies with aerial, cottony mycelium and irregular margins develop after 7 days on oatmeal agar. The colony develops a gelatinous, salmon-colored center covered by dark brown or black chlamydospores at 14 days. Slightly curved, single-celled conidia (6.4 × 2.5 µm) develop on single conidiophores born directly on hyphae.

Disease Cycle, Epidemiology, and Management

M. paspali survives unfavorable conditions as dormant mycelium in infected leaves and plant debris. Development of sparse leaf patch and leaf blight is favored by high humidity, heavy dew, frequent rainfall, and excessive soil moisture when temperatures are 25–32°C (77–90°F). The pathogens spread from leaf to leaf via mycelium. They are moved long distances by free water and by maintenance equipment, which disseminates conidia and mycelium in infected leaves and stems.

Sparse leaf patch was first reported on *P. vaginatum* golf turf in the tropical climate of Hainan province, China. In the United States, the disease has been observed throughout the Gulf Coast states in early and late summer; however, symptoms may persist throughout the growing season if left unchecked.

Fig. 257. Leaf blight of *Agrostis stolonifera*, caused by *Microdochium poae*. (Courtesy J. Liang—© APS)

Leaf blight of *P. pratensis* and *A. stolonifera* occurs during the summer months when the grasses are exposed to prolonged periods of heat greater than 30°C (86°F). This disease was first reported on golf courses in Beijing and Zhengzhou, China.

Little is known about control measures because these diseases were only recently described. Epidemiological factors suggest that mowing should be suspended until the turf dries. Putting greens showing symptoms should be mowed last to prevent movement of the pathogen to unaffected greens. Successful control with fungicides has been reported only for sparse leaf patch.

Selected References

Liang, J., Li, G., Zhao, M., and Cai, L. 2019. A new leaf blight disease of turfgrasses caused by *Microdochium poae*, sp. nov. Mycologia 111:265-273.

Tomaso-Peterson, M. 2010. Curative fungicide applications for control of Microdochium patch on a seashore paspalum green, 2009. Plant Dis. Manage. Rep. 4:T046.

Zhang, W., Nan, Z., Tian, P., Hu, M., Gao, Z., Li, M., and Liu, G. 2015. *Microdochium paspali,* a new species causing seashore paspalum disease in southern China. Mycologia 107:80-89.

(Prepared by M. Tomaso-Peterson)

Mollicute Diseases

Mollicutes are prokaryotic organisms that have no cell walls and include mycoplasmas, phytoplasmas (mycoplasma-like organisms), and spiroplasmas. Few reports implicate mollicutes as turfgrass pathogens. However, these organisms cause diseases of grasses used for food and feed, and the turfgrass environment appears conducive to the survival of these pathogens. Therefore, it is likely that mollicute diseases occur in turfgrasses but go largely undetected. This section describes diseases of turfgrasses in which mollicutes have been implicated, although the grasses may not have been maintained as turf at the time the diseases were identified. The overall importance of mollicutes in turfgrasses remains to be determined.

Symptoms

Mollicutes cause aster yellows of *Agropyron* and *Lolium* spp. in the Northern Hemisphere, white leaf of *Poa annua* in Italy and possibly white leaf of *Cynodon* spp. in Southeast Asia, and yellow dwarf of *Agrostis* and *Zoysia* spp. in Japan.

Aster yellows of mature plants is characterized by a general yellowing or reddening of leaves, chlorotic blotches, and stunting. Symptoms appear on *L. multiflorum,* but infected *L. perenne* plants are symptomless. Chlorotic blotches of unknown cause are also common in *Agrostis* spp. in North America during the summer.

White leaf of *P. annua* is caused by Bermudagrass white leaf phytoplasma, which belongs to the rice yellows dwarf phytoplasma and aster yellows phytoplasma groups and is found in Asia, Africa, and Europe. Infected plants lose chlorophyll and turn white, and the disease results in a white speckling of turf. A similar condition has been observed in *Agrostis* and *Poa* spp. in North America.

Yellow dwarf is a growing problem in *Agrostis* and *Zoysia* spp. in Japan. It is caused by an undescribed agent believed to be a mollicute. Symptoms include leaf yellowing, stunting, and tufting.

Causal Agents

Mollicutes lack cell walls, are bundled by a unit membrane, and reproduce by budding and by binary fission. They occur in various shapes, from spherical to ovoid to filamentous. High-resolution microscopy is required for accurate observation of mollicutes.

The aster yellows phytoplasma is spherical and tiny (0.3–0.5 µm in diameter) but increases in size during reproduction.

Disease Cycle and Epidemiology

Mollicutes generally are introduced into the phloem cells of plants during feeding by stylet-bearing insect vectors, such as leafhoppers. Insect-transmitted mollicutes do not require high humidity for efficient transfer. Mollicutes multiply in the phloem and may induce various symptoms, including chlorosis, smaller or stunted leaves, excessive tillering (witches'-brooms or tufting), and a general decline in turf vigor.

Management

Control of vectors is the primary means of controlling the spread of mollicutes. Mollicutes are temporarily suppressed but not killed by treatments with tetracycline antibiotics.

Selected References

Lee, I. M., Pastore, M., Vibio, M., Danielli, A., Davis, R. E., and Bertaccini, A. 1997. Detection and characterization of a phytoplasma associated with annual blue grass (*Poa annua*) white leaf disease in southern Italy. Eur. J. Plant Pathol. 103:251-254.

Marcone, C., Ragozzino, A., and Seemüller, E. 1997. Detection of Bermuda grass white leaf diseases in Italy and characterization of the associated phytoplasma by RFLP analysis. Plant Dis. 81:862-866.

Tani, T., and Beard, J. B. 1997. Color Atlas of Turfgrass Diseases. Ann Arbor Press, Chelsea, MI.

(Prepared by R. W. Smiley, P. H. Dernoeden, and B. B. Clarke)

Nigrospora Blight

Nigrospora blight occurs in *Lolium perenne, Festuca rubra,* and *Poa pratensis* under hot, humid conditions in summer. In *Stenotaphrum secundatum,* both foliar and stolon rot symptoms have been documented. Symptoms of this disease can easily be confused with phases of Ascochyta leaf blight, dollar spot, brown patch, Pythium blight, and Leptosphaerulina leaf blight. Furthermore, conidia of *Nigrospora* spp. are commonly found on necrotic tissues and in thatch during the summer.

Symptoms and Signs

In cool-season grasses, turf becomes uniformly blighted over large areas or may appear patchy because of local pockets of more heavily infected grass (Fig. 258). Plants may be affected in well-defined patches, 10–20 cm in diameter. The dieback generally starts at the leaf tip. Uniform lesions may extend all the way down to the leaf sheath, resulting in the death of entire leaves. Less commonly, leaves and sheaths become water soaked and may have an oily appearance similar to that of turf affected by *Pythium* spp. Affected tissues quickly develop tan, necrotic lesions, which in some cultivars are surrounded by purple to reddish brown margins (Fig. 259). Lesions often completely girdle leaf blades. The leaf tissue above the lesion becomes entirely blighted and twisted and turns tan. In some *P. pratensis* cultivars, the dying leaf turns deep purple.

In warm, very humid weather, the pathogen produces abundant, fluffy, white mycelium on blighted leaf blades. The mycelium can easily be confused with signs of dollar spot, brown patch, or Pythium blight.

In *S. secundatum,* foliar symptoms have been described as brown, circular leaf spots with yellow to rust brown margins. Dark lesions have been observed on infected stolons during periods of hot, dry weather, causing tillers to wilt, turn yellow, and die. Symptoms from a distance may manifest as a general thinning or patchiness of the infected turf.

Causal Agents

Nigrospora spp. are facultative parasites that normally persist as saprophytes on organic litter. Conidia are the most distinct diagnostic feature of *Nigrospora* spp.; they are shiny, black, smooth-walled spheres (12–18 µm in diameter) that are easily visible with a hand lens or dissecting microscope (Fig. 260). The conidia are produced atop hyaline conidiogenous cells arising directly from mycelium or from pale brown, branching conidiophores (Fig. 249).

N. sphaerica is the most common species associated with turfgrasses and has been observed on both cool- and warm-season species. *N. osmanthi* has been documented in *Stenotaphrum* spp. in China, and *N. oryzae* has been reported in *P. pratensis* in Ontario, Canada. Preliminary identification of species can be made based on the size of conidia. *N. sphaerica* conidia are mostly 16–19 µm in diameter, whereas *N. osmanthi* and *N. oryzae* conidia are typically 13–16 µm. Definitive identification is best carried out through sequencing of internal transcribed spacer rDNA or other loci.

Disease Cycle

N. sphaerica appears to survive as spores and mycelium in host debris. It sporulates in or on infected tissue. Conidia germinate in warm, wet weather and are apparently splashed, blown, or carried to leaves, stems, and stolons. The fungus produces abundant aerial mycelium under very humid conditions, and a new infection is initiated at the spot on the leaf where

Fig. 258. Nigrospora blight of *Poa pratensis*. The disease is sparse in the left half of the plot, where fungicides were routinely applied, and severe in the right half, where fungicides were not applied. (Courtesy R. W. Smiley—© APS)

Fig. 259. Leaf lesions on *Poa pratensis* caused by Nigrospora blight. (Courtesy R. W. Smiley—© APS)

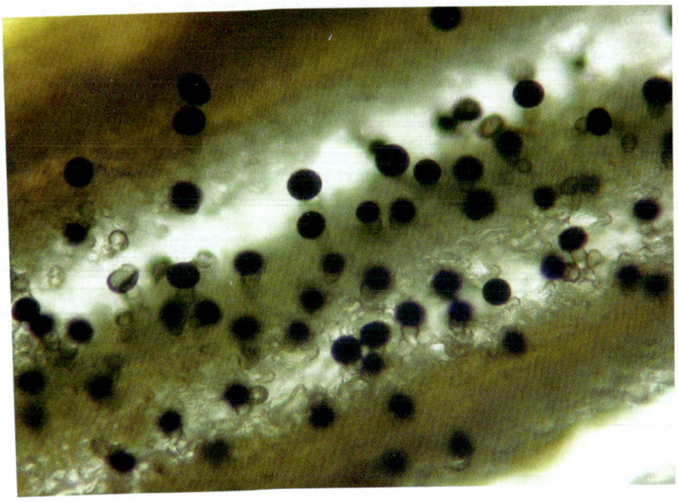

Fig. 260. Conidia of *Nigrospora* sp. on a *Cynodon* sp. leaf. (Courtesy M. Tomaso-Peterson—© APS)

aerial mycelium contacts healthy leaf tissue. Symptoms suggest that the pathogen also colonizes leaf tips.

Epidemiology

Periods of warm, humid weather are most conducive to Nigrospora blight in cool-season grasses. The most severe outbreaks occur when nights are warm and the air is saturated with moisture from dense fog or light rain.

In *S. secundatum,* Nigrospora blight has been observed from spring to summer, with foliar symptoms developing during warm, humid weather and stolon rot observed under hot, dry conditions. The disease is more severe in turf affected by St. Augustine decline or take-all root rot.

In all grass species, the disease occurs only in plants stressed by poor soil conditions, such as excessive dryness or infertility.

Management

Accepted turfgrass management practices help to prevent severe stress to turf. Adequate amounts of complete fertilizers should be applied to encourage healthy turf growth. Turf should be irrigated thoroughly (deeply) and as infrequently as possible to avoid moisture stress. Turf should not be watered in the late afternoon or evening. Grass should not be mowed when it is wet. Applying herbicides or installing new sod during or just before an extended period of hot, humid weather should be avoided.

Cultivars of *Festuca, Lolium,* and *Poa* spp. vary in susceptibility to Nigrospora blight. If the disease is severe and recurring, fungicides may be used to control it. Chlorothalonil, iprodione, and thiophanate-methyl have demonstrated activity against *Nigrospora* spp. in field or laboratory experiments.

Selected References

Mei, S. S., Wang, Z. Y., Zhang, J., and Rong, W. 2019. First report of leaf blight on *Stenotaphrum secundatum* caused by *Nigrospora osmanthi* in China. Plant Dis. 103:1783.

Thompson, D. C., Fowler, M. C., and Smiley, R. W. 1982. Recognizing Nigrospora blight. Plant Dis. 66:265.

Wang, M., Liu, F., Crous, P. W., and Cai, L. 2017. Phylogenetic reassessment of *Nigrospora:* Ubiquitous endophytes, plant and human pathogens. Persoonia 39:118-142.

Zheng, L., Shi, F., Kelly, D., and Hsiang, T. 2012. First report of leaf spot of Kentucky bluegrass (*Poa pratensis*) caused by *Nigrospora oryzae* in Ontario. Plant Dis. 96:909.

(Prepared by R. W. Smiley, P. H. Dernoeden, and B. B. Clarke; revised by L. P. Tredway)

Physoderma Leaf Spot and Leaf Streak

Physoderma leaf spot and leaf streak, previously known as Physoderma brown spot, occurs infrequently in turfgrasses but has been reported in North and South America, Japan, and Europe. Hosts of the pathogen include species of *Agropyron, Agrostis, Cynodon, Festuca, Paspalum, Phleum,* and *Poa.*

Symptoms and Signs

Small, yellowish spots or stripes may appear on foliage, which gradually turns brown and becomes ashen when dead. The spots are oval to somewhat elongated and may coalesce. Lesions may occur in bands across the leaf blade. When the disease is severe, basal rotting of the culm and wilting and shredding of leaves can occur. Microscopic observation reveals the presence of yellow brown masses of fungal sporangia embedded in leaf tissue. Diseased plants are dwarfed and have yellow, stiff, erect leaves; these symptoms may be confused with those of yellow tuft (downy mildew).

Causal Agents

Physoderma graminis and *P. paspali* are fungi belonging to the class Chytridiomycetes. They produce sporangia in pustules on infected foliage. The sporangia are golden brown, spherical to subspherical, about 20–40 μm in diameter, and smooth walled, each with a circular cap or lid at one end. Upon germination, sporangia release numerous zoospores, each about 4 × 6 μm with a single, posterior whiplash flagellum. Zoospores germinate to produce fine, fibrous hyphae that are coenocytic (nonseptate).

Disease Cycle

Thick-walled resting sporangia survive the winter embedded in leaves. When conditions favor germination, zoospores are released and move in water films along leaf surfaces and to the crown and lower leaves of a plant. Zoospores germinate to produce infectious hyphae that penetrate young leaves (usually coleoptiles) and spread from cell to cell through the host tissue. Internode elongation is curtailed, which may cause plants to change from a decumbent to an erect growth habit. Sporangia are produced in infected foliar tissue. The disease cycle can be repeated every 2–3 weeks as long as favorable conditions persist.

Epidemiology and Management

Infection occurs only when a film of moisture is present at high temperatures (23–30°C [73–86°F]). Control studies have not been reported. If the disease is present, management of watering practices during the day and improvement of water drainage through the soil should be helpful. Control may be attempted with fungicides that are effective against diseases caused by *Pythium* spp.

Selected References

Karling, J. S. 1950. The genus *Physoderma* (Chytridiales). Lloydia 13:29-71.

Smith, J. D., Jackson, N., and Woolhouse, A. R. 1989. Fungal Diseases of Amenity Turf Grasses. E. & F. N. Spon, London.

Sparrow, F. K. 1962. *Urophlyctis* and *Physoderma*. Trans. Mycol. Soc. Jpn. 3:16-18.

(Prepared by R. W. Smiley, P. H. Dernoeden, and B. B. Clarke)

Primitive Root-Infecting Fungus-like Organisms

Numerous primitive fungus-like organisms infect the roots of turfgrasses. These organisms have been subjected to extensive taxonomic revision, and most are currently classified in the kingdom Fungi or kingdom Chromista (see Diseases and Their Causal Agents). The organisms described in this section are obligate parasites that can damage root systems. They may infect roots without causing discernible damage. However, when environmental conditions are unfavorable for the growth of the plant, the once harmonious balance between roots and parasites can quickly shift in favor of the parasites, and plant vigor is further reduced as a result. Several of these organisms also serve as vectors of plant-pathogenic viruses.

Symptoms

These obligate parasites evolved in such a manner that they seldom kill or extensively damage root systems. Under very wet or dry conditions or cold or hot conditions, however, they may use more plant-derived compounds than the plant generates, thereby causing chlorosis, stunting, and susceptibility to attack by other root and foliage pathogens. Specific symptoms are not easily recognized.

Causal Agents

Primitive root-infecting fungus-like organisms include *Olpidium brassicae, Rhizophydium graminis, Lagenocystis radicicola,* and *Polymyxa graminis.* Among these, only *P. graminis* is common in roots of turfgrasses, but its importance as a turf pathogen is unknown. These organisms can be seen only when roots are cleared, stained, and observed under a microscope (Figs. 261 and 262).

Lagenocystis, Olpidium, and *Rhizophydium* spp. produce very little mycelium. *L. radicicola* forms spherical oospores 10–25 µm in diameter. *O. brassicae* produces sporangia 8–25 µm in diameter (Fig. 263). *R. graminis* produces sporangia up to 100 µm in diameter and thick-walled resting spores 6–12 µm in diameter. Oospores and sporangia of these organisms can easily be confused with those of *Pythium* spp. (oospores 10–25 µm and sporangia 10–50 µm in diameter) and possibly *Sclerophthora* spp. (oospores 50–75 µm and sporangia 50–80 µm in diameter).

P. graminis forms distinctive resting spores in roots (Fig. 263). The spores are about 5 µm in diameter and are produced in clumps resembling bunches of grapes. Clumps of spores are packed into widely scattered cells that have been infected by zoospores. Several strains of this fungus appear to occur. In studies in Canada, *P. graminis* strains did not infect *Poa pratensis, Lolium multiflorum, Festuca* spp., or *Agrostis* spp. and varied in pathogenicity to *Poa annua.* Other studies and observations have indicated the presence of *P. graminis* in roots of *Agrostis* spp.

Life Cycle

These organisms must infect living roots in order to complete their life cycles. Tiny, motile zoospores are borne in sporangia. Zoospores swim to new sites on the same root or adjacent roots, where they encyst, germinate, and infect root cells. Resting spores, sporangia, or oospores are then formed in the infected cells. Resting cells of *P. graminis* release zoospores, which can swim to and infect root hairs. An amorphous plasmodium is formed and produces additional zoospores or new resting cells.

These parasites are found on apparently healthy as well as diseased turf. Their pathogenic capabilities in turfgrasses are uncertain.

Epidemiology

These obligate parasites require wet soil to move from root to root. They may be transported over longer distances in running water or by the movement of soil (for example, during cultivation). Infections are most prevalent in low-lying, poorly drained, or overirrigated areas when the soil is cool. In grasses other than turf, viruses transmitted by these organisms are

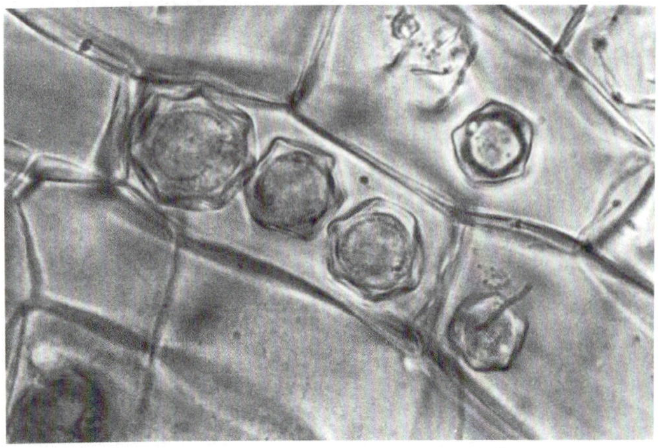

Fig. 261. Resting spores of *Olpidium brassicae* in root epidermal cells. (Courtesy M. J. Adams—© APS. Reproduced, by permission, from Bockus, W. W., et al., eds. 2010. Compendium of Wheat Diseases and Pests, 3rd ed. American Phytopathological Society, St. Paul, MN.)

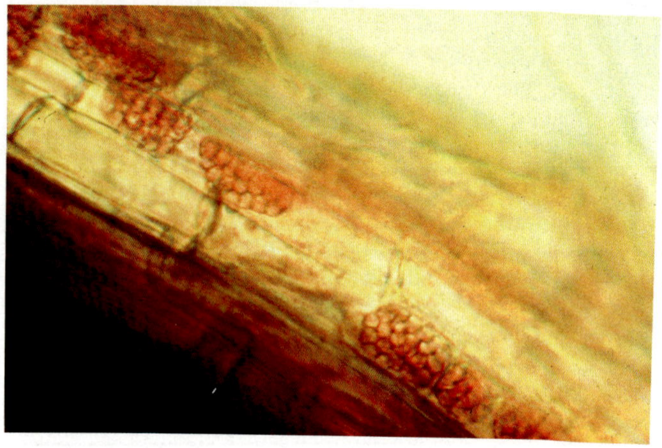

Fig. 262. Resting spores of *Polymyxa graminis* in root cells. (Courtesy T. E. Freeman)

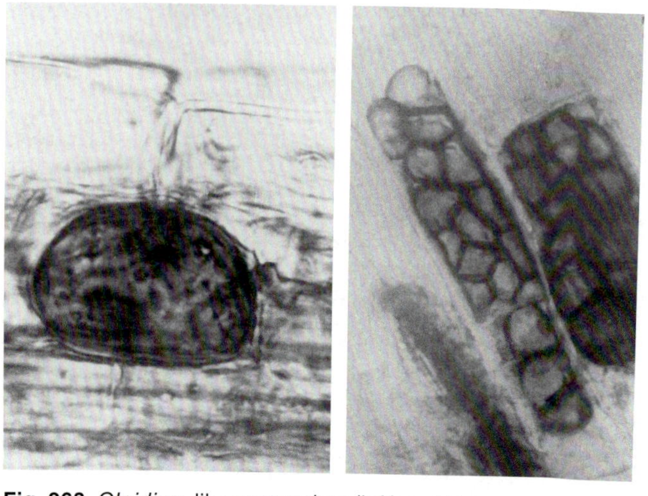

Fig. 263. *Olpidium*-like sporangium (left) and *Polymyxa graminis*-like resting spores (right) in root cortical cells. (Courtesy R. W. Smiley—© APS)

known to cause reduced plant vigor in wet areas. Reduced vigor predisposes plants to infections by other pathogens and to damage during periods of extreme temperature or moisture.

Management

If infections are numerous, improvements in soil drainage and irrigation practices are recommended.

Selected References

Adams, M. J. 1990. Epidemiology of fungally-transmitted viruses. Soil Use Manage. 6:184 189.

Barr, D. J. S. 1979. Morphology and host range of *Polymyxa graminis, Polymyxa betae,* and *Ligniera pilorum* from Ontario and some other areas. Can. J. Plant Pathol. 1:85-94.

Britton, M. P., and Rogers, D. P. 1963. *Olpidium brassicae* and *Polymyxa graminis* in roots of creeping bent in golf putting greens. Mycologia 55:758-763.

Dale, J. L., and Murdoch, C. L. 1969. *Polymyxa* infection of Bermuda grass. Plant Dis. Rep. 53:130-131.

Karling, J. S. 1968. The Plasmodiophorales. Hafner Publishing, New York.

Macfarlane, I. 1970. *Lagena radicicola* and *Rhizophydium graminis,* two common and neglected fungi. Trans. Br. Mycol. Soc. 55:113-116.

Salt, G. A. 1979. The increasing interest in "minor" diseases. Pages 281-312 in: Soil-Borne Plant Pathogens. B. Schippers and W. Gams, eds. Academic Press, New York.

(Prepared by R. W. Smiley, P. H. Dernoeden, and B. B. Clarke)

Pseudoseptoria Leaf Spot

Pseudoseptoria leaf spot (previously known as Selenophoma leaf spot, halo spot, red eye spot, and frog-eye) occasionally causes damage to cool-season turfgrasses.

Symptoms

Leaf lesions (Fig. 264) are elliptical, usually numerous, and only 1–4 mm long. Active lesions are purple brown; the color fades with age. As a lesion enlarges, the center becomes bleached and surrounded by a purplish margin. The necrotic center of the lesion becomes dotted with tiny, black pycnidia of the pathogen. When the disease is extensive, large numbers of leaf spots can form, giving individual plants or turfgrass areas an overall purplish appearance.

Causal Agents

Species associated with grass hosts include *Pseudoseptoria stomaticola, P. donacis,* and *P. everhartii.* These fungi produce dark brown to black, globose pycnidia, 40–150 μm in diameter (Fig. 254). The conidia are crescent shaped, and their sizes serve to differentiate the species: *P. everhartii* conidia are 1–1.5 × 10–15 μm; *P. stomaticola,* 1.6–3.1 × 11–20 μm; and *P. donacis,* 2–4.5 × 18–35 μm.

Fig. 264. Pseudoseptoria leaf spot of bromegrass (*Bromus inermis*). (Courtesy A. L. Scharen; photo by K. Clough)

Disease Cycle

These fungi survive stressful periods as pycnidia or possibly as mycelia in plant debris. When environmental conditions are favorable, conidia are released from pycnidia and are splashed or carried by wind to uninfected leaves. Infection results in the development of leaf lesions and the production of new pycnidia.

Epidemiology and Management

Pseudoseptoria leaf spot occurs during cool, wet periods in early spring. Little is known about the disease because of its infrequent occurrence at damaging levels, and control measures have not been reported. Fungicides effectively control the disease in grasses grown for seed production.

Selected References

Bissett, J. 1982. Description of *Pseudoseptoria stomaticola.* Fungi Canadenses, no. 238. National Mycological Herbarium, Agriculture Canada, Ottawa.

Punithalingam, E., and Waller, J. 1973. *Selenophoma donacis.* Descriptions of Pathogenic Fungi and Bacteria, no. 400. Commonwealth Mycological Institute, Kew, England.

Smith, J. D., Jackson, N., and Woolhouse, A. R. 1989. Fungal Diseases of Amenity Turf Grasses. E. & F. N. Spon, London.

(Prepared by R. W. Smiley, P. H. Dernoeden, and B. B. Clarke)

Ramularia Leaf Spot

Ramularia leaf spot has been reported in cool-season grasses in temperate climates throughout the world. It is usually of minor importance in mowed turfgrasses.

Symptoms

Leaf spots are elliptical to round and tan, brown, or dark red, and they often are surrounded by yellow halos. The lesions generally are not extensive on most hosts; but under ideal conditions for the pathogen, most of the leaf surface may become necrotic.

Causal Agents

Ramularia pusilla and *R. holci-lanati* cause this leaf spot. Conidiophores are hyaline and often twisted or bent. Conidia are hyaline, ellipsoidal to ovate, mostly aseptate, and 2–10 × 7–27 μm (Fig. 249). Conidia and conidiophores are sparse on infected tissue, and the conidiophores are difficult to see even with a low-power microscope.

Disease Cycle, Epidemiology, and Management

The pathogens cause most damage during the autumn and spring and apparently survive the winter and summer as stromata in leaves or clippings. Control measures have not been reported.

Selected References

Braun, U. 1990. Taxonomic problems of the *Ramularia/Cercosporella* complex. Stud. Mycol. 32:65-75.

Smith, J. D., Jackson, N., and Woolhouse, A. R. 1989. Fungal Diseases of Amenity Turf Grasses. E. & F. N. Spon, London.

(Prepared by R. W. Smiley, P. H. Dernoeden, and B. B. Clarke)

Septoria and Stagonospora Leaf Spots

Species of *Septoria, Stagonospora, Parastagonospora,* and *Neottiosporina* cause similar diseases in many Festucoideae, Eragrostoideae, and Panicoideae grasses throughout the world. Some species can attack actively growing tissues, but many are weak pathogens or saprophytes that colonize and decompose dead or senescing leaves of turfgrasses and cereal grains. On turfgrasses, leaf spots caused by pathogenic species of these fungi typically occur during early spring and autumn, but they rarely cause extensive damage. In cereals such as wheat (*Triticum aestivum*), diseases associated with *Septoria* and *Stagonospora* spp. are often referred to as leaf or glume blotch.

Symptoms and Signs

Small lesions occur mainly as stripes, spots, or blotches near leaf tips, especially those that have been cut during mowing. The lesions vary in color from gray to grayish green, light or deep brown, or dark purple (Fig. 265). They can enlarge to a length of 12.7 mm or more and typically fade to a straw color with age. Leaf tips often become chlorotic and mottled. From a distance, areas of affected turf may appear scorched. Tiny, golden brown to dark brown or black pycnidia of the pathogen develop in older lesions. The pycnidia are sometimes difficult to see without the aid of magnification.

Causal Agents

Several species initially described as *Septoria* are now classified in the genera *Zymoseptoria, Stagonospora, Parastagonospora,* or *Neottiosporina.* These fungi produce pycnidia (Fig. 265) that vary from golden or light brown to black and are mostly 40–200 μm in diameter (Fig. 254). Conidia are needle-like (rod shaped to filiform), straight or curved, nonseptate or one to five septate, 1–9 μm in diameter, and 5–85 μm long. The color and dimensions of these structures are different for each species.

The sexual state has been observed in some *Stagonospora* spp. Pseudothecia of *S. nodorum* are light brown to black and 150–200 μm in diameter. Asci are hyaline, club shaped, cylindrical to curved, and bitunicate. Each ascus contains eight ascospores. At maturity, ascospores are hyaline to pale brown, fusoid (19.5–22.5 μm long and 4 μm in diameter), and nonseptate or one to three septate, with constrictions at the septa. Ascocarps of *S. avenae* are brown to black and 100–220 μm in diameter. Asci are hyaline, club shaped, straight to slightly curved, and bitunicate. Each ascus contains eight ascospores. Ascospores are hyaline, biseriate, fusoid (16–28 μm long and 4–6 μm in diameter), three septate, and constricted at the septa.

Fig. 265. Pycnidia of *Septoria macropoda* in blighted leaves of *Poa pratensis.* (Courtesy R. W. Smiley—© APS)

Disease Cycle

These pathogens survive unfavorable periods in winter and summer as mycelia or pycnidia in infected debris. Conidia are produced and released from pycnidia during cool, wet periods in spring and autumn. Conidia are splashed or carried to healthy leaf tissue by equipment, people, or animals and may lie dormant on leaves for long periods when temperatures are below 10°C (50°F). At ambient air temperatures of 16–24°C (61–75°F), conidia germinate and produce mycelia, which penetrate leaves to initiate infection. Secondary infection occurs as long as cool, moist weather persists. During warmer periods, conidia are short-lived, although *S. nodorum* may be more tolerant of higher temperatures than the other species. Pycnidia are produced in older, necrotic areas of leaves killed by the pathogen.

Ascospores of *Zymoseptoria tritici* have been identified as a major source of inoculum in small grains, but the importance of this or any other sexual state in the infection process of turfgrasses remains unknown.

Epidemiology

Infection occurs only when a film of water is present on leaf surfaces. Penetration often occurs at the cut tips of leaf blades. Periods of cool, wet weather, especially in spring and autumn, cause rapid increases in the numbers of spores produced and liberated. Infrequently mowed, nutrient-deficient turf growing in shaded areas is most commonly affected by these leaf spots. Some species are seedborne. Biochemical and molecular techniques, particularly ELISA, PCR, and sequence analysis, have been used to differentiate species of these fungi isolated from cereals and other gramineous hosts.

Management

Little is known about the management of these diseases in turfgrasses. Specific turfgrass management practices for controlling these diseases have not been reported, and variations in the resistance of turfgrass cultivars to these fungi have rarely been recorded. Turf growing in nitrogen-deficient soil and turf treated with plant growth regulators appear to be more susceptible to infection. Adequate and balanced fertility, as determined by soil tests, should be maintained; foliage should be allowed to dry before mowing; and mower blades should be sharpened to reduce shredding of leaf tips.

Selected References

Bissett, J. 1982. Description of *Stagonospora avenae.* Fungi Canadenses, no. 239. National Mycological Herbarium, Agriculture Canada, Ottawa.

Bissett, J. 1982. Description of *Stagonospora nodorum.* Fungi Canadenses, no. 240. National Mycological Herbarium, Agriculture Canada, Ottawa.

Bissett, J. 1983. Description of *Septoria elymi.* Fungi Canadenses, no. 242. National Mycological Herbarium, Agriculture Canada, Ottawa.

Bissett, J. 1983. Description of *Septoria tritici.* Fungi Canadenses, no. 244. National Mycological Herbarium, Agriculture Canada, Ottawa.

Bockus, W. W., Bowden, R. L., Hunger, R. M., Morrill, W. L., Murray, T. D., and Smiley, R. W., eds. 2010. Compendium of Wheat Diseases. 3rd ed. American Phytopathological Society, St. Paul, MN.

Castellani, E., and Germano, G. 1977. Le stagonosporae graminicole. Ann. Fac. Sci. Agrar. Univ. Torino 10:1-135.

Cherepanova, N. P., and van Tchank, N. 1980. Key for the identification of *Septoria* Sacc. spp. on Gramineae in the Leningrad region. Nov. Sist. Nizshikh Rast. 17:100-102.

Cunfer, B. M., and Ueng, P. P. 1999. Taxonomy and identification of *Septoria* and *Stagonospora* species on small-grain cereals. Annu. Rev. Phytopathol. 37:267-284.

Khashnobish, A., and Shearer, C. A. 1996. Phylogenetic relationships in some *Leptosphaeria* and *Phaeosphaeria* species. Mycol. Res. 100:1355-1363.

Richardson, M. J., and Noble, M. 1970. *Septoria* species on cereals—A note to aid their identification. Plant Pathol. 19:159-163.
Smith, J. D., Jackson, N., and Woolhouse, A. R. 1989. Fungal Diseases of Amenity Turf Grasses. E. & F. N. Spon, London.
Sprague, R. 1950. Diseases of Cereals and Grasses in North America. Ronald Press, New York.

(Prepared by R. W. Smiley, P. H. Dernoeden and B. B. Clarke; revised by E. L. Butler and J. P. Kerns)

Spermospora Leaf Spot

Spermospora leaf spot has been recorded in *Agrostis, Festuca, Lolium,* and *Poa* spp. in cool, oceanic climates of the world.

Symptoms

Affected leaves of grasses develop discrete, straw-colored, grayish brown, or reddish brown, elliptical lesions with darker brown borders and surrounding chlorotic halos. On *Poa* spp., the lesions are straw colored with light brown borders. Lesions on all species may enlarge rapidly to cover much of the leaf surface, causing a dark brown scald or bleaching (Fig. 266) and resulting in a general blighting of foliage (Fig. 267).

Causal Agents

Spermospora ciliata, S. lolii, and *S. poagena* cause Spermospora leaf spot. These fungi produce hyaline to slightly yellowish mycelia, which form dark brown to black stromata on leaf surfaces. Short, hyaline to yellow conidiophores arise from the stromata and bear abundant conidia, which are hyaline and fusiform, tapering gradually from the basal septum to the hilum and then into a narrow, elongate, whiplike apical cell (Fig. 249). Conidia of *S. ciliata* are two to four septate and mostly 4.5–5.5 × 34–48 µm, including the whiplike apical cell, which is 5–16 µm long. Conidia of *S. lolii* are two to six septate and 3.5–4.7 × 40–70 µm. Conidia of *S. poagena* are three to seven septate and 3.5–4.5 × 45–95 µm.

Fig. 266. Spermospora leaf spot of *Festuca arundinacea.* (Courtesy B. B. Clarke—© APS)

Fig. 267. Foliar blight of *Festuca arundinacea* incited by *Spermospora lolii.* (Courtesy B. B. Clarke—© APS)

Disease Cycle, Epidemiology, and Management

Little is known about the disease cycle, epidemiology, and control of Spermospora leaf spot. The disease occurs primarily from autumn to spring and is favored by oceanic climates in cool temperate regions. There are no known reports of fungicides being used for the control of this disease.

Selected References

Deighton, F. C. 1968. *Spermospora.* Trans. Br. Mycol. Soc. 51:41-49.
MacGarvie, Q. D., and O'Rourke, C. J. 1969. New species of *Spermospora* and *Cercosporella* affecting grasses and other hosts in Ireland. Ir. J. Agric. Res. 8:151-167.
Smith, J. D., Jackson, N., and Woolhouse, A. R. 1989. Fungal Diseases of Amenity Turf Grasses. E. & F. N. Spon, London.

(Prepared by R. W. Smiley, P. H. Dernoeden, and B. B. Clarke)

Tar Spot

Tar spot, also called black leaf spot, occurs in many grass species but is of minor importance in turfgrasses. The disease is widely distributed throughout the world.

Symptoms and Signs

Small, black, circular to oval spots form on abaxial and/or adaxial leaf surfaces (Fig. 268). In the smallest spots, the black area may be surrounded by a light, chlorotic halo, but the halo usually disappears as the spot enlarges. Heavily infected areas appear mottled yellow green to bright yellow with black spots. As leaves mature and senesce, a "green island" symptom often develops, in which the area around the black spots remains green longer than the healthy tissue.

Causal Agents

Many species of *Phyllachora* cause tar spot of grasses in the family Poaceae. The black spots they cause (Fig. 269) look like lesions but are actually clypei—crusty, shieldlike structures covering perithecia embedded in the leaf tissue. Multiple perithecia are formed in the leaf mesophyll under a single clypeus.

Fig. 268. Black clypei of *Phyllachora graminis,* a tar spot pathogen. (Courtesy Z. K. Punja—© APS)

Fig. 269. Tar spot of *Paspalum vaginatum* caused by *Phyllachora paspalicola.* (Courtesy A. D. Martínez-Espinosa—© APS)

Clypei vary from circular to elliptical to linear and, in most species, measure 0.1–1 × 0.3–2 µm. *P. graminis* differs in producing black clypei up to 5 mm long, which tend to coalesce to form long, shiny, black stripes. Ascospores of *Phyllachora* spp. are hyaline, ovoid to elliptical, unicellular, and mostly 4–7 × 9–14 µm (Fig. 254).

Disease Cycle

Phyllachora spp. are obligate parasites, requiring living plant tissues to complete their development. Perithecia are formed in infected leaves and release ascospores to initiate new infections. Little is known about the specific conditions that are conducive to disease development.

Epidemiology and Management

Tar spot occurs in many grasses and is especially pronounced in wet, shaded sites. No control measures have been recommended.

Selected References

Hao, J., Ding, X., Liao, L., Rong, W., Bai, C., Raymer, P., Chen, Z., and Wang, Z. 2018. First report of *Phyllachora* sp. causing tar spot on seashore paspalum in South China. Plant Dis. 102:2661.

Martínez-Espinoza, A. D., Martinez Uribe, O. A., and Kim, D. 2012. Identification and characterization of tar spot on seashore paspalum in Georgia. Phytopathology 102(suppl.):S4.76.

Parbery, D. G. 1967. Studies on graminicolous species of *Phyllachora* Nke. in Fckl. V. A taxonomic monograph. Aust. J. Bot. 15:271-375.

Smith, J. D., Jackson, N., and Woolhouse, A. R. 1989. Fungal Diseases of Amenity Turf Grasses. E. & F. N. Spon, London.

Sprague, R. 1950. Diseases of Cereals and Grasses in North America. Ronald Press, New York.

(Prepared by R. W. Smiley, P. H. Dernoeden, and B. B. Clarke; revised by L. P. Tredway)

Viral Diseases

Viruses are rarely associated with economically important diseases of grasses maintained as turf. Viral infections of turf may go unnoticed or unreported because they do not usually lead to the death of the host plant and foliar symptoms often are mild. There are important exceptions, such as the decline of *Stenotaphrum secundatum* (St. Augustine decline) caused by Panicum mosaic virus and mosaic of ryegrass (*Lolium* spp.) caused by ryegrass mosaic virus. Additionally, a severe dieback of the *S. secundatum* cultivar Floratam has been associated with sugarcane mosaic virus in Florida. The etiology of this viral disorder, referred to as "lethal viral necrosis," is still unproved; however, a second virus, related to Panicum mosaic virus, Bermuda grass latent virus, also likely plays a key role. The less debilitating viral diseases may increase the severity of disease complexes, but these interactions have not been studied carefully. Viral diseases may be of some importance in turf species grown for seed production, especially under environmental stress, because they have been shown to be problematic for seed production of other grasses. Select viruses that have been identified in turfgrasses are described in this section; however, the grasses may not have been maintained as turf at the time the viruses were identified.

Symptoms

Viral disease symptoms in turf are often mild or imperceptible, but some viruses are associated with severe physiological damage or death (Fig. 270). In general yellowing, mottling, streaking, and mosaic of leaves are common symptoms of viral infections (Figs. 271 and 272). Other observable symptoms include stunting to various degrees, excessive tillering, and leaf necrosis. Symptoms in plants infected with two or more viruses may be significantly worse than those in plants infected with one virus alone. Different strains of a virus may cause different syndromes, and a mild strain of one virus may cross-protect plants from an otherwise severe strain of the same or a related virus.

St. Augustine decline is a viral disease whose biology is not completely known. It initially appears as a chlorotic mottle or mosaic on leaf blades (Fig. 273). At this stage, the symptoms may resemble those of iron or zinc deficiency, mite feeding damage, or downy mildew infection. The chlorosis gradually becomes more extensive until the affected turf appears uni-

Fig. 270. Severe damage to *Stenotaphrum secundatum* caused by mechanical transmission of lethal viral necrosis. (Courtesy P. F. Harmon—© APS)

Fig. 271. *Stenotaphrum secundatum* with symptoms of viral infections. (Courtesy P. F. Harmon—© APS)

formly chlorotic (Fig. 274), lacks vigor, and becomes thin as stolon growth is retarded. The disease can kill an entire lawn in as little as 3 years after the first appearance of symptoms. The longer the turf is infected, the more severe the symptoms become and the greater the possibility of decline and damage from a combination of stresses, such as insect and nematode damage, shade, drought, and cold. St. Augustine decline is thought to have a stress-related component, because virus-infected plants with severe symptoms in the field have been shown to recover to a uniform mild mosaic after a few months in a greenhouse.

Panicum mosaic virus also causes a mosaic disease in *Eremochloa ophiuroides* in North America (Fig. 275). Decline and death have not been associated with this disease, but the viral infection may be an additive stress factor.

Lethal viral necrosis has been reported only on the *S. secundatum* cultivar Floratam, which is resistant to St. Augustine decline. In mixed genotype swards, cultivars of *S. secundatum* other than Floratam develop mosaic disease symptoms caused by sugarcane mosaic virus and no additional symptoms (Fig. 276). Lethal viral necrosis symptoms progress from mosaic to a systemic necrosis of all aboveground plant parts attached to infected stolons (Fig. 277). Symptoms progress when

Fig. 272. *Stenotaphrum secundatum* with symptoms of viral infections. (Courtesy P. F. Harmon—© APS)

Fig. 273. St. Augustine decline of *Stenotaphrum secundatum.* (Courtesy G. E. Holcomb—© APS)

Fig. 274. *Stenotaphrum secundatum* leaves with symptoms of St. Augustine decline. (Courtesy J. L. Dale—© APS)

Fig. 275. Mosaic of centipedegrass (*Eremochloa ophiuroides*), caused by Panicum mosaic virus. (Courtesy G. E. Holcomb—© APS)

Fig. 276. Lethal viral necrosis of *Stenotaphrum secundatum* cv. Floratam on the left compared with an unknown *S. secundatum* cultivar on the right with mosaic symptoms only. (Courtesy P. F. Harmon—© APS)

Fig. 277. Lethal viral necrosis of *Stenotaphrum secundatum* cv. Floratam. (Courtesy P. F. Harmon—© APS)

Floratam is coinfected with Bermuda grass latent virus and when the coinfected turfgrass is then exposed to cool evening temperatures in the fall. Systemic necrosis can also be triggered by incubation at 18°C (64°F).

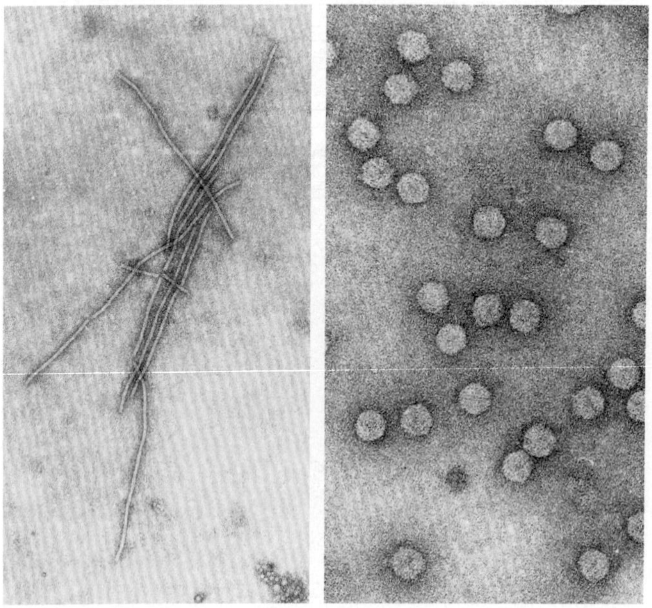

Fig. 278. Flexuous, threadlike particles of *Maize dwarf mosaic virus* (left, 40,000×), a member of the family *Potyviridae,* and polyhedral particles of *Barley yellow dwarf virus* (right, 275,000×), a member of the family *Luteoviridae.* (Courtesy H. W. Israel and W. F. Rochow—© APS)

Causal Agents

Viruses are submicroscopic agents that require living host cells in which to multiply. They are grouped by characteristics such as size, shape (e.g., spheres, rods, and long filaments) (Fig. 278), nucleic acid type (RNA or DNA), host range, and means of transmission. Some viruses that may cause disease in turfgrasses are listed in Table 6. These virus groups are the taxonomic equivalent of genera or families, as noted in the table. Viruses that are common in grasses represent at least 11 different groups.

Some strains of Panicum mosaic virus are associated with satellite viruses, that is, small viruses that cannot infect plants without the parent virus as a helper. These satellites have been shown to modulate symptoms caused by the parent Panicum mosaic virus, but their epidemiological role is not known.

Lethal viral necrosis is likely the result of a synergistic coinfection between sugarcane mosaic virus, which has been known to cause mosaic on *S. secundatum* since the 1950s, and a more recently described virus related to Panicum mosaic virus, Bermuda grass latent virus. Bermuda grass latent virus was discovered in, but is not known to cause symptoms on, *Cynodon* spp. Additional research into this disorder is needed to confirm this hypothesis; however, maize lethal necrosis is a well-documented example in which a similar severe symptom results from coinfection of maize with sugarcane mosaic virus and a virus similar to Bermuda grass latent virus.

Disease Cycle

Viruses are not motile, and therefore they can be transmitted from plant to plant only by other organisms or agents, by seed or pollen, or by the mechanical transfer of plant sap. Most viruses have vector relationships with one or a few species of aphids, leaf beetles, mites, leafhoppers, or planthoppers that feed on foliage; stylet-bearing nematodes that feed on roots;

TABLE 6. Viral Agents Reported in Turfgrasses[a]

Virus Species[b]	Virus[c,d]	Turf Host Genera	Vector or Method of Transmission[d]	Distribution
Agropyron mosaic	*Potyviridae*	*Agropyron, Festuca, Lolium, Poa*	Mite, sap	Northern Hemisphere
Barley stripe mosaic	*Hordeivirus*	*Agropyron, Lolium*	Seed, pollen	Worldwide
Barley yellow dwarf	*Luteoviridae*	*Agropyron, Cynodon, Festuca, Lolium, Phleum, Poa*	Aphid	Worldwide
Bermuda grass etched-line	*Marafivirus*	*Cynodon*	Leafhopper	Morocco, Iran
Bermuda grass latent	*Tombusviridae*	*Cynodon*	Sap	North America
Brome mosaic	*Bromoviridae*	*Agropyron, Agrostis, Lolium, Phleum, Poa*	Sap	Northern Hemisphere
Cocksfoot mild mosaic	*Sobemovirus*	*Festuca, Lolium*	Aphid	Europe
Cocksfoot streak	*Potyviridae*	*Lolium, Paspalum*	Aphid	Europe
Cynosurus mottle	*Sobemovirus*	*Lolium*	Beetle	Worldwide
Maize chlorotic dwarf	*Sequiviridae*	*Pennisetum*	Leafhopper	North America
Oat necrotic mottle	*Potyviridae*	*Agrostis, Festuca, Lolium, Poa*	Sap	North America
Oat sterile dwarf	*Reoviridae*	*Lolium, Poa*	Planthopper	Europe
Panicum mosaic	*Tombusviridae*	*Dactylotenium, Eremochloa, Pennisetum, Stenotaphrum*	Sap	North America
Poa semilatent	*Hordeivirus*	*Agropyron, Lolium, Phleum, Poa*	Sap	Northern Hemisphere
Rice dwarf	*Reoviridae*	*Poa*	Leafhopper	East Asia
Ryegrass chlorotic streak	. . .	*Lolium*	. . .	New Zealand
Ryegrass cryptic	*Alphacryptovirus*	*Lolium*	Mite	Australia
Ryegrass mosaic	*Potyviridae*	*Agrostis, Festuca, Lolium, Poa*	Mite	Northern Hemisphere
Soil-borne wheat mosaic	*Furovirus*	*Agropyron*	Fungus, sap	Northern Hemisphere
Sugarcane mosaic	*Potyviridae*	*Cynodon, Paspalum, Pennisetum, Stenotaphrum*	Aphid, sap	Worldwide
Western ryegrass mosaic	. . .	*Festuca, Lolium*	. . .	North America
Wheat American striate mosaic	*Rhabdoviridae*	*Lolium*	Leafhopper	North America
Wheat spindle streak mosaic	*Potyviridae*	*Agropyron, Lolium, Phleum, Poa*	Fungus, sap	Northern Hemisphere
Wheat streak mosaic	*Potyviridae*	*Agropyron, Lolium, Poa*	Mite, sap	Northern Hemisphere
Wheat yellow leaf	*Closteroviridae*	*Agropyron, Lolium*	Aphid	Japan
Zoysia dwarf	. . .	*Zoysia*	. . .	Japan
Zoysia mosaic	*Potyviridae*	*Zoysia*	. . .	Japan

[a] Prepared by R. W. Smiley, P. H. Dernoeden, and B. B. Clarke; revised by P. Harmon and B. Hillman—© APS.

[b] Italicized names are species recognized by the International Committee on Taxonomy of Viruses (ICTV). Names not italicized are not recognized by the ICTV.

[c] Virus family (*-viridae*) or genus (*-virus*).

[d] . . . = Unknown.

or primitive fungi that infect roots. Transmission from plant to plant may occur within a few minutes of feeding, as with Agropyron mosaic virus, or may take days, as with rice dwarf virus.

Insect-vectored viruses may be generally distributed across large areas or concentrated along fringe areas close to other infected plants that serve as feeding sites for the insect vector. Diseases caused by viruses transmitted by soilborne fungi (mainly *Polymyxa graminis*) or nematodes (species of *Xiphinema, Longidorus,* or *Trichodorus*) often occur in patterns reflecting the low-lying, wet areas favored by these vectors. Loose, sandy soils favor the movement of soilborne vectors.

Several viruses of turfgrass (e.g., Panicum mosaic virus, Poa semilatent virus, and brome mosaic virus) have no known insect vectors and are transmitted only by physical means, such as mowing, cultivating, or transplanting infected sod. Barley stripe mosaic virus is efficiently transmitted through seed and in plant sap. Viruses with wide host ranges, including sugarcane mosaic virus, can survive between turfgrass renovations in weedy alternative hosts such as *Digitaria* spp., in asymptomatic hosts, or in neighboring lawns. Lawns that fail because of lethal viral necrosis and are then reestablished with Floratam become infected from this reservoir of inoculum and fail again.

Because viruses are transmitted in many ways, it is likely that they are more common than currently recognized in turfgrasses. The damage inflicted on turfgrasses by most of these viruses, however, is not well documented.

Detection and Diagnosis

The identification of viral disease agents is a complex and expensive procedure. Often, sufficient clues to the identities of viral agents are provided by the host species affected, symptomatology, and characteristics of the disease cycle. Since viral diseases of turf are rarely major threats, standardized tests for most viral agents affecting turf are not commercially available. Tests that include serological techniques (such as ELISA), nucleic acid-based probes (such as dot blot hybridization), and molecular genetic-based tests (such as PCR or reverse transcriptase PCR) can be used for consistent identification of particular viruses. Such tests have been developed for viruses that pose serious threats and are used for testing potentially resistant turf cultivars in specific situations, but most are not yet commercially available.

Lethal viral necrosis is diagnosed in Floratam when symptoms become more severe than leaf blade mosaic, and the sugarcane mosaic virus is detected by double antibody sandwich ELISA or by a commercially available potyvirus group-specific lateral flow immunoassay. No reliable detection method is available for Bermuda grass latent virus to date.

Epidemiology and Management

Viruses cannot be controlled directly in grass plants, but they can be managed indirectly by excluding them or by slowing or stopping their spread. The transmission of a vectored virus from plant to plant depends on conditions favorable for the survival and movement of the vector when the virus is present. Insect-vectored viruses are thus less likely to become problems on golf greens that are frequently treated with insecticides. Nematode-vectored viruses are also less likely to occur where turf is regularly treated with nematicides or where soil conditions are not conducive to the survival of *Xiphinema, Longidorus,* or *Trichodorus* spp. Wet, poorly drained soils favor the activity of *P. graminis* and the viruses that it transmits. The role of mechanical sap transmission of viruses in turf is poorly understood but has been shown to be a factor in St. Augustine decline. Presumably, sap transmission is most likely to occur in grasses that are mowed when wet.

The spread of St. Augustine decline may be reduced by mowing when the turfgrass canopy is dry and by planting sprigs of resistant cultivars. Watering to avoid drought stress, the use of complete fertilizers, and fungal disease control reduce stress-induced damage in virus-infected turfgrasses. Virus-resistant turf cultivars should be planted in areas where viral diseases are already established or where they pose a particular threat. Cultivars resistant to Panicum mosaic virus tend to crowd out susceptible cultivars showing symptoms of St. Augustine decline. Furthermore, practices that enhance turf growth may reduce symptoms of this disease.

The spread of lethal viral necrosis from lawn to lawn in affected communities has followed patterns consistent with sap transmission by mowing equipment and in a similar manner observed with St. Augustine decline. Introductions of lethal viral necrosis into new communities and observations of infrequent long-distance transport events suggest an inefficient and infrequent insect vector (possibly aphids) also may play a role. Management options for lethal viral necrosis include resodding or interplanting turfgrass cultivars other than Floratam. Lawn maintenance companies and pest control operators are encouraged to allow plant sap to dry on equipment between accounts and, when possible, to use registered chemical sanitizers, bleach, or alcohol solutions when moving from communities or counties where lethal viral necrosis has been confirmed to unaffected areas.

Viral diseases were the first target of research in genetic engineering to produce disease-resistant plants. Stable resistance to ryegrass mosaic virus was demonstrated in fertile transgenic perennial ryegrass lines, the first demonstration of such a strategy in an agronomically important perennial monocot. Such a strategy may be useful against other viral diseases threatening turf, but such technologies have other regulatory and public acceptance hurdles yet to overcome.

Selected References

Alcalá-Briseño, R. I., Harmon, P., and Polston, J. 2017. Viral metagenomics unravels the etiology of lethal necrosis of St. Augustinegrass 'Floratam'. (Abstr.) Proc. Annu. Mtg. Am. Phytopathol. Soc. https://apsnet.confex.com/apsnet/2017/meetingapp.cgi/Paper/5990

Bockus, W. W., Bowden, R. L., Hunger, R. M., Morrill, W. L., Murray, T. D., and Smiley, R. W., eds. 2010. Compendium of Wheat Diseases and Pests. 3rd ed. American Phytopathological Society, St. Paul, MN.

Commonwealth Mycological Institute and Association of Applied Biologists. 1970–1988. Descriptions of Plant Viruses. CMI and AAB, Kew, England.

Holcomb, G. E., Liu, T. Z., and Derrick, K. S. 1989. Comparison of isolates of panicum mosaic virus from St. Augustinegrass and centipedegrass. Plant Dis. 73:355-358.

Mahuku, G., Lockhart, B. E., Wanjala, B., Jones, M. W., Kimunye, J. N., Stewart, L. R., Cassone, B. J., Sevgan, S., Nyasani, J. O., Kusia, E., Kumar, P. L., Niblett, C. L., Kiggundu, A., Asea, G., Pappu, H. R., Wangai, A., Prasanna, B. M., and Redinbaugh, M. G. 2015. Maize lethal necrosis (MLN), an emerging threat to maize-based food security in sub-Saharan Africa. Phytopathology 105:956-965.

McCoy, N. L., Toler, R. W., and Amador, J. 1969. St. Augustine decline (SAD)—A virus disease of St. Augustine grass. Plant Dis. Rep. 53:955-958.

Pyle, J. D., Mandadi, K. K., and Scholthof, K. B. G. 2019. *Panicum mosaic virus* and its satellites acquire RNA modifications associated with host-mediated antiviral degradation. mBio 10. doi:10.1128/mBio.01900-19

U.S. Department of Agriculture. 1977. Plant Pests of Importance to North American Agriculture: Index of Plant Virus Diseases. U.S. Dep. Agric., Agric. Handb. 307.

van Regenmortel, M. H. V., Fauquet, C. M., Bishop, D. H. L., Carstens, E. B., Estes, M. K., Lemon, S. M., Maniloff, J., Mayo, M. A., McGeoch, D. J., Pringle, C. R., and Wickner, R. B. 2000. Virus Taxonomy. 7th Rep. International Committee on Taxonomy of Viruses, Academic Press, San Diego, CA.

Xu, J., Jörg, S., and Alpeter, F. 2001. Dissection of RNA-mediated ryegrass mosaic virus resistance in fertile transgenic perennial ryegrass (*Lolium perenne* L.). Plant J. 26:265-274.

(Prepared by R. W. Smiley, P. H. Dernoeden, and B. B. Clarke; revised by B. I. Hillman and P. F. Harmon)

White Blight

White blight, also known as Melanotus white patch, is primarily a summer disease of *Festuca* spp., but it also occurs in *Agrostis, Lolium,* and *Poa* spp. It is most commonly reported in the southeastern and transition zones of the United States, but it probably also occurs in other warm, humid regions.

Symptoms and Signs

The disease first appears as blighted, circular patches, 8–12 cm in diameter (Fig. 279), which may enlarge to about 1 m in diameter. Patches may coalesce to form large, irregularly shaped areas of blighted turf. The patches are distinctly white and often have pink to salmon-colored borders. Blighted leaf blades often are matted and appressed to the soil surface. Matted leaves are covered by a network of grayish white mycelium of the causal fungus. Leaf blades are killed first at the tip, and blighting proceeds from the tip toward the sheath. The pathogen does not infect stems, and therefore plants are not directly killed. Stalkless basidiocarps (fruiting bodies) of the pathogen form on blighted grass blades (Fig. 280).

Causal Agent

Melanotus phillipsii causes white blight of *Festuca* spp. in the southeastern United States. It produces basidiocarps on leaf blades, where they are clearly visible. The basidiocarps have no central stipe, are grayish white, and are 2–8 mm in diameter. They appear first as small, round balls on leaf blades but then open at the bottom to expose their "gills." Basidiospores are smooth and 5.7–6.3 × 2.7–3.3 µm. The importance of basidiospores in the dispersal of the pathogen is unknown.

Fig. 279. White blight of *Festuca arundinacea,* caused by *Melanotus phillipsii.* (Courtesy N. Jackson—© APS)

Fig. 280. Basidiocarps of *Melanotus phillipsii,* the causal agent of white blight. (Courtesy L. T. Lucas—© APS)

Disease Cycle and Epidemiology

M. phillipsii causes more severe damage during hot, humid weather than during cooler periods. Blighting occurs during summer and early autumn, mostly on fully sunlit sites, but it may occur in partially shaded areas. The most severe damage occurs when daytime temperatures exceed 29°C (84°F) and nighttime temperatures remain above 21°C (70°F). Heat stress and low soil moisture render turf more susceptible to the disease. White blight is more severe in immature stands of *Festuca* spp. than in mature turf. Leaf blighting may be extensive, but affected turfs normally recover rapidly with the advent of cooler, less humid conditions.

Management

Irrigating deeply but infrequently helps to avoid drought stress. Autumn applications of nitrogen stimulate growth and hasten recovery of the turf. There are no specific chemical control measures, but fungicides effective against basidiomycetes may reduce the severity of white blight.

Selected Reference

Wick, R. L. 1982. Occurrence of Melanotus white patch on red and Chewings fescue. Plant Dis. 72:268.

(Prepared by R. W. Smiley, P. H. Dernoeden, and B. B. Clarke)

Yellow Ring

Yellow ring is a disease of *Poa pratensis* characterized by distinctive yellow rings. There are no other known hosts.

Symptoms and Signs

The causal agent is a basidiomycete fungus that lives saprophytically in thatch. In cool weather, it infects epidermal root cells, and leaves become chlorotic. The disease may be evident from spring to autumn. Distinct lemon-colored or golden yellow rings develop (Figs. 281 and 282), and large amounts of white mycelium can be seen in the thatch. Thatch permeated with the fungus has a "mushroom" odor. The rings are ephemeral, and yellowed leaves regain their green color in time.

Causal Agent

Trechispora alnicola causes yellow ring. The pathogen grows most rapidly and causes disease at 20–25°C (68–77°F). It produces conidia in culture. Basidiospores are 2.5–3.0 × 2.5–3.5 µm, and each usually contains an oil drop. The importance of basidiospores in the dispersal of the pathogen is unknown.

Fig. 281. Yellow ring in *Poa pratensis* sod. (Courtesy H. T. Wilkinson—© APS)

Fig. 282. Yellow ring in *Poa pratensis* sod. (Courtesy B. B. Clarke—© APS)

Disease Cycle and Epidemiology

Yellow ring generally develops during cool, moist periods in midspring and autumn when temperatures are 20–25°C (68–77°F), but rings may also appear during moist periods in summer. Symptoms develop only when massive amounts of mycelium accumulate in thatch. *T. alnicola* produces basidia and basidiospores on a hymenial layer but does not produce basidiocarps. The disease is most likely to occur in stands of *P. pratensis* that are less than 2 years old. *T. alnicola* can infect roots and crowns, but it does not cause extensive cellular death or necrosis of leaves. There are no root lesions or necrosis, but infected roots may appear light brown or tan. Symptoms dissipate as daytime temperatures increase to above 30°C (86°F).

Management

Yellow ring causes no permanent damage to turf, and fungicide use is generally not warranted. Vertical cutting and core aeration disrupt the fungus and reduce thatch and thus may speed the disappearance of the yellow rings. Fungicides that specifically target basidiomycetes may reduce the severity of the disease.

Selected References

Fermanian, T. W., Shurtleff, M. C., Randell, R., Wilkinson, H. T., and Nixon, P. L. 2003. Controlling Turfgrass Pests. 3rd ed. Prentice Hall, Upper Saddle River, NJ.

Wilkinson, H. T. 1987. *Trechispora alnicola* associated with yellow ring on *Poa pratensis*. Can. J. Bot. 65:150-153.

Wilkinson, H. T. 1987. Yellow ring on *Poa pratensis* caused by *Trechispora alnicola*. Plant Dis. 71:1141-1143.

(Prepared by R. W. Smiley, P. H. Dernoeden, and B. B. Clarke)

Part II. Noninfectious Agents

Biotic Agents of Noninfectious Diseases

Turfgrasses are often damaged by the feeding activity of insects and by competition from weeds, moss, and algae. None of these pests are considered to cause true disease, but the disorders they cause may mimic disease. Insects that feed on crowns, roots, or stem bases or extract plant sap from foliage may cause plants to appear diseased. Careful diagnosis is required to assess the true cause of the damage, which can be difficult if infectious agents become established in wounds created by insects or mechanical damage. In such cases, the primary cause of the disorder may rapidly become masked. Weeds, moss, and algae are usually quick to occupy voids that remain after the turf has been thinned by infectious diseases or environmental stresses. Competition from these invaders can slow turf recovery even after the primary problem has been corrected. Integrated pest management practices can prevent or minimize competition from most noninfectious pests.

(Prepared by L. P. Tredway)

Algae

Algae are a diverse group of organisms, encompassing both eukaryotic and prokaryotic lineages and multicellular and single-celled growth habits. Although not related by common ancestry, algae share the ability to photosynthesize using chlorophyll but lack the vascular system and complex vegetative structures characteristic of higher plants.

As a group, algae are adapted to a wide range of habitats, including both terrestrial and aquatic systems. Both eukaryotic and prokaryotic algae have been documented in turfgrasses, but the most common inhabitants are prokaryotic cyanobacteria, also known as "blue-green algae" because of their bluish-green appearance afforded by chlorophyll a and phycobilin pigments. The genera *Lyngbya, Nostoc, Oscillatoria,* and *Phormidium* are observed most frequently in turfgrass systems.

Blue-green algae produce individual cells, typically 1–3 × 1–40 µm, that divide by binary fission. Cells are arranged in filaments called "trichomes," with neighboring cells connected via microplasmodesmata. Trichomes may be branching or unbranched and can resemble fungal hyphae but are distinguished by the presence of photosynthetic pigments. One or more trichomes are typically surrounded by a mucilaginous envelope or sheath to form filaments of varying sizes and shapes. Trichomes may fragment into individual cells called "hormogonia" that can be disseminated through various means. Certain species produce differentiated cells such as heterocytes, which fix atmospheric nitrogen, or akinetes, thick-walled resting cells capable of surviving harsh conditions. Many cyanobacteria, such as *Oscillatoria* spp., can move up and down in the turf canopy through a unique mechanism called "gliding motility." Some species, including common turf inhabitants in the genus *Nostoc,* form macroscopic colonies, evident as slimy growth in the turf canopy or on soil surfaces.

Algal growth is most aggressive in warm, humid climates, in poorly drained areas, and where mowing heights are low (Fig. 283). Excessive algal growth may cause indirect injury to turfgrasses by slowing air and water movement, creating conditions unfavorable for turf growth. The mucilaginous material produced by blue-green algae may bind soil particles together or clog soil pores, further reducing soil aeration and drainage. For example, blue-green algae have been implicated in the development of black layer by contributing to the development of anaerobic conditions.

Blue-green algae may also cause direct injury to turfgrasses by producing metabolites with herbicidal properties. Blue-green algae are known to reduce plant root growth in other systems, but this effect has not been studied in turfgrasses. *Phormidium* and *Oscillatoria* spp. have been implicated as possible causes of yellow spot of creeping bentgrass and black staining of Kentucky bluegrass foliage in the United States and South Korea, respectively.

In addition to their more direct effects on turf, algae are common secondary invaders, colonizing areas of thin turf and slowing turf recovery or reestablishment (Fig. 284). Mechanical methods such as spiking or verticutting combined with sand topdressing may be necessary to physically disrupt established mats of algal growth and facilitate turf recovery.

Algae can be controlled by improving soil drainage and aeration, avoiding overirrigation, maintaining proper nutritional levels, increasing the mowing height, and removing trees and

Fig. 283. Algal growth evident in the canopy of a *Poa agrostis* putting green. (Courtesy L. P. Tredway—© APS)

Fig. 284. Thick algal crust on the soil surface impeding recovery of the turfgrass. (Courtesy L. P. Tredway—© APS)

shrubs as needed to increase air movement and sunlight penetration. Blue-green algae are favored by soil pH greater than 7, so overapplication of lime should be avoided. The fungicides chlorothalonil, mancozeb, potassium phosphite, and copper hydroxide can prevent or control algal growth in turfgrass stands. Hydrated lime has been recommended in the past but can encourage growth of blue-green algae by increasing soil pH. Irrigation water is often implicated as a source of algal infestations in turf, but a survey in the U.S. state of Mississippi found very low similarity in algal populations in irrigation water compared with those in golf course putting greens.

Selected References

Baldwin, N. A., and Whitton, B. A. 1992. Cyanobacteria and eukaryotic algae in sports turf and amenity grasslands: A review. J. Appl. Phycol. 4:39-47.

Hodges, C. F. 1992. Interaction of cyanobacteria and sulfate-reducing bacteria in subsurface black-layer formation in high-sand content golf greens. Soil Biol. Biochem. 24:15-20.

Maddox, V. L., Krans, J. V., and Sullivan, M. J. 1997. Survey of algal and cyanobacterial species on golf putting greens in Mississippi, USA. Int. Turfgrass Soc. Res. J. 8:495-505.

Tredway, L. P., Stowell, L. J., and Gelertner, W. D. 2006. Yellow spot and the potential role of cyanobacteria as turfgrass pathogens. Golf Course Manage. 74:83-86.

(Prepared by L. P. Tredway)

Black Layer

Black layer is a symptom associated with anaerobic (lack of oxygen) conditions in the soil. Under these conditions, black layers or pockets 0.6–5 cm thick form in the soil profile (Fig. 285). These formations are composed of metal sulfides (primarily ferrous sulfide), which are not directly harmful to turfgrasses but indicate the activity of sulfur-reducing bacteria such as *Desulfovibrio* spp. In the absence of oxygen, these bacteria convert elemental sulfur to hydrogen sulfide gas, which is directly toxic to turfgrass roots and contributes to turf decline (Fig. 286).

Although black layer occurs in native soils, it is especially noted in artificially constructed sites with high sand content, such as golf course putting greens. Black layer can form at any depth in the profile but is most common in the upper 10 cm. Affected turf initially develops a yellow, bronzed, or wilted appearance, followed by thinning and death. Symptoms are often most severe in areas subjected to other stresses such as shade, poor air movement, or mechanical wear. Root growth in affected areas is typically sparse or nonexistent, and dark deposits of metal sulfides are observed on the surfaces of belowground tissues (Fig. 287). The noxious, rotten-egg odor of hydrogen sulfide is often detected in affected soils.

Anaerobic soil conditions are prerequisite for black layer formation. The most common causes are excessive rainfall or irrigation, poor surface or subsurface drainage, layering in the soil profile, excessive organic matter, or combinations of these factors. Any other factor that reduces soil oxygen levels is likely to contribute to black layer formation. For example,

Fig. 285. Black layer in the soil profile beneath an ultradwarf *Cynodon* sp. putting green. (Courtesy L. P. Tredway—© APS)

Fig. 286. Chlorosis and decline of Bermudagrass turf caused by black layer formation in the soil profile. (Courtesy M. Tomaso-Peterson—© APS)

Fig. 287. Deposits of metal sulfides evident on Bermudagrass roots and stolons caused by black layer formation. (Courtesy M. Tomaso-Peterson—© APS)

the blue-green algae *Oscillatoria* and *Nostoc* spp. are known to facilitate black layer formation, likely by slowing water and gas exchange and producing organic compounds readily digested by sulfur-reducing bacteria.

Maintenance of adequate soil aeration through proper design, construction, and maintenance is essential to prevent black layer. Surface and subsurface drainage should be provided to reduce waterlogging during periods of wet weather, and irrigation heads should be spaced to provide uniform coverage. Routine aerification and sand topdressing are recommended to reduce organic matter accumulation and increase soil porosity. Sand particles should be properly sized and consistent over time to prevent development of layers. Application of elemental sulfur, sulfate fertilizers, or organic forms of nitrogen can increase the risk of black layer formation and should be used with caution on susceptible sites. Spoon-feeding with nitrate forms of nitrogen has been shown to prevent or alleviate black layer by temporarily increasing the soil redox potential. Preventing or controlling algae invasions (see above) can assist with black layer management as well.

Selected References

Berndt, W. L. 2016. Redox potential and black layer: By understanding black layer and its causes and employing appropriate cultural practices, superintendents can reduce the potential for its occurrence. Golf Course Manage. 84:86-91.

Berndt, W. L., and Vargas, J. M., Jr. 2010. The nature and control of black layer. Golf Course Manage. 78:104-108.

Hodges, C. F. 1992. Growth of *Agrostis palustris* in subsurface black-layered sand induced by cyanobacteria and sulfate-reducing bacteria. Plant Soil 142:91-96.

Hodges, C. F. 1992. Interaction of cyanobacteria and sulfate-reducing bacteria in subsurface black-layer formation in high-sand content golf greens. Soil Biol. Biochem. 24:15-20.

(Prepared by L. P. Tredway)

Fairy Ring

Fairy ring is one of the most widespread diseases of turfgrass, occurring on both cool- and warm-season species worldwide. It is also, without question, the most complex turfgrass disease. Fairy ring is not caused by a single pathogen but rather by many thatch- and soil-inhabiting basidiomycetes that vary in their appearance and behavior. Fairy ring is also unique in that the causal fungi rarely infect or cause direct harm to turfgrass plants. Instead, fairy ring fungi colonize the thatch and soil and alter the physical and/or chemical properties of the turfgrass root zone, which in turn elicits a change in the health or appearance of the turf. Effective management of fairy ring, therefore, requires not only control of the causal fungus but also remediation of soil conditions to restore turfgrass health.

Symptoms and Signs

Fairy ring symptoms can occur at any time of year on both cool- and warm-season turfgrasses. Rings, semicircles, or arcs of darker green or faster-growing grass often appear in the spring to summer when soil moisture is plentiful (Fig. 288). These bands of lush turf may be 5–20 cm wide. Some rings may have an inner band of thin or dead turf, and occasionally a second zone of stimulated grass occurs inside the zone of plant death (Fig. 289). During periods of dry weather, especially in summer, the outer ring of lush grass may be missing, leaving a ring of dead or severely damaged turf (Fig. 290). After rain or heavy irrigation during mild weather, often in late summer and autumn, the causal fungus may produce mushrooms or puffballs in the outer zone of lush grass (Fig. 291).

Thatch and soil infested by fairy ring fungi often have a distinct "mushroom" odor. White mycelium may also be evident in the field or after incubation in a moist chamber for 2–3 days (Fig. 292). Infested thatch may appear orange to reddish brown, evidence of organic acid accumulation as a by-product of prolific fungal growth (Fig. 293). The soil profile beneath dead or damaged turf is often extremely dry because of the development of water repellency (hydrophobicity) at the surface.

Rings or arcs caused by fairy ring fungi may be very small when young. They may grow indefinitely, disappear unexpectedly, or be disrupted when intercepted by a flower bed, a sidewalk, or another ring. Rings on slopes are not closed on the lower side, often forming arcs of lush green or dead turf. Rings seldom overlap or cross. As the advancing edges of rings merge, they stop and coalesce to produce a scalloped effect (Fig. 294).

Three types of fairy ring symptoms have been distinguished in turfgrasses. Type I symptoms appear as rings or arcs with a zone of severely wilted, necrotic, or dead grass (Fig. 290). Type I symptoms may be accompanied by basidiocarps and outer or inner zones of stimulation (Fig. 289). Type II symptoms appear as rings or arcs of darker green stimulated grass that may include the production of basidiocarps (Fig. 295). Type III symptoms appear as basidiocarps formed in a circular or semicircular arrangement with no visible effect on the growth of the grass (Fig. 296). In all cases, basidiocarps are produced only briefly, when environmental conditions in the soil stimulate their production.

Fig. 288. Type II fairy ring symptoms on a creeping bentgrass fairway with no ring of dead grass. (Courtesy M. A. Fidanza—© APS)

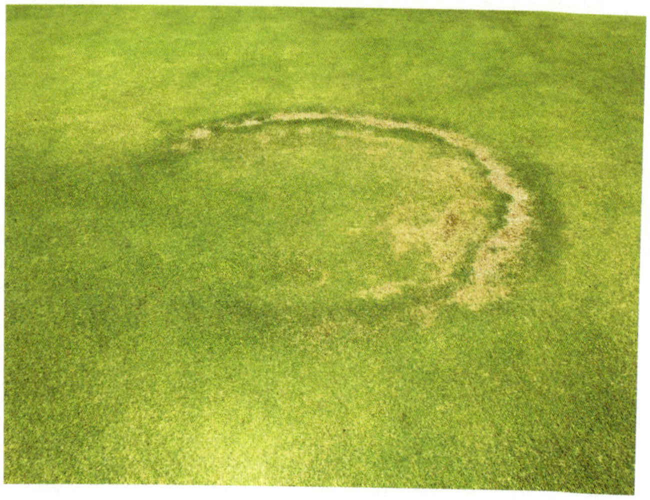

Fig. 289. Type I fairy ring symptoms with inner and outer zones of stimulated turfgrass growth on a Bermudagrass putting green. (Courtesy L. P. Tredway—© APS)

Fig. 290. Type I fairy ring symptoms on a creeping bentgrass putting green. (Courtesy L. P. Tredway—© APS)

Fig. 291. Type II fairy ring symptoms with basidiocarps growing within the zone of stimulation. (Courtesy G. L. Miller, Jr.—© APS)

Fig. 292. Mycelial growth of a fairy ring fungus in the soil profile of a creeping bentgrass putting green after incubation in a moist chamber. (Courtesy L. P. Tredway—© APS)

Fig. 293. Orange coloration of a creeping bentgrass thatch layer associated with fairy ring development. Note the dry condition of the sand on the right resulting from lack of water infiltration. (Courtesy L. P. Tredway—© APS)

Fig. 294. Type II fairy rings on a *Poa agrostis* putting green that developed a scalloped effect as they expanded and merged. (Courtesy B. B. Clarke—© APS)

Fig. 295. Type II fairy ring with basidiocarps on a creeping bentgrass putting green. (Courtesy L. P. Tredway—© APS)

Causal Agents

Many basidiomycetes in the order Agaricales cause fairy ring. Most of these fungi produce basidiocarps of either mushrooms (i.e., toadstools) (Fig. 296) or puffballs (Fig. 297). The most commonly cited are species of *Agaricus, Agrocybe, Arachnion, Bovista, Calvatia, Chlorophyllum, Clitocybe, Coprinus, Hygrocybe, Lepiota, Lycoperdon, Marasmius, Panaeolus, Scleroderma, Tricholoma,* and *Vascellum.* Research has been conducted mostly on fairy ring caused by *Marasmius oreades* and species of *Agaricus, Lycoperdon,* and *Vascellum.* Accurate identification of most fairy ring fungi is based on characteristics of the mature basidiocarp and basidiospores. Molecular primers have been developed to aid in the identification of some of these fungi. Of note, the basidiocarps of several fairy ring fungi are poisonous and must not be eaten.

Disease Cycle

Whether fairy ring is initiated by a bit of fungal mycelium or by the germination of a basidiospore is not known. From the point at which the fungus becomes established, it grows outward in a radial fashion, colonizing the thatch and the upper root zone. Generally, the highest concentration of mycelia is found along the outer perimeter of the ring where symptoms are evident. Fairy ring fungi may grow deep into the soil, with a dense, white, threadlike network of mycelia penetrating to depths of 20 cm or more. For example, *M. oreades* colonizes dead litter at the soil surface before growing downward into the thatch and underlying soil. The mycelium may penetrate to a depth of 30 cm in light-textured soils but seldom goes deeper than 5 cm in fine-textured clay soils.

Fairy ring symptoms occur in three zones corresponding to distinct stages of the disease as the fungus spreads outward in a radial fashion (Fig. 289). An "outer stimulated zone" develops along the outer perimeter of the ring, where plant-available nitrogen is released as the fairy ring decomposes organic matter in newly colonized soil. The "inner stimulated zone" forms where older fungal mycelium is degraded by other microbes, releasing additional nitrogen that stimulates turf growth.

When type II fairy ring symptoms are evident, the inner and outer stimulated zones coincide. Development of a "dead zone" containing wilted, dead, or severely damaged turf between the inner and outer stimulated zones leads to the appearance of type I symptoms. The dead zone is most often caused by hydrophobic (water-repellent) conditions in the thatch or soil (Fig. 298); the turf ultimately succumbs to drought stress or accumulation of ammonium to toxic levels. Other mechanisms include direct parasitization of turfgrass roots or the release of toxic levels of ethylene or hydrogen cyanide in the turfgrass root zone.

Fig. 296. Type III fairy ring symptoms of the pleated inkcap (*Coprinus plicatilis*) in a Kentucky bluegrass lawn. (Courtesy M. A. Fidanza—© APS)

Fig. 297. Puffball basidiocarp typical of fairy ring fungi in the genera *Arachnion, Bovista, Lycoperdon,* and *Vascellum.* (Courtesy L. P. Tredway—© APS)

Epidemiology

Fairy ring is most prevalent and damaging where fertility is low and there is insufficient soil moisture. Typically, fairy ring symptoms are more numerous and severe on lightly watered and underfertilized golf turf, sports fields, and lawns compared with well-irrigated and fertilized sites. Turfs with thick thatch layers and turfs grown in sand are especially vulnerable to fairy ring development and subsequent damage. Fewer rings are generally found in high-rainfall areas than in drier regions, but this is not always the case globally. A preponderance of microorganisms antagonistic to fairy ring fungi has been proposed to account for the lower level of damage in moist, fertile turfs. Consistent and uniform soil water content may also suppress fairy ring development, whereas environmental conditions that produce repeated wet–dry cycles have been associated with persistent fairy ring occurrence.

Management

Sound agronomic practices provide effective control of fairy ring by minimizing the establishment of the causal fungi and, once they are established, reducing the severity of turfgrass symptoms. Deep and infrequent irrigation providing thorough wetting of the root zone is recommended, along with the use of wetting agents to facilitate a uniform and consistent root zone volumetric water content. Practices to reduce thatch and organic matter accumulation, such as aerification, dethatching, or topdressing, help to reduce the growth of fairy ring fungi, improve turfgrass rooting, and slow the development of hydrophobic conditions. Sufficient fertilization to maintain optimal turfgrass growth minimizes the appearance of type II fairy

Fig. 298. Water droplet test demonstrating the hydrophobic condition of the soil profile beneath fairy ring symptoms. Water droplets were placed either 2 days (white arrow) or immediately (yellow arrow) before photo was taken. (Courtesy G. L. Miller, Jr.—© APS)

ring symptoms (Fig. 299), but excessive fertilization could increase thatch accumulation and increase the severity of type I symptoms.

Fungicides can be useful tools for prevention of fairy ring in turfgrasses but are often unsuccessful in completely controlling the disease on their own. Many fungicides are rather water-insoluble and are difficult to distribute uniformly through the entire soil profile, especially if it is already colonized by fairy ring mycelium. Therefore, fungicides are most effective when used in conjunction with a soil surfactant, which helps the active ingredient to penetrate deeper and more uniformly into the soil profile. Wetting agents and fungicides can be applied separately or together as a tank mixture and should be followed with irrigation to move the active ingredients into the soil profile.

Once fairy ring symptoms are evident, curative fungicide applications often provide poor results unless steps are first taken to remediate soil properties and facilitate turf recovery. Type II symptoms can often be disguised or "masked" with application of moderate amounts of nitrogen or iron fertilizer at appropriate times of the year. Application of a turf colorant or pigment may also help to mask type II symptoms. Thorough irrigation and application of a soil wetting agent, if necessary, are often helpful to mitigate any hydrophobicity that may be present and dilute potentially harmful accumulations of ammonium.

Alleviating type I fairy ring symptoms requires rewetting the soil profile to facilitate turf recovery. It is generally recommended to core aerate or spike affected areas, apply a soil wetting agent, and follow with heavy irrigation. Tree root watering wands, water forks, or hydro-aerification can be useful to introduce water directly into hydrophobic areas. Regular irrigation should be continued to maintain consistently moist to wet soil conditions, which facilitate turf recovery and may stimulate microorganisms that antagonize or parasitize the fairy ring fungus. Application of a fungicide at this time can help to prevent or delay the recurrence of fairy ring symptoms and protect from other disease activity.

Where type III symptoms occur (basidiocarps with no zone of stimulated grass), the fruiting structures can be simply mowed or raked off and removed. This does not weaken or control the fungus, but it eliminates unsightly basidiocarps.

Other approaches to controlling fairy ring require the destruction of the existing turfgrass. These methods include soil fumigation, soil replacement, and soil cultivation and mixing.

Soil fumigation requires the removal of existing turf and cultivation of the underlying soil. Thorough soil preparation is necessary to enable the fumigant to be completely distributed through infested soil aggregates. Procedures for soil fumigation are discussed in Part III (see Sanitation, Exclusion, and Avoidance). The efficacy of currently available soil fumigants against fairy ring fungi is unknown.

Fig. 299. Fairy ring symptoms are more evident in underfertilized turfgrass (foreground) than in well-fertilized turfgrass (background). (Courtesy L. P. Tredway—© APS)

Soil replacement is a labor-intensive method for rapid eradication of the causal agents. The sod and underlying soil in a band of about 50 cm on either side of the advancing ring are excavated to a depth of 20–75 cm, depending on the extent of ring development. Infested soil and sod are hauled away or composted. The resulting hole should be filled with clean, preferably sterilized, topsoil similar in texture to the soil removed. Reestablishing the turf with seed is preferable to using sod, because seed poses less risk of reintroducing fairy ring fungi. This method is impractical in most situations and is generally cost prohibitive.

Soil cultivation and mixing have also been effective in eradicating or reducing the severity of fairy ring. The turf is cut as sod and removed from the site, or, less desirably, the grass is killed with a nonselective herbicide such as glyphosate. The soil then is cultivated repeatedly with a heavy-duty rotary cultivator to mix infested and noninfested soil. The cultivated area is soaked with water, with or without a wetting agent, to promote microbial antagonism against the fairy ring fungus. The soil should be kept bare for a period of time, such as through the winter, before a new grass stand is established.

Selected References

Blenis, P. V., Chow, P. S., Duncan, I., and Knowles, N. R. 2004. Cyanide levels near fairy rings affect the growth of grasses and fungi. Can. J. Bot. 82:1324-1329.

Elliott, M. L., Hickman, R. B., and Hopkins, M. 2002. Flutolonil for control of necrotic fairy ring on bermudagrass putting greens. HortTechnology 12:656-657.

Fidanza, M. 2009. Fairy ring 101: The curious, elusive, but troublesome fairy ring. USGA Green Sect. Rec. 47:8-10.

Fidanza, M., Cisar, J. L., Kostka, S. J., Gregos, J. S., Schlossberg, M. J., and Franklin, M. 2007. Preliminary investigation of soil chemical and physical properties associated with type-I fairy ring symptoms in turfgrass. Hydrol. Processes 21:2285-2290.

Fidanza, M. A., Colbaugh, P. F., Engelke, M. C., Davis, S. D., and Kenworthy, K. E. 2005. Use of high-pressure injection to alleviate type-I fairy ring symptoms in turfgrass. HortTechnology 15:169-172.

Miller, G. L., Grand, L. F., and Tredway, L. P. 2011. Identification and distribution of fungi associated with fairy rings on golf putting greens. Plant Dis. 95:1131-1138.

Miller, G. L., Soika, M. D., and Tredway, L. P. 2012. Evaluation of preventive fungicide applications for fairy ring control in golf putting greens and in vitro sensitivity of fairy ring species to fungicides. Plant Dis. 96:1001-1007.

Shantz, H. L., and Piemeisel, R. L. 1917. Fungus fairy rings in eastern Colorado and their effects on vegetation. J. Agric. Res. 11:191-245.

Smith, J. D. 1978. Control of *Marasmius oreades* fairy rings: A review of methods and new approaches to their elimination. J. Sports Turf Res. Inst. 54:106-114.

Smith, J. D. 1980. Fairy rings: Biology, antagonism and possible new control methods. Pages 81-86 in: Advances in Turfgrass Pathology. P. O. Larsen and B. G. Joyner, eds. Harcourt Brace Jovanovich, Duluth, MN.

Smith, J. D. 1980. Is biologic control of *Marasmius oreades* fairy rings possible? Plant Dis. 64:348-354.

Stevenson, D. R., and Thompson, D. J. 1976. Fairy ring kinetics. J. Theor. Biol. 58:143-163.

(Prepared by M. A. Fidanza and L. P. Tredway)

Superficial Fairy Ring

Superficial fairy ring, also referred to as "white patch" or "superficial white patch," is caused by numerous thatch-inhabiting basidiomycetes. Fungi causing the disease have been reported in turfgrasses in North America, the United Kingdom, Europe, New Zealand, and Australia. Most of these fungi appear to be saprophytes, but some appear to be facultative parasites that can colonize weakened living tissues (see Introduction, Ecological Groupings of Pathogens and Parasites). Superficial fairy ring is most prevalent on golf greens but is likely to occur in any turf.

Yellow ring, caused by *Trechispora alnicola*, is a disease of *Poa pratensis* with symptoms that may appear similar to those of superficial fairy ring, but yellow ring is covered elsewhere because root infection has been documented as a primary cause of its symptoms (see Part I, Infectious Diseases: Other, Less Commonly Observed Diseases).

Symptoms and Signs

Superficial fairy ring (Fig. 300) appears as rings or patches of felted white mycelium in the foliage of *Agrostis* or *Poa* spp. on golf greens during cool or warm, humid weather. The disease also occurs in closely mowed *Festuca* spp. in the United Kingdom. The mycelium is most dense in a band 2–5 cm wide at the margin of the ring and less dense to sparse within the ring. Occasionally, mycelium is diffuse throughout the thatch, and well-defined patches may not be present. Mycelium is confined mostly to the thatch and the surface of the soil, and infested thatch has a strong "mushroom" odor. Older leaves of the turf senesce prematurely, and young roots may be stunted. Aerial mycelium may colonize older leaf blades and sheaths, blocking sunlight and causing premature senescence and thinning of the turf. Turf in the center of a patch may be sunken and less dense and may or may not appear healthy. There are no dark green rings of grass with stimulated growth, actual dead zones, or basidiocarps. Mycelium in thatch may become creamy white and mealy with age. Rings or patches of this description have been reported to be 5–35 cm in diameter in most regions and up to 1 m in Australia.

Causal Agents

Several basidiomycete fungi causing superficial fairy ring have been described. *Trechispora cohaerens* is recognized as a causal agent of superficial fairy ring in the United Kingdom. *T. farinacea* has been observed in *Agrostis* spp. and *Poa pratensis* in the Netherlands. *Coprinopsis kubickae* has been associated with superficial fairy ring in *A. stolonifera* in Canada. There are likely many more causal agents of superficial fairy ring that remain undescribed. In general, fungi that cause superficial fairy ring colonize the thatch with mycelium consisting of hyaline, septate hyphae with clamp connections. Under certain conditions, they may form hymenial layers bearing basidia and basidiospores, appearing as creamy white, mealy structures in the thatch. The importance of basidiospores in the dispersal of these fungi is unknown. There are only a small number of formal reports in which causal agents of superficial fairy ring have been identified. The host ranges of these fungi are unknown, but it is unlikely that they are as restricted as reported above.

Fig. 300. Superficial fairy ring in *Agrostis stolonifera* on a golf green. (Courtesy P. H. Dernoeden—© APS)

Disease Cycle and Epidemiology

Superficial fairy ring fungi produce abundant aerial mycelia under moist conditions and grow radially from the primary locus. In warm or dry weather, they continue to grow slowly as sparse vegetative mycelia in thatch. In the United Kingdom, superficial fairy ring fungi grow rapidly in cool weather from autumn to spring, but in the United States the disease is most prevalent during warm, humid periods in summer. The disease affects primarily *Agrostis* and *Poa* spp. grown on golf greens and tees. It is more common on sand-based greens but is occasionally observed on greens, tees, and fairways on native soil.

Superficial fairy ring is sometimes associated with the use of fungicides. This phenomenon is believed to be a result of the suppression of organisms that compete with or are antagonistic to the pathogen. A similar observation was made in 1-year-old turf planted in fumigated soil.

Management

Control measures specific for superficial fairy ring have not been reported. Mechanical thatch removal, frequent aeration, and topdressing to reduce the substrates on which the pathogens subsist are likely beneficial. Fungicide evaluations for control of superficial fairy ring are very limited, but fungicides that specifically target similar basidiomycetes (i.e., species such as *Laetisaria* and *Rhizoctonia*) may reduce the severity of the disease.

Selected References

Jackson, N. 1972. Superficial invasion of turf by two basidiomycetes. J. Sports Turf Res. Inst. 48:20-24.

Kackley, K. E., Dernoeden, P. H., and Grybauskas, A. P. 1989. Effect of fungicides on the occurrence and growth *in vitro* of basidiomycetes associated with superficial fairy rings in creeping bentgrass. Plant Dis. 73:127-130.

Redhead, S. A., and Smith, J. D. 1981. A North American isolate of *Coprinus kubickae* associated with superficial fairy ring. Can. J. Bot. 59:410-444.

Smith, A. M., Stynes, B. A., and Moore, K. J. 1970. Benomyl stimulates growth of a basidiomycete on turf. Plant Dis. Rep. 54:774-775.

(Prepared by R. W. Smiley, P. H. Dernoeden and B. B. Clarke; revised by M. A. Fidanza)

Moss

Mosses are primitive plants (bryophytes) that lack a vascular system. They are rapid invaders of areas where turf density is low, especially when surface moisture is high because of frequent rainfall, overirrigation, and/or poor soil drainage (Figs. 301 and 302). Although more appropriately classified as weeds, mosses share some similarities with fungal diseases—they are capable of long-distance spread as sexual spores or locally via small propagules (vegetative fragments, bulbils, and protonema). Once established, mosses form a dense, perennial colony of vegetative growth that competes effectively with the surrounding turf. They are capable of surviving extended periods of heat or drought in a dehydrated and dormant state, resuming growth when conditions become favorable.

Many species of moss are found in high-cut turf, particularly in wet and shady areas where turfgrass growth is reduced. More recently, silvery thread moss (*Bryum argenteum*) has become a severe problem on golf course putting greens in temperate regions, likely because of reduced mowing heights and frequent applications of irrigation and foliar nutrients, all factors shown to favor its establishment and growth.

Cultural practices can be used to favor turfgrass growth over moss. Because of their lack of a true root system, mosses are favored by poor surface infiltration, light and frequent irrigation,

Fig. 301. Severe infestation of silvery thread moss (*Bryum argenteum*) on a *Poa agrostis* golf green. (Courtesy L. P. Tredway—© APS)

Fig. 302. Colony of silvery thread moss in the canopy of an *Agrostis* sp. golf green. (Courtesy L. P. Tredway—© APS)

and foliar applications of nitrogen. Regular topdressing and cultivation, deep and infrequent irrigation, and soil-directed fertilizer applications have been shown to reduce moss infestations. Improved sunlight penetration and air movement promote turfgrass health, and mowing at recommended heights for the turfgrass species being managed is recommended. Moss is more competitive at low soil pH levels (5–6), so application of lime based on soil test results may be beneficial.

Because of their lack of a vascular system, mosses are sensitive to osmotic stress. Applications of iron sulfate, calcium or potassium bicarbonate, quaternary ammonia, and certain soaps provide control by burning the moss foliage. Certain herbicides and fungicides are also available for control. They are most effective when applied to actively growing moss in the spring and fall. These measures typically control the green foliar tissue but leave behind a dense layer of protonema underneath, from which the colony may regrow. Cultural practices to physically break up the colony and encourage turfgrass growth can improve the long-term efficacy of a moss control program.

Selected References

Boesch, B. P., and Mitkowski, N. A. 2005. Chemical methods of moss control on golf course putting greens. Appl. Turfgrass Sci. 2. doi:10.1094/ATS-2005-1006-01-RV

Borst, S. M., McElroy, J. S., Breeden, G. K., and Flessner, M. 2009. Cultural practices and silvery-thread moss control: Topdressing and nitrogen can enhance the efficacy of a herbicide on silvery-thread moss. Golf Course Manage. 77:137-140.

Raudenbush, Z., Keeley, S. J., and Stark, L. R. 2015. A review: Establishment, dispersal, and management of silvery-thread moss (*Bryum argenteum* Hedw.) in putting greens. Crop, Forage Turfgrass Manage. 1. doi:10.2134/cftm2014.0094

Thompson, C., Fry, J., and Kennelly, M. 2011. Evaluation of conventional and alternative products for silvery-thread moss control in creeping bentgrass. Appl. Turfgrass Sci. 8. doi:10.1094/ATS-2011-1018-01-RS

Yelverton, F. H. 2005. Managing silvery thread moss in creeping bentgrass greens. Golf Course Manage. 73:103-107.

(Prepared by L. P. Tredway)

Slime Molds

Slime molds do not cause disease, but their sudden appearance as white, yellow, gray, pink, purple, or brown patches on grasses may cause alarm. These organisms use the leaves and stems of grasses to support their reproductive structures and ultimately facilitate their dispersal to surrounding areas. They feed on bacteria and other microbes and are not parasitic to plants. However, they may shade individual grass blades to the extent that they are weakened. Some yellowing of leaves may occur when slime mold reproduction is heavy. They are primarily a cosmetic problem on turfgrasses.

The plasmodial stage of a slime mold feels slimy and may appear as a mucilaginous mass giving rise to sporangia. Large numbers of pinhead-sized sporangia (fruiting bodies) of slime molds may suddenly appear on grass blades and stems in circular to irregularly shaped patches 2–60 cm in diameter (Figs. 303 and 304). Sporangia of *Fuligo* spp. are often cushion shaped and may cover an area up to 60 cm in diameter and 3 cm thick, particularly when they are produced in organically rich soil or on wood and bark mulch. Sporangia are typically white, yellow, gray, or purplish brown, but other colors have also been reported. Affected grass does not normally die or turn yellow, and signs of slime molds usually disappear within 1–2 weeks. These organisms normally reproduce in the same general location each year. *Mucilago crustacea, Didymium squamulosum,* and species of *Physarum* and *Fuligo* are slime molds affecting turfgrasses. These organisms are myxomycetes, taxonomically classified as protists rather than fungi. Reports of slime molds often refer to *Physarum cinereum,* but this is only one of about 100 species with sporangia of similar size, shape, and color. The sporangia of these genera are highly variable, but all become filled with dark masses of spores that are produced in very large numbers on the leaf surface. The spores are easily dislodged from the sporangia and are dispersed by wind, water, equipment, people, or animals. They are deposited in soil or

Fig. 303. Slime mold sporulating on an *Agrostis* sp. putting green. (Courtesy L. P. Tredway—© APS)

Fig. 304. Sporangia of slime mold on leaf blades of an *Agrostis* sp. (Courtesy L. P. Tredway—© APS)

thatch and lie dormant until conditions are favorable for germination. Sporangia are the only readily visible stage in the life cycle of these organisms. Slime molds may be spread to new locations as sporangia transported with grass seed from seed production fields.

Cool, wet, humid weather favors the discharge of swarm cells, and warm, wet, overcast weather favors the movement of plasmodia onto grass sheaths and leaves. An abundance of thatch also favors slime molds by providing sources of food, both directly, in the form of organic litter, and indirectly, in the form of higher microbial populations. Control measures are unnecessary because the slime molds are more unsightly than harmful. Spore masses may be removed from foliage by washing the leaves with a stream of water or by raking or brushing them. If leaves are growing rapidly, frequent mowing also removes spore masses. Cultivation practices that reduce thatch may decrease the incidence of slime molds. Fungicides containing mancozeb are labeled for control but are generally not necessary because of the ephemeral nature of slime mold outbreaks.

Selected References

Alexopoulos, C. J., Mims, C. W., and Blackwell, M. 1996. Introductory Mycology. 4th ed. John Wiley & Sons, New York.

Martin, G. W., and Alexopoulos, C. J. 1969. The Myxomycetes. University of Iowa Press, Iowa City, IA.

(Prepared by R. W. Smiley, P. H. Dernoeden, and B. B. Clarke; revised by M. Tomaso-Peterson)

Insect Pests

Feeding by, and occasionally the mere presence of, certain insects can often be confused with symptoms of turfgrass diseases. When turfgrass problems are diagnosed, the potential role of insect pests should be considered along with diseases or other abiotic stresses. Following is a description of insect pests and the damage they cause that is most often confused with symptoms of turfgrass diseases, with an emphasis on sampling techniques and diagnostic features. The selected references at the end of the section provide a more comprehensive treatment of insect pests and their management in turfgrasses.

Insects that feed on turfgrasses can be placed into one of three categories based on their habitats or zone of feeding (soil-inhabiting insects, stem- and thatch-inhabiting insects, and foliage feeders). Damage caused by these groups as well as by non-turfgrass feeding or "nuisance" pests of turf can be readily distinguished from plant disease if the insect itself or its remains (e.g., fecal material or exoskeletons) are detected. Selecting the appropriate sampling technique or techniques is required before an accurate diagnosis can be made. This often involves coupling plant symptoms with knowledge of the pest's ecology prior to selecting a sampling method.

Soil-Inhabiting Insects

Soil-inhabiting pests comprise the greatest number and diversity of turfgrass insects. Examples include pests of cool-season (e.g., craneflies) and warm-season (e.g., mole crickets) turfgrasses as well as insects that are capable of feeding on both turfgrass groups (e.g., white grubs). With the exception of ground pearls (*Margarodes* or *Eumargarodes* spp.), all root-feeding insects in turfgrass have chewing mouthparts. Damage caused by immature stages includes root feeding and pruning, which may initially cause the turf to appear chlorotic before taking on a wilted appearance. Additionally, both adult and immature stages may damage turf stands when they burrow through the soil or to the surface. Burrowing and feeding can sever the root system, causing the turf to be spongy underfoot or to separate the aboveground plant material from the root system.

White grubs. The larvae of scarab beetles, commonly referred to as "white grubs," are pests of turfgrasses worldwide (Fig. 305). Several exotic species that are problematic in North America produce a single generation per year (annual white grubs). Larvae hatch from eggs deposited in midsummer, feed vigorously on roots until mid to late fall, and then move lower into the soil profile to overwinter. Larval feeding damage may initially appear as wilt or chlorosis and may be exacerbated during periods of high heat and extended drought (Fig. 306). Populations tend to be patchily distributed, which may lead to an incorrect diagnosis of disease caused by *Pythium* spp. or a patch disease like summer patch or take-all root rot.

White grub infestations can be distinguished from diseases through passive (i.e., visual) and active sampling techniques. Active sampling techniques are required to distinguish early grub damage from a disease outbreak. A golf course cup cutter is the most common white grub sampling device, allowing the turfgrass manager to rapidly sample the turf in question and surrounding areas. The core (approximately 10 cm in diameter) is taken only from the top 8–10 cm of the soil during the period when larvae feeding may be confused with disease. The core is then placed on a tray and quartered. Larvae that have fed on the root system usually are exposed at this time, although further investigation may be required. Since grubs are patchily distributed, multiple samples should be taken from an affected area.

Fig. 305. Scarab beetle larvae, commonly referred to as "white grubs," in a turfgrass root zone. (Courtesy B. A. McGraw—© APS)

Vertebrate predators (e.g., raccoons, skunks, and armadillos) may dig up the turf in search of larvae. Bird probing or increased activity in the damaged areas in late summer and fall may also indicate the presence of subterranean larvae. It is possible for birds and vertebrate predators to find populations of white grubs in areas that would otherwise not be symptomatic. In these instances, the secondary damage caused by predators may be more severe than the insect infestation.

Crane flies. Two European species of crane flies, *Tipula oleracea* and *T. paludosa,* can be pests of cool-season turfgrasses in northeastern North America and the Pacific Northwest. In warm-season grasses, only *T. umbrosa* has been associated with damage. Crane flies are large (1.75–2.5 cm long) but delicate flies that tend to be problematic in moist, poorly drained areas. Their larvae (maggots) are called "leatherjackets" and live in burrows by day and feed on turfgrass roots (Fig. 307). However, they come to the soil surface at night or on cloudy days and also feed on foliage. Leartherjacket larvae are legless, which enables them to be distinguished from caterpillars that occur coincident with them in turfgrass. Damage may appear in the spring (*T. paludosa*) or fall (*T. oleracea*) or when the insects are vigorously feeding prior to pupation (Fig. 308). Large populations of leatherjackets may often go unnoticed in excessively moist and thatchy areas. Damage may be more apparent when the turf dries, appearing as wilt or a general thinning on high heights-of-cut areas to isolated chlorotic or sunken spots on greens, which become brown over time. The initial chlorosis on putting greens may be confused with anthracnose, although as the damage progresses, it may appear more like damage caused by dollar spot.

Crane fly larvae can be detected in the same manner as white grubs. However, they are not as readily seen as white grub larvae in soil samples and can often be overlooked. Multiple cup cutter samples may be required to detect larvae in a suspected area before crane flies may be ruled out. When applied to small areas as a soil drench, an insecticide irritant such as carbaryl drives larvae to the soil surface where they can be counted. Leatherjackets are expelled from the soil in a short period of time (less than 15 minutes) and are often seen hanging at the distal end of the turfgrass plant. Crane fly larvae pupate at the soil surface, and these pupal remains can be detected in short-mowed grass if it is sampled before mowing.

Mole crickets. Three species of introduced mole crickets (*Neoscapteriscus* spp.) are common pests of warm-season turfgrass, mainly in the southern and southwestern areas of the United States. Of these, tawny mole crickets (*N. vicinus*) and southern mole crickets (*N. borellii*) are the most widespread and damaging. Two native species, *Neocurtilla hexadactyla* and *Gryllotalpa major,* are occasionally reported as pests of both cool- and warm-season turfgrasses. Adults are large, 3.0–3.5 cm long, with elongated bodies and conspicuous fossorial forelegs (Fig. 309). Nymphs are smaller and lack fully grown wings but otherwise resemble adults and co-occur in the same

Fig. 306. Severe turf damage caused by the feeding activity of white grub larvae. (Courtesy L. P. Tredway—© APS)

Fig. 308. Severe turfgrass damage caused by crane fly larvae and the foraging activity of vertebrates. (Courtesy B. A. McGraw—© APS)

Fig. 307. Crane fly larva, commonly referred to as a "leatherjacket" because of its leathery appearance. (Courtesy B. A. McGraw—© APS)

Fig. 309. Adult tawny mole cricket. (Courtesy B. A. McGraw—© APS)

habitats. Adult and nymphal stages create extensive subterranean tunnels in which they live and develop. These tunnels can erupt to the surface, which can excessively aerate the root zone causing a droughtlike appearance or problems with ball roll on putting greens (Fig. 310). Damage from tunneling can occur in the spring, late summer, and fall. Tunneling is most extensive where soils are naturally sandy or where native soil profiles have been modified with sand. When surface eruptions are not present, tunneling may produce a wilting or thinning of turfgrass in spring and fall that could be mistaken for patch diseases or dollar spot.

Mole crickets are most commonly detected by using a soap disclosing solution or soap flush technique. Briefly, 30 ml of dish soap is diluted in 4 L of water. The soap solution is poured over a turf area approximately 0.5 × 0.5 m and causes adult and immature mole crickets to crawl out of the moist soil to the surface where they can be identified. The success of this technique is dependent on soil moisture. It works best where there is adequate soil moisture and in areas near fresh tunnels.

Ground pearls. Ground pearls are a primitive group of subterranean scale insects associated with roots of all warm-season grasses. Two species, *Margarodes meridionalis* and *Eumargarodes laingi,* attack turfgrass. The "pearls" of ground pearls are the rounded cyst stage (0.5–2 mm in diameter) attached to turfgrass roots (Fig. 311). Cysts can occur year round but can easily be mistaken for root-knot nematode females or even fertilizer in the soil. Adults (wingless, pinkish red females or winged, gnatlike males) are the only stage found outside the cysts. Females lay eggs that produce nymphs. These nymphs are exposed temporarily as they settle on roots. They attach to the root by feeding and then cover themselves with a waxy substance, eventually forming the pearls.

Ground pearls attack a range of warm-season turfgrasses, but damage to centipedegrass and certain cultivars of Bermudagrass are most common. As nymphs feed, they either remove fluid or possibly inject salivary toxins, causing infested turfgrass to develop an unhealthy, yellow appearance that eventually leads to local dead patches (Fig. 312). Delayed green-up, spring dead spot, take-all root rot, and other patch diseases are common misdiagnoses for ground pearl damage. Excavation of the soil with a spade or cup cutter should reveal cysts, if present; however, cysts can occur as deep as 30 cm.

Fig. 310. Injury to *Cynodon* sp. turf caused by the tunneling activity of mole cricket nymphs and adults. (Courtesy L. P. Tredway—© APS)

Fig. 311. Ground pearls feeding on roots of *Eremochloa ophiuroides.* (Courtesy C. P. Freeman—© APS)

Stem- and Thatch-Inhabiting Insects

Insects that feed on the crown or stem of the turfgrass plant pose great challenges to early detection and control. The annual bluegrass weevil, several billbug species, and chinch bugs are some of the most misdiagnosed turfgrass insect pests, and the damage they cause is often confused with that of several foliar and root diseases. Damage caused by stem borers (weevils) may differ from that caused by insects that draw fluids from the plant (chinch bugs).

Stem-boring weevils. The annual bluegrass weevil (*Listronotus maculicollis*) is a pest of close-cut (less than 0.13 cm) turfgrass found on golf course tees, fairways, and greens in eastern North America. The insect completes its development on annual bluegrass (*Poa annua*) and creeping bentgrass (*Agrostis stolonifera*). Overwintering adults (Fig. 313) emerge in the spring and search for short-mowed host plants where the mated female chews a hole in the stem and oviposits. Developing larvae are initially stem borers, feeding within the plant. Older larvae become too large for the tiller, exit the plant, and feed externally on the crown. Damage is not usually expressed until crown-feeding stages are present. The damage begins as chlorosis and, given the host plants, mowing heights, and time of the year, is most often confused with anthracnose (Fig. 314). First-generation larval feeding damage appears in mid-May to mid-June in most areas of the insect's distribution. The number of generations a population undergoes varies with northern latitude, and thus the reoccurrence of damage throughout the growing season also varies.

Several species of billbugs cause damage to cool- and warm-season grasses (Fig. 313). The hunting billbug (*Sphenophorus venatus vestitus*) and bluegrass billbug (*S. parvulus*) are the most widely distributed species in North America, although several species may cause considerable damage regionally. Generally, billbugs cause more damage to high-mown turf than does the annual bluegrass weevil. However, the insect utilizes the host in much the same manner. Given that these insects attack turfs with higher heights of cut, the diseases that billbug damage is often mistaken for also differ. Early summer billbug damage may be confused with dollar spot, delayed green-up,

Fig. 312. Damage caused by ground pearls in *Eremochloa ophiuroides.* (Courtesy M. C. Martin)

or spring dead spot in warm-season grasses. Midsummer infestations may produce wiltlike symptoms or be misdiagnosed as summer patch. Late summer to early fall damage in warm-season sod, mainly evident as compromised tensile strength, could be misdiagnosed as various root dysfunctions.

Annual bluegrass weevils and billbugs may be detected by the same techniques. Visual examination of droughty areas (billbugs) followed by shallow soil sampling of the turf with a knife may reveal late-instar larval infestations. Droughty-looking turf that currently or recently harbored billbug larvae breaks off at the crown when pulled. This procedure, known as the "tug test," often reveals the billbug's sawdust-like frass (fecal material) within hollowed-out stems.

It may be advantageous to sample turf before young larvae are present or before advanced damage occurs. It is common to combine soil core sampling with saltwater extraction. A golf course cup cutter or smaller tubular turf plugger may be used to extract samples from the suspected area. Soils are manually separated from plant material and visually examined for late-instar larvae. Unlike white grubs, early-instar larvae are located within the plant and must be extracted or irritated from the plant. Soil and plant material may be placed in a beaker filled with lukewarm water saturated with table or agricultural-grade salt. Stages that were not detected in the soil float to the water's surface within 15 minutes. Stages inside of the plant may take 45 minutes to an hour to surface. These stages may require the use of a hand lens to see.

Fig. 313. Hunting billbug (left) and annual bluegrass weevil (right) adults. (Courtesy B. A. McGraw—© APS)

Fig. 314. Early-season damage caused by the annual bluegrass weevil, which can resemble symptoms of anthracnose. (Courtesy B. A. McGraw—© APS)

Chinch bugs. Chinch bugs are minute (3–4 mm long), sap-sucking insects that cause damage to lawns, golf course surrounds, and native areas (Fig. 315). These insects use a needlelike mouthpart (stylet) to withdraw fluids from the plant and sometimes also to transfer toxins in their saliva to the plant. Large populations of feeding nymphs and adults can cause the turf to appear wilted, although this may not be apparent until an additional stressor (e.g., heat or drought) occurs.

Two species are problematic in North America. The hairy chinch bug (*Blissus leucopterus hirtus*) occurs mainly in the Great Lakes, Mid-Atlantic, and northeastern portions of the United States and is problematic on Kentucky bluegrass and fine leaf fescues. The southern chinch bug (*B. insularis*) attacks St. Augustine grass and zoysiagrass in the southeastern and southwestern portions of the United States. Despite geographic and host plant differences, both species have multiple generations per year, overwinter as adults, and have overlapping life stages (eggs, nymphs, and adults) from late spring to fall.

Damage from hairy chinch bugs is typically not observed until August or September in the northeast. A warm and dry spring promotes the buildup of local populations in thatchy areas such as home lawns, bunker faces, and fine leaf fescue native areas on golf courses. These populations may go unnoticed if ample moisture is provided or precipitation is abundant. However, hot and dry conditions at the end of summer may cause infested turf to appear affected by drought or confused with damage caused by dollar spot (localized damage) or brown patch (turf with higher heights of cut) (Fig. 316). Damage from southern chinch bugs is typically isolated to thatchy areas of lawns in full sun. Heavy populations of southern chinch bugs first create the appearance of drought, but the damage persists in spite of ample moisture. Southern chinch bugs are aggressively managed because lawns susceptible to damage can be killed entirely. In southern lawns, chinch bug damage may be misdiagnosed as take-all root rot (*Gaeumannomyces graminis*) in St. Augustine grass.

Adult chinch bugs, though relatively small, can readily be seen in a large infestation merely by teasing apart the turf. Additional measures such a vacuum sampling or water flotation may be employed to get density estimates. A water flotation can is a hollow, metal cylinder that is driven in the ground. Water

Fig. 315. Long-winged (left) and short-winged (right) chinch bug adults. (Courtesy B. A. McGraw—© APS)

is added to the cylinder and replaced as it enters the ground. Eventually, the soil becomes saturated and the water remains relatively level. Chinch bugs float to the surface of the water in a relatively short time (less than 5 minutes) where they can be counted.

Foliage Feeders

Several species of foliage-feeding mites cause damage that can be confused with disease, herbicide injury, or drought symptoms. Banks grass mite (*Oligonychus pratenis*), for example, is associated with drought stress or dry, sunny areas of turfgrass. Feeding by this species on St. Augustine grass causes stippling that can be mistaken for the mottled appearance of foliage associated with the St. Augustine decline virus. An infestation of Banks grass mites is usually accompanied by webbing on the grass blades. Winter grain mite (*Penthaleus major*) injury to cool-season grasses occurs in the fall to early spring. In the southeastern United States, overseeded ryegrass infested with winter grain mites may produce foliar symptoms similar to those of drought. Both winter grain and Banks grass mites are visible to the unaided eye. Zoysiagrass and Bermudagrass mites (*Eryiophes* spp.), however, are not visible without a microscope and do not have the ticklike form typical of mites. These mites have a larval form and are usually found under the leaf sheaths. Eriophyid mite populations can increase rapidly under warm, dry conditions, initially causing chlorosis that progresses into deformed growth and possibly localized dead patches. A tufted appearance, or witches'-broom, is common in infested Bermudagrass (Fig. 317), while a straplike or rolled appearance occurs in zoysiagrass. Both types of damage can be mistaken for herbicide injury.

Mites in turfgrass can be difficult to sample because they are small and relatively fast moving. An alcohol wash or extraction is a useful method to isolate mites. Symptomatic grass from the field or a submitted sample can be submerged in 70% ethanol and shaken vigorously. The liquid portion can then be checked under a dissecting microscope for mites that were dislodged from the plant material.

Fig. 316. Severe damage to *Festuca* sp. caused by chinch bug feeding. (Courtesy B. A. McGraw—© APS)

Fig. 317. Tufted growth (witches'-brooms) caused by Bermudagrass mite activity. (Courtesy L. P. Tredway—© APS)

Nuisance Pests

Several insects found in or around turfgrass sites may garner pest status through their presence rather than because of any damage they may cause to plants. Several ant species and wasps provide beneficial ecosystem services in turfgrass. However, they may indirectly damage turf through their activity (e.g., mounding or burrowing) or create human health issues (e.g., stings). The turfgrass ant (*Lasius neoniger*) is a predator of several major turfgrass insect pests in temperate regions of North America. The ant is most often a nuisance around short-mown tees and putting surfaces on golf courses where their mounds can disrupt the playing surface or interfere with mowing. Occasionally, small, circular, bleached areas of turf resembling dollar spot are seen near mounds, possibly caused by the formic acid secreted by ants in the subfamily Formicinae as a defense mechanism.

Selected References

Brandenburg, R. L., and Freeman, C. P. 2012. Handbook of Turfgrass Insects. 2nd ed. Entomological Society of America, Lanham, MD.

Potter, D. A. 1998. Destructive Turfgrass Insects: Biology, Diagnosis, and Control. Ann Arbor Press, Chelsea, MI.

(Prepared by B. A. McGraw and D. W. Held)

Thatch Collapse

Thatch collapse is a disease of fine turf mowed to a height of 1.3 cm or lower. The disease affects most turfgrass species and has been identified throughout the United States, the United Kingdom, Australia, and New Zealand.

Symptoms and Signs

In both cool- and warm-season grasses, symptoms include depressed or sunken patches of dark green turf 8–46 cm or more in diameter (Fig. 318A). A yellow ring may be present at the outer edge of an active patch. A cross-section of the soil profile reveals profuse white to fawn-colored mycelium in the top 2.5 cm of the thatch (Fig. 318B–D). As the organism moves through the soil, organic matter is degraded, leaving behind healthy, white plant tissue. Under field conditions, basidiocarps (fungal fruiting bodies) occasionally form near the crowns of affected turfgrasses and may extend down into the thatch (Fig. 318F). Although the disease does not kill affected turfgrasses, it can cause surfaces to become sunken and can negatively impact uniformity, firmness, and playability of golf course turf (Fig. 319).

Causal Agent

Sphaerobolus stellatus is a wood-rot fungus often found in mulch commonly used in home landscapes. Colonies grown on potato dextrose agar at 25°C (77°F) produce concentric rings of creamy white mycelia that may be aerial or appressed. Clamp connections (Fig. 318E) are a distinguishing morphological characteristic of the mycelia, as with many basidiomycetes. If present, basidiocarps are white (immature) to creamy orange (mature) and measure 1–3 mm in diameter. When mature,

Fig. 318. Thatch collapse in *Agrostis stolonifera*. **A,** Sunken area created by degraded organic matter; **B and C,** fawn-colored decomposing thatch, which emits a mushroom odor; **D,** profuse mycelium visible within the upper 2.5 cm of the soil–thatch interface; **E,** clamp connections in mycelium; and **F,** basidiocarps within the thatch and turfgrass canopy. (Courtesy A. M. Baetsen-Young—© APS)

Fig. 319. Negative impact on the surface of a golf course putting green caused by thatch collapse. (Courtesy A. M. Baetsen-Young—© APS)

sticky spore masses (glebae) within basidiocarps measure 734 × 686 μm and can be forcefully ejected up to 6 m. Individual basidiospores average 7.6 × 4.8 μm.

Disease Cycle

The causal organism survives unfavorable conditions as hardened glebae and/or mycelia within the soil profile. Glebae are highly tolerant of environmental extremes and can survive for several years. When conditions are favorable for growth, *S. stellatus* spreads radially through the thatch. Although the causal organism does not directly infect turfgrass plants, it aggressively degrades organic matter as it moves laterally. Under optimal conditions, as much as 20% of the surface organic matter may be lost in as little as 6 weeks. The degradation of organic matter results in the sunken, dark green areas characteristic of the disease.

Epidemiology

Thatch collapse is most severe where thatch accumulation is excessive. Optimal temperatures for the growth of *S. stellatus* are 21–30°C (70–86°F), but the fungus can grow at temperatures as low as 5°C (41°F). Depending on location, symptoms may appear all year long but are generally more severe from late spring to early fall. Disease is favored by extended dry periods followed by heavy rains or periodic wet–dry cycles. Since glebae can be forcefully discharged several meters from an individual infection center, they likely play a significant role in the spread of the disease. Although glebae develop most rapidly at 21°C (70°F), they can develop over a wide range of temperatures (10–25°C [50–77°F]).

Management

Little information is available regarding suppression of thatch collapse symptoms. Cultural recommendations include core cultivation to minimize the buildup of organic matter, application of nitrogen to mask symptoms, and judicial application of sand topdressing to smooth the surface of sunken patches. Fungicide recommendations generally include those specified for the suppression of fairy ring.

Selected References

Baetsen-Young, A. M., Kaminski, J. E., Kasson, M. T., and Davis, D. D. 2015. Insights into the biology of *Sphaerobolus stellatus* as a causal agent of thatch collapse in golf turfs. Crop Sci. 55:2342-2351.

Baetsen-Young, A. M., Kaminski, J. E., and Tien, M. 2017. Lignocellulose degrading capabilities of *Sphaerobolus stellatus* in creeping bentgrass. Int. Turfgrass Res. Soc. J. 13:145-152.

Brantley, E. A., Davis, D. D., and Kuhns, L. J. 2001. Biological control of the artillery fungus, *Sphaerobolus stellatus,* with *Trichoderma harzianum* and *Bacillus subtilis*. J. Environ. Hortic.19:21-23.

Dernoeden, P. H., Kaminski, J. E., and Haspell, C. 2011. Thatch collapse: A disease of fine-leaf fescue. Golf Course Manage. 79:88-91.

(Prepared by J. E. Kaminski)

Abiotic Agents of Noninfectious Diseases

Turfgrasses may be harmed directly by various chemical, mechanical, and physical conditions in the plant environment. Chemical causes of abiotic damage include pesticides, animal urine or salts, saline or sodic soils, excessive or deficient soil fertility, plant growth regulators, gaseous pollutants, and chemicals released in spills. Physical agents include extremes of air temperature and availability of water, compaction, thatch, shading, excessively wet soil, and roots of ornamentals and trees in competition with the turf. Mechanical agents include mower shredding or scalping of leaves, frost heaving or bruising of frozen plant tissues, and abrasive wear. These abiotic factors often interact with infectious agents, making diagnosis difficult.

(Prepared by R. W. Smiley, P. H. Dernoeden, and B. B. Clarke; revised by E. H. Reasor and M. Tomaso-Peterson)

Chemical Agents

Pesticides

Plant health aids should always be applied precisely according to the label directions. Some pesticides may have nontarget effects on the turfgrass ecosystem (see Nontarget Effects of Pesticides in Part III). Excessive applications, especially of herbicides, may kill turfgrass plants. Several herbicides and fungicides have growth-regulating properties that may adversely modify grass growth when they are not used according to the label directions. Any pesticide that is intended to be washed off the foliage must be applied to achieve maximum efficiency and avoid killing or damaging foliage, especially on warm, dry days.

Symptoms of pesticide damage often appear in the patterns in which a pesticide has been applied, such as broad swaths, narrow streaks, or other regular patterns. Specific symptoms are highly variable, including leaf speckling (Fig. 320), chlorosis, burn, droughty appearance, and death. Symptoms may occur soon after application or may be delayed for several weeks.

Animal Urine and Salts

Animal urine contains soluble salts, urea, and other compounds. It causes little damage to turf if it is uniformly dispersed and has a low soluble salt content or if the soil is moist and fertile. Local stimulation of leaf growth and dark green coloration may be apparent from nitrogen released from the urea. However, if the salt content is high and the urine is deposited in one spot, the affected turf may be killed (Fig. 321), especially in dry or infertile soil. A margin of dark green, rapidly growing grass may surround the dead area. Other sources of salts, such as cooling water from homemade ice cream, fertilizer spills, and materials spread on highways and sidewalks to melt snow and ice, can also kill turf. Only animal urine causes dead patches with marginal areas exhibiting dark green, stimulated growth. Heavy watering helps to disperse salts and lower their concentration.

Patches caused by salt damage can be confused with several patch-type diseases caused by infectious agents.

Fertilizers

Excessive use of fertilizers can damage grasses by disrupting nutrient balances in the plants and inducing excessive growth, which may be subject to scalping during mowing. Many types of fertilizers applied to the foliage of wet grass can cause salt burn (Fig. 322). Watering turf thoroughly after fertilizer application is recommended, particularly after application of water-soluble nitrogen fertilizers with a high salt index. Natural and many synthetic organic fertilizers generally have a low burn

Fig. 320. Phytotoxic response to nightly frosts in turf treated with the herbicide oxadiazon. (Courtesy R. W. Smiley—© APS)

Fig. 321. Damage caused by fox urine. (Courtesy M. J. Goatley—© APS)

Fig. 322. Salt burn appearing in footprints and wheel tracks following fertilizer application with a rotary spreader. (Courtesy M. Tomaso-Peterson—© APS)

potential. Fertilizers should be applied uniformly and during the proper season for the type of turf being grown.

Nutrient Deficiencies

Like the chemical excesses discussed above, nutrient deficiencies also cause direct damage to plants. Some soils are deficient in phosphorus and potassium. Micronutrient deficiencies can occur in turf grown in sandy soils or sand mixes used on golf greens or athletic fields. Lime, sulfur, or acidifying sources of nitrogen, such as ammonium sulfate, should be applied according to soil test recommendations to maintain a pH of 6–7. Sulfur should be used with caution, because it can be phytotoxic, especially on golf greens. A soil test should always be used for guidance in alleviating nutrient deficiencies and extremes in soil alkalinity or acidity.

Plant Growth Regulators

Plant growth regulators are natural or synthetic chemicals applied to golf courses, athletic fields, and home lawns to yield a desired growth response. Most turfgrass plant growth regulators are applied to decrease vertical shoot growth to reduce mowing requirements; however, other benefits may include increased shoot density, turfgrass color, and abiotic stress tolerance and decreased seedhead production. Many plant growth regulators inhibit gibberellic acid biosynthesis either early or late in the metabolic pathway. Certain herbicides can also be used as plant growth regulators without resulting in plant death. Examples of commonly applied plant growth regulators are trinexapac-ethyl, prohexadione-calcium, paclobutrazol, flurprimidol, and ethephon.

Plant growth regulator programs typically include repeat applications at user-selected intervals. Proper application interval is critical to maximize the desired effect while not sacrificing overall plant health. The metabolism rate of the active ingredient is the basis of selecting an accurate application interval. If plant growth regulators are applied too frequently, the active ingredient can accumulate, leading to turfgrass "overregulation," which can increase disease susceptibility, reduce traffic tolerance, and hinder turfgrass competition against weed or algae infestations. Symptoms of overregulation include bronzing or chlorosis and an overall thinning of the turfgrass stand.

Fig. 323. Damage to a *Zoysia* sp. caused by a phytotoxic topramezone herbicide, which was spilled and then spread by foot traffic. (Courtesy M. Tomaso-Peterson—© APS)

Chemical Spills

Spills of chemical fuels or lubricants, pesticides, soaps, or cleaning products cause severe and often long-lasting damage to grasses (Fig. 323). Injury appears as dead patches, for example, where a fuel or sprayer filter or line was changed, where a fuel tank was filled, or where a household floor cleaner was discarded. Streaks may also occur where hydraulic fluids, oils, or fuels escape from moving equipment. Some contaminants can be difficult to recognize, such as salts in topsoil, persistent herbicides or other chemicals in soils, and pesticides transported by water to low-lying areas. Damage can often be minimized when spills are detected quickly and the area is thoroughly washed with detergents and then treated with an absorbent substance, such as activated charcoal.

Selected References

Beard, J. B. 2002. Turfgrass Management for Golf Courses. Ann Arbor Press, Chelsea, MI.

Dernoeden, P. H. 2013. Creeping Bentgrass Management. 2nd ed. CRC Press, Boca Raton, FL.

Turgeon, A. J., and Kaminski, J. E. 2019. Turfgrass Management 1.0. Turfpath, State College, PA.

Walker, C. 1971. Turf Culture. New Zealand Institute for Turf Culture, Palmerston North, New Zealand.

(Prepared by R. W. Smiley, P. H. Dernoeden, and B. B. Clarke; revised by E. H. Reasor and M. Tomaso-Peterson)

Physical Agents

Temperature

Extremely high or low temperatures can directly damage grasses. The overall effect of temperature extremes depends on the physiological status of the plants at the time of exposure.

Winter-kill may be indirect (desiccation) when unfrozen plants are in frozen soil without a protective cover of snow, turf covers, or mulch (Fig. 324). The degree of exposure of a turf area to winter winds and sun can affect desiccation. Direct tissue damage and death can occur if ice crystals form and disrupt cell walls and membranes. Ice encasement can be especially damaging. The extent of the damage depends on the rate and frequency of freezing and thawing, the hydration status of the plant tissues, and the soil type.

Damage caused by high air or soil temperatures is more likely to occur in cool-season grasses than in warm-season grasses. High-temperature damage to grasses may result from

Fig. 324. Winter-kill on the north-facing slope of a *Cynodon dactylon* fairway. (Courtesy M. Tomaso-Peterson—© APS)

direct solar radiation; engine exhaust (a combination of heat and gaseous pollutants); contact with objects such as containers of hot liquids, motors, automotive floor mats, glass, or metal pails in full sunlight; and lightning strikes (Fig. 325). Heat is most detrimental to grass when the plants are unable to cool themselves by transpiration. Transpirational cooling is reduced under very high atmospheric humidity, under dry soil conditions, or when grass blades are completely covered by a hot object. Direct solar radiation also increases the respiratory activities of plant roots and soil microorganisms, which may lead to an accumulation of toxic gases or lethal temperatures in the root zone, especially in wet and compacted soils. Turfgrass root growth is slowed or stopped when soil temperatures are too high, generally above 18°C (64°F) for cool-season grasses or above 30°C (86°F) for warm-season grasses.

Water and Ice

Extremes in the availability of water probably cause more problems than any other single abiotic agent, but the effects of water and temperature are difficult to separate. The importance of transpirational cooling and other interactions were mentioned above. Lack of water restricts plant growth and makes plants more susceptible to mechanical injury and heat damage. Drought stress is often more pronounced next to sidewalks, driveways, and buildings; at the tops of ridges and slopes; and in areas where objects such as large rocks, concrete and other construction debris, or septic tanks are buried. In some turfs, hydrophobic (water-repellent) soil conditions develop because of microbial activity. Areas where such conditions develop are called "localized dry spots."

Excessive water, combined with poor drainage, creates an unfavorable soil environment for root growth. The lack of oxygen, the accumulation of toxic gases, and supraoptimal temperatures in soil may weaken or kill plants directly. Excessive humidity reduces transpiration and can lead to increases in leaf temperatures, which may be harmful to enzymatic reactions necessary for growth. Turfgrasses inundated by water on hot, sunny days are subjected to extreme heat; the resulting injury is called "scald."

Prolonged ice cover can cause suffocation (anoxia) of turfgrass, although respiratory processes are greatly reduced during very cold weather. Under the ice, oxygen is depleted by microbes and plants, and an anaerobic condition develops as a result. Anoxia kills plants directly or predisposes them to freeze injury. Permeable covers are used on golf greens in cold regions to protect the turf from desiccation and freeze injury. Excessive soil wetness can be controlled by proper surface drainage, installation of subsurface drainage systems, and proper irrigation.

Fig. 325. Damage caused by a lightning strike to an *Agrostis stolonifera* putting green. (Courtesy M. Tomaso-Peterson—© APS)

Shallow Soil

Soil overlying shallow rock outcroppings or buried objects dries rapidly, and plants growing in it may show drought stress much sooner than those in nearby, deeper soils. Patches or streaks of wilted turf may appear on gravel or rocky soils. Turf on one residential lot may become dry much sooner than that on neighboring properties if the depths of excavation and topsoil refill differ significantly. The cause of such patches or patterns can be determined by probing the soil profile. Concrete, bricks, lumber, or large rocks buried during construction projects should be removed to alleviate persistent problems.

Soil Compaction

Heavily trafficked turfgrass areas, especially those on overly wet or fine-textured soils, can become so compacted that normal gas exchange and root growth are restricted. Such areas include footpaths, animal paths, and places where heavy mowers or vehicles always follow the same routes.

Compaction is difficult to correct without major renovation. The goals of corrective measures are to alleviate causal factors, to improve gas exchange, and to improve penetration of the soil by water and roots. Traffic over affected areas should be redirected or restricted. The soil can be modified to include a higher proportion of sand, as specified by a soil particle analysis. Effective soil modification is difficult to achieve and generally is expensive. Aerification machines that remove many soil cores to a depth of 5–10 cm or more are particularly useful in alleviating compaction.

Thatch

Thatch is a layer of plant litter (an accumulation of dead plant roots, crowns, rhizomes, and stolons) that develops above the soil surface. Thatch is not markedly influenced by the removal or return of leaf clippings to the turf during mowing. Clippings decompose rapidly and recycle nutrients back to living plants and to microorganisms that decompose plant litter. Thick layers of thatch reduce the movement of water, air, fertilizer, and pesticides into the soil and the movement of carbon dioxide and other gases out of the soil. Stems and roots tend to grow in thatch rather than down into the underlying soil, and this increases the risk of damage to plant tissues caused by drought and high temperature.

Thatch depths may be reduced by using vertical cutting (dethatching) and core aerification equipment when the turf is actively growing and by applying recommended topdressing sands or soils. A long-term program of managed microbial decomposition, including precise control of water, pesticides, soil pH, and fertilizers, helps to reduce the rate of thatch development.

Trees and Shrubs

Ornamental plants and trees with shallow roots or dense canopies compete with turfgrasses for water, nutrients, and light. Damage can be reduced by rearranging species of woody ornamentals, turf, or both. Water competition can be reduced by irrigating the turf more deeply and frequently and by pruning tree roots. In general, turf growing in shaded sites should be mowed higher and requires less nitrogen fertilizer than turf grown in the full sun. Trees should be fertilized with special root-feeding equipment and products. Problems caused by competition for sunlight can be reduced by planting shade-tolerant grass species or cultivars, increasing the mowing height in shaded areas, selective pruning or removal of trees and shrubs, or exchanging turfgrass for more shade-tolerant ground covers.

Selected References

Beard, J. B. 2002. Turfgrass Management for Golf Courses. Ann Arbor Press, Chelsea, MI.

Dernoeden, P. H. 2013. Creeping Bentgrass Management. 2nd ed. CRC Press, Boca Raton, FL.

Turgeon, A. J., and Kaminski, J. E. 2019. Turfgrass Management 1.0. Turfpath, State College, PA.

(Prepared by R. W. Smiley, P. H. Dernoeden, and B. B. Clarke; revised by E. H. Reasor and M. Tomaso-Peterson)

Mechanical Agents

Turfgrasses are injured when mowing equipment is dull or poorly adjusted; when the mowing height is too low or mowing is too infrequent; when the grass on mounds or on uneven areas is cut too low; when people, animals, or equipment move across drought-stressed, excessively wet, or frozen turf; and when the turf is subjected to excessive use.

Mower Injury

Dull mower blades tear or shred grass leaves rather than cut them (Fig. 326). When a rotary mower with dull blades is used, vascular bundles are left protruding from grass leaves, giving the turf a grayish cast. This shredding causes the tips of injured leaves to turn straw brown and die within a few days. Such injuries also allow infectious agents to penetrate plants to cause disease. Dull blades or poorly adjusted mowers can cause abraded bands across leaf blades. Mower blades should be sharpened and adjusted frequently to prevent excessive injury to grass leaves.

Fig. 326. Shredded leaf blades caused by a dull rotary mower. (Courtesy R. C. O'Knefski—© APS)

Scalping

Scalping occurs where grass is mowed so short that yellow or brown stem tissue is exposed. It may occur where a mower wheel drops into a depression, on embankments or other uneven areas, or in grass that is not mowed frequently enough. Such damage also occurs when the mowing height is lowered too rapidly.

Scalping can occur when weedy patches of stoloniferous grasses become established in turf cut at a normal residential lawn height. A patch of weedy grass becomes stringy and fluffy and periodically is lifted and severed by the mower. This leaves a brown area in an otherwise healthy turf.

Rhizomatous or stoloniferous grasses with scalping injuries often recover on their own, but it may be necessary to reseed damaged areas in bunchgrasses such as *Festuca* (especially fine-leaf fescues) and *Lolium* spp. Patches of undesirable grasses in lawns should be controlled and reseeded or sodded with desirable grasses.

Leaf and Crown Bruises

When turfgrass plants are frozen or dormant during dry weather, the leaf and crown tissues become brittle and can be damaged by pressure from the wheels of mowers or vehicles and from the feet of people or animals. When plant growth resumes, the damaged leaves may turn brown and fail to grow (Fig. 327). The damage often affects only exposed leaves, but crown damage and plant death can also occur. These injuries occur in recognizable patterns, such as brown, footprint-shaped marks that develop after someone walks across turf with heavy frost or turf that is drought stressed. Brown wheel tracks, snowmobile or ski tracks, and a dog's or cat's favorite path around the yard also appear in frozen, drought-stressed, or dormant turf. To remedy this problem, traversing dry turf or frosted turf without snow cover should be avoided.

Fig. 327. Leaf blades and crowns killed where equipment was operated on dry, dormant grass. Symptoms appear when growth resumes after rain or watering. (Courtesy R. W. Smiley—© APS)

Abrasive Injury

Green grass blades may be injured by excessive use. Such injuries can occur around the bases in backyard ball fields, in the goal areas of soccer and lacrosse fields, in the middle of football fields, at the front lines of a volleyball court, and in the push-off areas under swings. Mowing the peripheral areas of golf greens or collars during the summer, particularly during wet periods, causes thinning and sometimes death of the turf. Grass blades in high-use areas are subject to abrasion, which damages the leaf epidermis. The internal tissues dry quickly, giving the foliage a dull, gray green appearance that resembles drought stress. The grass fades to a brown or bleached color within several days. The damage may be permanent if the crown is killed or if the turf is completely worn away. Rotating activities to different areas or mowing the peripheral areas of golf greens less frequently promotes regrowth and prevents permanent damage.

Selected References

Beard, J. B. 2002. Turfgrass Management for Golf Courses. Ann Arbor Press, Chelsea, MI.

Dernoeden, P. H. 2013. Creeping Bentgrass Management. 2nd ed. CRC Press, Boca Raton, FL.

Turgeon, A. J., and Kaminski, J. E. 2019. Turfgrass Management 1.0. Turfpath, State College, PA.

(Prepared by R. W. Smiley, P. H. Dernoeden, and B. B. Clarke; revised by E. H. Reasor and M. Tomaso-Peterson)

Physiological Disorders of Turfgrasses

Turfgrasses are sometimes grown in environments to which they are not well adapted or managed in ways that are detrimental to their health. In such cases, the turf stand often declines under adverse weather conditions. Symptoms may appear suddenly or gradually and typically include foliar chlorosis, thinning of the canopy, root dieback, and eventual death of the turfgrass. In a few cases, these problems have been sufficiently common and widespread to be described as decline syndromes: summer bentgrass decline, summer Bermudagrass decline, and centipedegrass decline.

Early research on decline syndromes focused on identification of potential pathogens and characterization of their pathogenicity. Fungi and bacteria are often observed in turf that is declining because of a physiological disorder, and in some cases, application of fungicides was shown to sustain turfgrass health or slow its decline. Pathogens including *Gaeumannomyces, Pythium, Colletotrichum,* and *Rhizoctonia* species have been associated with decline syndromes, but no singular pathogen has been consistently associated or shown to duplicate field symptoms.

As our understanding of turfgrass physiology has advanced, abiotic stresses including heat, shade, anaerobic soils, and nutrient deficiencies are increasingly recognized as primary causes of decline syndromes. These stresses limit turfgrass growth and metabolism, rendering the plant more prone to pathogen infection, which further accelerates the decline process. Selection of a well-adapted turf species and cultivar, proper site design and construction, and sound agronomic practices are the most effective and sustainable ways to manage decline syndromes. Application of biostimulants or synthetic chemistries to enhance abiotic stress tolerance and protect from pathogen infection may also be effective in some circumstances.

(Prepared by L. P. Tredway)

Summer Bentgrass Decline

Summer bentgrass decline is a common occurrence on *Agrostis* spp. putting greens in tropical and subtropical climates. It may also occur in more-temperate regions during prolonged periods of hot, humid weather. Summer bentgrass decline is rarely observed where mowing heights are greater than 5 mm, highlighting the importance of low mowing as a contributing factor.

Symptoms

Symptoms of summer bentgrass decline are variable, but they often begin with a rapid dieback of the root system. Root hairs and fibrous roots are typically absent. Epidermal and cortical tissues slough away from coarse secondary roots, leaving behind bare vascular cylinders. Above ground, chlorosis begins with the older leaves and progresses to the youngest, and eventually the foliage turns yellow to orange, tan, or brown as it collapses (Fig. 328). Symptoms usually appear in irregular areas but may occur in discrete patches corresponding to *Agrostis* genotypes that are less vigorous.

Summer bentgrass decline is most severe where air movement is restricted or in high-traffic areas, mower cleanup passes, and poorly drained areas. Turf growing in recent aerification holes often remains healthy but may eventually succumb if unfavorable conditions persist.

Known pathogens of *Agrostis* spp., including *Colletotrichum cereale, Magnaporthe poae,* and *Gaeumannomyces avenae,* are often observed in declining turf and may serve as primary or secondary causes. *Pythium* spp. are frequently observed in the roots and/or foliage, but species range from aggressive pathogens to nonpathogenic. Weakly pathogenic or nonpathogenic fungi such as *Fusarium* spp. or ectotrophic root colonizers may also be encountered. This presents a significant challenge to diagnosticians working with samples from *Agrostis* spp. putting greens. A thorough review of field symptoms, weather conditions, agronomic practices, and product applications is often necessary to formulate an accurate diagnosis.

Epidemiology

Heat and humidity are the predominant factors that favor summer bentgrass decline. *Agrostis* spp. are cool-season grasses well adapted to temperate or Mediterranean climates. However, because of its density and fine leaf texture, *A. stolonifera* became a popular choice for golf course putting greens in subtropical environments such as the transition zone of the United States, where hot and humid conditions prevail for much of the summer.

Air temperatures above 32°C (90°F) and soil temperatures above 25°C (77°F) are most detrimental to creeping bentgrass health, especially when humidity is high. Under these conditions, transpiration rates are limited, photosynthetic efficiency declines, and root respiration increases, leading to energy deprivation in the plant. The root system dies back, reducing the plant's ability to absorb water and nutrients from the soil, further limiting transpiration and normal metabolic functioning. The plant becomes more susceptible to diseases and abiotic stresses and has little recuperative potential.

These weather conditions are also highly conducive to pathogens including *C. cereale, M. poae,* and *Pythium* spp. While heat is thought to be the primary incitant of summer bentgrass decline, these pathogens are also involved in many cases.

Other factors including mowing heights below 3.8 mm, restricted air movement, poor soil aeration, excessive organic matter accumulation, and traffic or mechanical wear also favor the development of summer bentgrass decline.

Management

Certain agronomic practices have been shown to limit the development of summer bentgrass decline, including (1) increasing

Fig. 328. Summer bentgrass decline on an *Agrostis stolonifera* putting green. Note that turf growing in recent aerification holes remains healthy. (Courtesy L. P. Tredway—© APS)

mowing height to 3.8 mm or higher during periods of summer stress; (2) reducing mowing frequency to every 2 days (lightweight rolling can be employed instead of mowing on alternate days); (3) increasing air movement through tree removal or installation of high-powered fans; (4) providing sufficient nitrogen fertilization throughout the growing season (a study in North Carolina demonstrated that at least 195 kg N/ha was needed to maintain acceptable creeping bentgrass quality); (5) regular aerification and topdressing to limit organic matter accumulation and maintain soil aeration; and (6) use of soil surfactants to improve water infiltration and maintain uniform soil moisture.

Tank mixtures of a phosphonate fungicide, a pigment, and a contact fungicide such as chlorothalonil or mancozeb provide effective prevention of summer bentgrass decline when combined with the above cultural practices. Addition of the plant defense activator acibenzolar-*S*-methyl provides further benefits. These mixtures increase creeping bentgrass tolerance to abiotic stresses such as heat and protect against many of the pathogens that can facilitate decline. See Disease Management with Synthetic Chemistries in Part III for more information on these chemistries and their influence on turfgrass health.

Application of the plant growth regulator trinexapac-ethyl has been shown to increase heat and drought tolerance and limit development of summer bentgrass decline. Application of seaweed extracts containing cytokinins has proved beneficial as well.

Selected References

Carrow, R. N. 2004. Surface organic matter in creeping bentgrass greens: Controlling excessive organic matter can lead to healthy greens in summer. Golf Course Manage. 72:96-101.

Dernoeden, P. H. 2013. Creeping Bentgrass Management. 2nd. ed. CRC Press, Boca Raton, FL.

Huang, B. 2004. Summer bentgrass decline and oxidative stress: Oxidative stress contributes to summer bentgrass decline. Golf Course Manage. 72:119-122.

Huang, B., and Gao, H. 2000. Growth and carbohydrate metabolism of creeping bentgrass cultivars in response to increasing temperature. Crop Sci. 40:1115-1120.

Huang, B., and Liu, X. 2009. Physiological responses of creeping bentgrass to heat stress affected by phosphonate fungicide application. Int. Turfgrass Soc. Res. J. 11:799-806.

Jespersen, D., and Huang, B. 2017. Effects of trinexapac-ethyl and Daconil Action (acibenzolar-*S*-methyl and chlorothalonil) on heat and drought tolerance of creeping bentgrass. Crop Sci. 57:S-138-S-146.

Lucas, L. T. 1994. Diseases and stresses involved in summer decline of bentgrass. Pages 41-42 in: Proc. Int. Golf Course Conf. Show, 65th. Golf Course Superintendents Association of America, Lawrence, KS.

Miller, G. L., and Brotherton, M. A. 2020. Creeping bentgrass summer decline as influenced by climatic conditions and cultural practices. Agron. J. 112:3500-3512.

Wang, K., Zhang, X., Goatley, M., and Ervin, E. 2014. Heat shock proteins in relation to heat stress tolerance of creeping bentgrass at different N levels. PloS One 9(7):e102914.

Xu, Y., and Huang, B. 2010. Responses of creeping bentgrass to trinexapac-ethyl and biostimulants under summer stress. HortScience 45:125-131.

Zhang, X., and Ervin, E. H. 2008. Impact of seaweed extract-based cytokinins and zeatin riboside on creeping bentgrass heat tolerance. Crop Sci. 48:364-370.

(Prepared by L. P. Tredway)

Summer Decline of Bermudagrass Greens

Summer decline of *Cynodon* spp. greens is a disorder triggered by abiotic and biotic stresses that, once set in motion, become difficult to remedy. Symptoms associated with summer Bermudagrass decline vary depending upon overriding factors. In general, symptoms may appear in solitary, irregular areas within a green. At onset, the turf canopy is off color to yellow to brown. Turf thins as symptoms progress. Stunted roots appear off color to brown and rotted (Figs. 329 and 330). Affected areas may coalesce.

Epidemiology

Mowing height is a stress on *Cynodon* spp. growth and development. The pursuit of ever-faster greens speeds has led to lower mowing heights. In the era of dwarf *Cynodon* spp. greens (e.g., the hybrids Tifdwarf and Tifgreen 328), mowing heights were approximately 4.3 mm. With the introduction of ultradwarf hybrids (Champion, TifEagle, and MiniVerde), mowing heights of 3.2 mm or less became possible. In some instances, this put demands on dwarf hybrids to be mowed at 3.2 mm, which those cultivars rarely tolerate. Even though ultradwarf cultivars are more tolerant of lower mowing heights, there are negative consequences to plant health, including reduced photosynthesis. Other abiotic stress factors include light, photoperiod, temperature, moisture, nutrient availability and uptake, oxygen in the root zone, and traffic. Diseases, nematodes, and insects are biotic stresses that contribute to Bermudagrass decline.

These stress factors, alone or in concert, impart harmful physiological effects on the Bermudagrass plant. When the

Fig. 329. Summer decline of the *Cynodon* spp. hybrid MiniVerde. (Courtesy M. Tomaso-Peterson—© APS)

Fig. 330. Off-color, stunted roots of the *Cynodon* spp. hybrid MiniVerde. (Courtesy M. Tomaso-Peterson—© APS)

Fig. 331. Loss of *Cynodon* spp. turf caused by summer decline that requires sodding and plugging for recovery. (Courtesy M. Tomaso-Peterson—© APS)

plant cannot compensate for the stress, an overall decline can be observed. Excessive heat (ambient and soil), reduced light caused by shade or excessive cloud cover, and diminished soil moisture exacerbate summer decline symptoms.

Cynodon spp. grow best in 100% full sunlight. The optimal temperature for *Cynodon* spp. hybrids is approximately 31°C (88°F). An increase in soil temperatures above 30°C (86°F) results in a gradual decline in root vigor. Photosynthetic capabilities of compromised plants are reduced; therefore, available carbohydrates are limited. The plant's decline predisposes it to attack by fungi, nematodes, and insects. The affected areas are weakened as they transition into winter dormancy. As a result, these areas will likely exhibit a delay in spring green-up when growth resumes. Fungal pathogens, particularly ectotrophic root-infecting fungi (see Take-all Root Rot and Bermudagrass Decline), *Pythium* spp. (see Pythium Root Rot), and nematodes (see Nematodes) are often associated with roots of declining plants.

Management

In severe cases of turf loss, plugging or strip sodding may be necessary (Fig. 331). Identifying and correcting the stress factors that caused the decline is key to turf recovery. Selective tree trimming or removal increases light intensity and photoperiods if shade is an issue. Irrigation heads must apply uniform water coverage on the green. Core aerification, spike, slice, or pencil tine practices open the root system to adequate oxygen levels. Soil and tissue analyses should guide fertilizer application. Wetting agents reduce areas of hydrophobicity, and vertical mowing or topdressing improve thatch health. Submitting samples to a diagnostic laboratory may help to identify pathogens or nematodes that are contributing to turf decline and aid in development of a fungicide and nematicide program to improve overall turf health.

Many factors are involved in plant health. A harmonious balance of the previously mentioned factors ensures plant health and vigorous growth, resulting in a putting green surface in which color retention and turf density are maintained at optimal levels.

Selected References

Chan, Z., and Shi, H. 2015. Improved abiotic stress tolerance of bermudagrass by exogenous small molecules. Plant Signaling Behav. 10(3):e991577.

Hodges, B. P., Baldwin, C. M., Stewart, B., Tomaso-Peterson, M., McCurdy, J. D., Blythe, E. K., and Philley, H. W. 2016. Quantifying a daily light integral for establishment of warm-season cultivars on putting greens. Crop Sci. 56:2818-2826.

Huang, B., and Gao, H. 2000. Growth and carbohydrate metabolism of creeping bentgrass cultivars in response to increasing temperatures. Crop Sci. 40:1115-1120.

McCurdy, J. D., Flournoy, E. T., Stewart, B. R., Philley, H. W., Reddy, K. R., Kreuser, W. C., Reasor, E., and Baldwin, C. M. 2020. Estimates of turf-type hybrid bermudagrass base and optimal growth temperatures. J. Miss. Acad. Sci. 65:258-266.

Stanford, R. L., White, R. H., Krausz, J. P., Thomas, J. C., Colbaugh, P., and Abernathy, S. D. 2005. Temperature, nitrogen, and light effects on hybrid bermudagrass growth and development. Crop Sci. 45:2491-2496.

Turgeon, A. J., and Kaminski, J. E. 2019. Turfgrass Management 1.0. Turfpath, State College, PA.

(Prepared by M. Tomaso-Peterson)

Centipedegrass Decline

Centipedegrass (*Eremochloa ophiuroides*) is a warm-season lawn turf, ideally suited for the southern U.S. Gulf Coast states. Centipedegrass is a stoloniferous, medium-textured, slow-growing, bright green grass that can be established with seed, sprigs, or sod. Its low fertility requirements make it suitable for home lawns. However, when centipedegrass is managed beyond the minimal requirements, the turfgrass suffers.

Symptoms and Signs

Centipedegrass decline is a disorder that appears in lawns older than 3 years. Slow spring green-up or successful green-up followed by a gradual decline into the summer is a primary characteristic of the disorder. The general decline of turf is associated with yellow or chlorotic leaves in irregular patches that gradually turn necrotic, resembling drought stress. When patches coalesce, the decline is more widespread throughout the lawn. Pathogens including *Gaeumannomyces graminis*, *Rhizoctonia* spp., *Curvularia* spp., and nematodes are often associated with centipedegrass decline but do not appear to play primary roles in its development.

Epidemiology

Centipedegrass decline is an abiotic disorder. Overfertilization, overirrigation, excessive thatch accumulation, high soil pH, and low mowing heights are the most common contributing factors. Herbicides containing 2,4-D or MSMA may also injure centipedegrass and contribute to turf decline. Only herbicides that are labeled for use on centipedegrass should be applied.

Management

Centipedegrass requires minimal fertilization, making it a desirable low-maintenance home lawn turf. No more than 1.0 lb N/1,000 ft^2 (48 kg N/ha) per year should be applied to centipedegrass. In sandy soils in climates where the growing season is prolonged, 2.0 lb N/1,000 ft^2 (96 kg N/ha) per year is appropriate. The leaves of healthy centipedegrass are light (lime) green. Dark green foliage indicates that too much nitrogen has been applied, which could lead to decline during the following growing season. Periodic soil testing is very important to maintain a healthy centipedegrass lawn. Soil pH of 5–6 is critical. Iron is unavailable in alkaline soil, and high phosphorous levels also reduce the availability of iron. Acidifying fertilizers are recommended when the soil pH exceeds 6.

A strong root system can withstand drought and winter injury. To promote healthy rooting of centipedegrass, deep, infrequent irrigation to deliver 2.5 cm of water per week is recommended. Centipedegrass's responses to desiccation initially appear as a change from the normal color to a light gray or silver and curling of the leaves. If drought stress is prolonged, the leaves turn necrotic and brittle. Poor root development may be attributed to soil compaction, improper pH, inadequate fertility, nematodes, or poor watering practices.

Thatch is the accumulation of organic matter at the soil surface. Thatch production is promoted by maintaining turf at mowing heights greater than 5 cm, excessive nitrogen applications, and allowing clippings to accumulate. Centipedegrass should be mowed at a height of 2.5–5 cm. Removal of clippings is not necessary if one-third or less of leaf length is removed during mowing. Growth slows in the autumn, and the mowing height should be increased by 0.6–1.2 cm. Excessive thatch reduces rooting into the soil, resulting in shallow roots and exposure of the roots and crowns to temperature extremes. Cold injury to the crowns and roots results in delayed spring green-up or nonuniform green-up within the lawn. When the turf is exposed to high temperatures, the shallow roots become desiccated, which may lead to excessive irrigation. The stolons of poorly rooted turf lift easily from the soil surface. Dethatching is recommended if the turf canopy is soft and spongy. Many fungal turf pathogens thrive in the humid thatch layer, so thatch management is also a good practice to reduce the incidence of fungal diseases.

Early in the spring, while centipedegrass leaves are still brown, all the dead leaves should be removed with a lawn mower and the dead plant debris should be collected. The mowing height should not be so low that it cuts the stolons because this damages the plants and delays green-up. Nonaggressive vertical mowing or core aerification followed by topdressing are recommended practices for thatch management when centipedegrass is actively growing.

These factors, along with nematodes, fungi, improper herbicide applications, and poor soils, may contribute to the decline of centipedegrass. When this grass is treated as a low-maintenance turf and the above recommendations are implemented, the likelihood that centipedegrass decline will occur is greatly reduced. Anthracnose, large patch, and take-all root rot can mimic centipedegrass decline; therefore, a diagnosis from a local disease diagnostician is warranted.

Selected References

Duble, R. L. n.d. Centipede Decline. Texas Cooperative Extension. https://aggie-horticulture.tamu.edu/plantanswers/turf/publications/Centipdecline.html

Martinez, A., and Waltz, C. 2017. Centipedegrass Decline. Univ. Ga. Coop. Ext. Circ. 1003. https://secure.caes.uga.edu/extension/publications/files/pdf/C%201003_3.PDF

(Prepared by M. Tomaso-Peterson)

Part III. Disease Management

Diseases of plants develop only when virulent pathogens, susceptible hosts, and the environment interact simultaneously over a given period of time. Disease development is driven to a large extent by the environment (e.g., moisture, temperature, relative humidity, and soil pH). The deleterious relationship between a susceptible host and an aggressive pathogen does not occur if the environment does not favor colonization and infection by the pathogen. Resistance to the pathogen by the host is also important and plays a major role in disease management. In general, severe disease outbreaks often develop when environmental conditions favor the pathogen and are less than optimal for the host.

From these basic considerations, four important disease control strategies arise: (1) to prevent pathogens from becoming established in a new turfgrass stand; (2) to modify the genetic composition of plants so that they resist attack by pathogens; (3) to alter the environment so that susceptible plants can overcome infection; and (4) to protect plants from disease with appropriate chemical treatments and cultural practices. The tactics used to implement these four strategies include seed and seedbed sanitation, planting disease-resistant species and cultivars, the application of cultural management practices, and treatment with biological controls or synthetic pesticide chemistries.

(Prepared by R. W. Smiley, P. H. Dernoeden, and B. B. Clarke)

Sanitation, Exclusion, and Avoidance

Sanitation is a disease management practice that aims to reduce pathogen populations or eliminate them from the soil. Soil fumigants—highly toxic gases that target nearly all living entities—are the most common sanitation method employed in turfgrasses. Because of the cost of materials and labor, fumigation is typically used only for golf courses, professional athletic fields, and high-value sod production. Fumigation is conducted before planting to reduce populations of weeds and plant pathogens, helping to ensure rapid establishment of the desirable turf. As a disease management practice, the benefits are often short lived because plant pathogens are quick to reinvade fumigated sites. Indeed, diseases such as take-all patch and fairy ring are often more severe in recently fumigated soils. Nematodes and fungal pathogens are often reintroduced into the site on sod or sprigs or through other means, either during or after establishment.

Methyl bromide was the most widely used soil fumigant from the 1960s until the early 2000s, but it is no longer used in turfgrasses because of its ozone-degrading properties. Products containing dazomet are labeled for application to turfgrasses and are used to control perennial weeds during turf renovations. Dazomet also has broad-spectrum activity against nematodes and fungal and bacterial pathogens. 1,3-Dichloropropene is used in sod production as a preplant soil fumigant to control plant-parasitic nematodes and may also be applied to established golf course turfgrasses with specialized application equipment (see Disease Management with Synthetic Chemistries).

Soil fumigants are applied as liquid or granular formulations and incorporated into the soil profile by shank injection, tilling, aerification, or heavy irrigation. The surface is then sealed with a plastic tarp to retain fumigant gases in the soil profile for a specified treatment time. Certain fumigants, such as dazomet or 1,3-dichloropropene, can be retained in the soil with light, frequent irrigation. Soil fumigants must be applied carefully and according to all label instructions to maximize performance and minimize risks to workers and the environment. In general, fumigants are most effective when applied to moist soils (approximately 25–50% of field capacity) and when the soil temperature is above 15°C (59°F).

Exclusion is a management practice that aims to prevent introduction of pathogen inoculum into a turfgrass stand. In the long term, this is difficult to achieve in managed turfgrasses where pathogens may be introduced by natural processes (e.g., wind and water) and by humans (e.g., foot traffic, maintenance equipment, or athletic equipment). However, exclusionary practices can help to ensure a rapid and successful establishment of a new turf stand. Care should be exercised to purchase plant materials that are as free of pathogens as possible, especially plant-parasitic nematodes, root-infecting pathogens, vascular wilt pathogens, and viruses. The most effective way to exclude these agents is to purchase high-quality material from a reputable dealer. Inexpensive seed, sprigs, and sod are usually of the lowest quality and purity, and paying a slightly higher initial cost may prevent expensive disease control efforts during and after establishment. Commercial seed treatment with one or more fungicides provides added assurance that the introduction of pathogens on the seed will be minimized. Seed treatment is especially important for preventing damping-off caused by *Pythium* spp., *Rhizoctonia* spp., and other pathogens.

Seedbed sanitation and pathogen-free seed or sod are compromised if contaminated soil or equipment is used to prepare a seedbed or to manage a new stand. When soil is introduced prior to establishment of a turf, it is best to use pasteurized soil or clean soil from a site where turfgrass has not been growing. Maintenance equipment should be washed thoroughly to remove all soil and plant debris before moving from infested to noninfested areas. Sanitation of equipment with bleach or peroxide-based materials may further reduce the risk of spreading certain pathogens. When possible, dedicated mowers should be used during grow-in to minimize introduction of pathogens from established sites.

Planting during periods most favorable for growth of the grass species helps to avoid seedling diseases. It is a good practice to establish cool-season grasses in late summer to early autumn and warm-season grasses in late spring. Planting at the proper time and implementing the other three disease control tactics provides the most effective long-term disease control program.

Selected References

Agrios, G. N. 2005. Plant Pathology. 5th ed. Academic Press, San Diego, CA.

Beard, J. B. 2002. Turf Management for Golf Courses. Ann Arbor Press, Chelsea, MI.

Lembright, H. W. 1990. Soil fumigation: Principles and application technology. J. Nematol. 22:632-644.

Unruh, J. B., Brecke, B. J., Dusky, J. A., and Godbehere, J. S. 2002. Fumigant alternatives for methyl bromide prior to turfgrass establishment. Weed Technol. 16:379-387.

(Prepared by L. P. Tredway)

Disease-Resistant Species and Cultivars

Turfgrass species and cultivars vary in their susceptibility to pathogens and tolerance to adverse environmental conditions. The priority of every disease control program is to establish turfgrasses that have high levels of resistance or tolerance to the most debilitating pathogens present in the area. Seeds of the most disease-resistant species and cultivars should be selected for planting in new seedbeds or for improving the composition of established turf. Herbicides and grass seed drills are available to help modify the cultivar or species composition of established turfs.

Because turfgrass cultivars and species differ in their adaptability to specific sites and in their susceptibility to diseases, it has become commonplace to use blends (combinations of different cultivars of the same species) and mixtures (combinations of two or more species). This tactic provides the highest probability of turfgrass survival with minimal maintenance costs. Blends and mixtures are more common than monostands (stands consisting of a single cultivar) for high-quality turf sites. Blends are particularly important when seeding Kentucky bluegrass because each cultivar represents a single, asexually produced clone. Diversifying the constituents of the blends or mixtures provides maximum resistance to diseases as well as broad adaptability to environmental stresses and variations in shade and soil texture. Guidelines for preparing mixtures and blends are available, but the composition of a mixture or blend should be based on the demonstrated performance of species and cultivars in the area in which the turf will be grown and on their resistance to the most damaging diseases of the region. Selecting a mixture or blend recommended by a state university turfgrass extension agent or specialist is the best approach if available. Another resource is the National Turfgrass Evaluation Program (ntep.org), which coordinates turfgrass cultivar performance trials across the United States.

Resistance to a disease may be conditioned by a single plant gene (monogenic resistance) or may be conditioned by several genes (polygenic resistance or tolerance). When resistance depends on a single gene, the level of resistance is unlikely to vary with environmental conditions. This type of disease resistance does not occur in most turfgrasses. Resistance in turfgrasses generally depends on the interaction of several or many genes. Polygenic resistance can be influenced by the environment, and considerable turf damage can occur if environmental conditions strongly favor pathogen growth and development. The level of resistance, therefore, depends heavily on cultural management and on selecting well-adapted species and cultivars. Local recommendations are very important, because turfgrass seeds are generally not labeled for use or sold only in the regions where they perform best. A cultivar that performs well in one geographic region may be barely adequate for the same use in another region.

Selected References

Bonos, S. A., and Huff, D. R. 2013. Cool-season turfgrass breeding. Pages 591-600 in: Turfgrass: Biology, Use and Management, vol. 56. J. C. Stier, B. P. Horgan, and S. A. Bonos, eds. Crop Science Society of America, Madison, WI.

Hanna, W., Raymer, P., and Schwartz, B. 2013. Warm-season grasses: Biology and breeding. Pages 543-590 in: Turfgrass: Biology, Use and Management, vol. 56. J. C. Stier, B. P. Horgan, and S. A. Bonos, eds. Crop Science Society of America, Madison, WI.

McCarty, L. B. 2004. Best Golf Course Management Practices. 2nd ed. Prentice-Hall, Upper Saddle River, NJ.

Meyer, W. A., and Funk, C. R. 1989. Progress and benefits to humanity from breeding cool-season grasses for turf. Pages 31-48 in: Contributions from Breeding Forage and Turf Grasses. Crop Science Society of America, Madison, WI.

Turgeon, A. J., and Kaminski, J. E. 2019. Turfgrass Management. TurfPath, State College, PA.

(Prepared by R. W. Smiley, P. H. Dernoeden, and B. B. Clarke; revised by S. A. Bonos and W. A. Meyer)

Endophytic Fungi

Endophytic fungi commonly infect many species of cool-season grasses in the subfamily Pooideae (see Table 1). These fungi have been studied most in the genera *Festuca* and *Lolium.* Infected pasture grasses can be toxic to some grazing animals; but for turfgrasses, the association with endophytes is a beneficial symbiosis. Plants that are hosts of endophytic fungi may possess greater resistance to or tolerance for insect pests, pathogenic fungi, nematodes, and environmental stresses and may exhibit improved growth and persistence.

Endophytic Fungi in Turfgrasses

Many fungi have endophytic stages, but those in the family Clavicipitaceae, in the genus *Epichloë,* are of particular interest because they form symbiotic relationships that range from pathogenic to mutualistic. *E. typhina* causes choke disease, in which a fungal stroma prevents the emergence of flower panicles. The degree of pathogenic expression varies greatly in each grass–fungus association. In some associations, nearly all panicles are "choked," while in others only a few are diseased. The white stroma consists only of mycelia and conidia (spermatia). Cross-fertilization of mating types occurs by transfer of spermatia from one mating type to another by symbiotic flies and results in the production of perithecia. Perithecia are orange tan and pyriform (100–129 µm at the base and 227–327 µm long). Asci are unitunicate and cylindrical (128 × 298 µm), and each ascus contains eight ascospores. At maturity, ascospores are hyaline, multiseptate, and filiform (128 × 298 µm).

There are other species of *Epichloë* that are asymptomatic and do not produce stromata or other obvious symptoms of infection in their hosts. Asymptomatic endophytes include *E. coenophiala,* which infects *F. arundinacea,* and *E. festucae* var. *lolii,* which infects *L. perenne.* Many other species of *Epichloë* infect *Agrostis, Bromus, Elymus, Festuca, Hordeum, Lolium, Phleum,* and *Poa* spp., among others.

Life Cycle of Endophytic Fungi

Epichloë spp. infect meristem tissues and grow intercellularly. If flowering occurs, they grow into the inflorescence and are disseminated by seeds. Some endophytic species do not produce conidia within plant tissue but form them only on stromata, on leaf surfaces, and in culture. The reniform to subulate conidia are borne at the tip of the conidiophore (Fig. 332) and in culture often appear in a T-shaped configuration. Endophytic mycelia are easily detected in basal leaf sheaths (Fig. 332) and in seeds by staining or by ELISA.

Asymptomatic endophytes (i.e., where no stromata are produced) are obligate parasites and may rarely infect new hosts naturally. There is increasing evidence, however, that some asymptomatic endophytes that produce conidia externally on leaf blades can spread to uninfected grasses and originate new infections.

Role of Endophytic Fungi in Pest Management

Four classes of alkaloids (pyrrolizidine, peramine, lolitrems, and ergot alkaloids) have been isolated from various infected grasses. These alkaloids are directly or indirectly toxic to insects and/or mammals. They occur in varying concentrations and combinations in different grass–fungus associations. A few associations apparently do not produce any alkaloids, while others produce one, two, or three.

Endophyte-infected grasses have exhibited enhanced resistance to the following important insect species: *Listronotus bonariensis* (Argentine stem weevil), *Sphenophorus* spp. (billbugs), *Parapediasia teterrella* (bluegrass webworm), *Spodoptera frugiperda* (fall armyworm), *Schizaphis graminum* (greenbug aphid), and *Blissus leucopterus hirtus* (hairy chinch bug).

Grasses associated with endophytes have also shown resistance to plant-pathogenic nematodes in the genera *Helicotylenchus, Paratrichodorus, Pratylenchus, Tylenchorhynchus,* and *Meloidogyne.*

Association with endophytes has also been connected with enhanced resistance to *Clarireedia* spp. (the dollar spot fungi) in *F. rubra* subsp. *fallax, F. rubra* subsp. *rubra,* and *F. brevipila*; to *Cladosporium phlei* (a leaf spot pathogen) and *Puccinia graminis* (the stem rust fungus) in *Phleum pratense*; and to *Rhizoctonia zeae* in greenhouse-grown *F. arundinacea* seedlings.

In addition, some grass–fungus associations result in improved plant growth, persistence, resistance to weed encroachment, and drought tolerance.

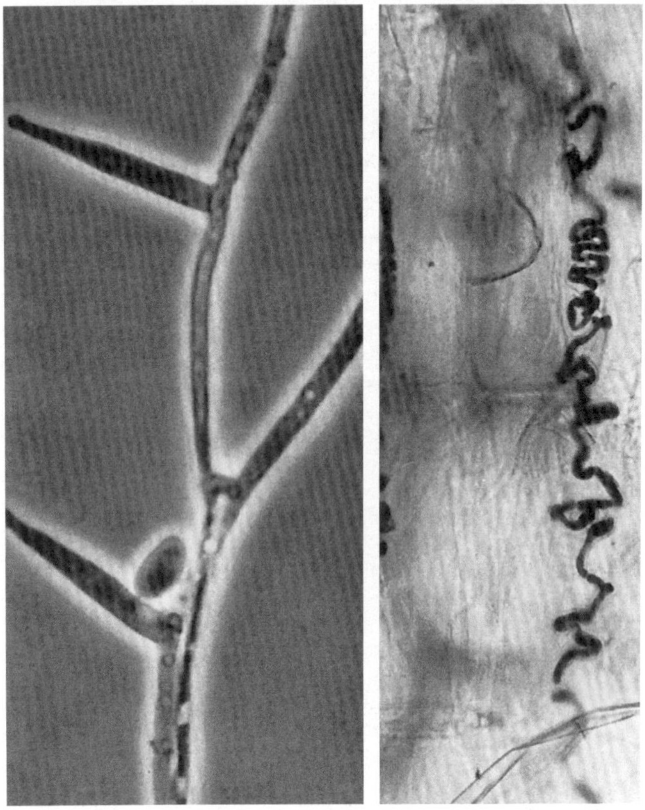

Fig. 332. Detached conidium at the base of a conidiophore of an endophytic species of *Epichloë* on agar medium (left) and convoluted intercellular hyphae of an *Epichloë* sp. in a pith-scraping preparation stained with aniline blue (right). (Courtesy J. F. White, Jr.—© APS)

There is considerable variation among biotypes or strains of endophytic fungi, and therefore the benefits to turfgrasses vary in different symbiotic associations.

The ability to introduce endophytic fungi into turfgrass germplasm offers an opportunity to significantly improve pest and stress tolerance in turfgrasses. New cultivars containing either partially ectophytic (i.e., low-disease-producing) types or non-choke-inducing endophytes have been developed for use as turf through maternal line selection and conventional breeding practices. Research continues into the use of tissue culture, seedling inoculation, and genetic engineering as methods of incorporating useful endophytes into turfgrasses. Studies have led to the development of endophytes that have beneficial properties but are of low toxicity to grazing mammals.

In the greenhouse, undesirable endophytes can be eradicated by the use of high rates of penetrant fungicides or by heat treatment of seed. There is no evidence, however, that fungicides applied in the field at labeled rates reduce endophyte infection in turfgrasses. Extended storage of seed destroys the viability of mycelium, but loss of viability is dependent on time, temperature, and the moisture content of the seed. To maintain viability, endophyte-infected seed should be stored under cold (0–5°C [32–41°F]), dry conditions.

Selected References

Bacon, C. W., and Siegel, M. R. 1988. Endophyte parasitism of tall fescue. J. Prod. Agric. 1:45-55.

Bacon, C. W., and White, J. F., Jr., eds. 2000. Microbial Endophytes. Marcel-Dekker, New York.

Clay, K. 1989. Clavicipitaceous endophytes of grasses: Their potential as biocontrol agents. Mycol. Res. 92:1-12.

Funk, C. R., Halisky, P. M., Ahmad, S., and Hurley, R. H. 1985. How endophytes modify turfgrass performance and response to insect pests in turfgrass breeding evaluation trials. Pages 137-145 in: Proc. Int. Turfgrass Res. Conf., 5th. F. Lemaire, ed. INRA Publications, Versailles, France.

Funk, C. R., White, R. H., and Breen, J. P. 1993. Importance of *Acremonium* endophytes in turf-grass breeding and management. Agric. Ecosyst. Environ. 44:215-232.

Leuchtmann, A., Bacon, C. W., Schardl, C. L., White, J. F., and Tadych, M. 2014. Nomenclatural realignment of *Neotyphodium* species with genus *Epichloë*. Mycologia 106:202-215.

Saikkonen, K., Young, C. A., Helander, M., and Schardl, C. L. 2016. Endophytic *Epichloë* species and their grass hosts: From evolution to applications. Plant Mol. Biol. 90:665-675.

Schardl, C. L. 1994. Molecular and genetic methodologies and transformation of grass endophytes. Pages 151-165 in: Biotechnology of Endophytic Fungi of Grasses. C. W. Bacon and J. F. White, eds. CRC Press, Boca Raton, FL.

Siegel, M. R., Latch, G. C. M., and Johnson, M. C. 1987. Fungal endophytes of grasses. Annu. Rev. Phytopathol. 25:293-315.

Tadych, M., Bergen, M. S., and White, J. F. 2014. *Epichloë* spp. associated with grasses: New insights on life cycles, dissemination and evolution. Mycologia 106:181-201.

Takach, J. E., and Young, C. A. 2014. Alkaloid genotype diversity of tall fescue endophytes. Crop Sci. 54:667-678.

White, J. F., Jr. 1987. Widespread distribution of endophytes in the Poaceae. Plant Dis. 71:340-342.

(Prepared by R. W. Smiley, P. H. Dernoeden, and B. B. Clarke; revised by J. F. White, Jr., and B. B. Clarke)

Management of Turfgrass Climate and Culture

The goal of turfgrass management is to maintain a functional or aesthetically pleasing ground cover with minimum effort and cost. To achieve this goal, factors that affect host resistance to pests must be taken into account when management practices are determined. These factors include species and cultivar adaptability, seeding density, age of plant tissue, mowing height and frequency, soil moisture and aeration, soil and air temperature, soil fertility and pH, shade intensity and air movement, and pesticides. Interactions among all these factors should also be considered. A great amount of literature is available describing the preferred methods of establishing and maintaining turfgrasses.

Other factors that must be considered in management decisions are those that affect pathogen activity and survival. These factors include plant nutrition; plant vigor, as influenced by cultural practices; soil and air temperature; soil aeration and moisture; the duration of leaf wetness; and pesticides. Only a brief introduction to these factors is provided here, but they are also discussed in the descriptions of diseases in Parts I and II.

The microclimate in the turfgrass canopy is governed mostly by temperature, humidity, and light intensity and quality. In some instances, these elements can be modified slightly by means of mowing, lightweight rolling, thatch control, and applications of fertilizer, soil surfactants, lime, sulfur, iron, water, or pesticides to reduce the disease-inducing potential of certain pathogens.

Temperature

Temperature has a direct effect on the growth rates of fungi and plants. Disease development is often affected by temperatures at which the growth rates of the host and pathogen are not matched. When the growth rate of turfgrasses is slowed by low temperatures, fungi capable of growing rapidly at these temperatures may incite disease. Similarly, fungi that grow best at high temperatures are able to cause disease when the growth of turfgrasses adapted to a cool, humid climate is restricted by high temperatures. High nighttime temperatures are particularly injurious to cool-season turfgrasses and can enhance the development of diseases caused by *Pythium, Pyricularia,* and *Rhizoctonia* spp. On the other hand, *Pythium* and *Rhizoctonia* spp. can incite disease on warm-season grasses as nighttime temperatures cool in the fall and remain cool through winter and spring.

The cardinal (minimum, optimum, and maximum) temperatures for growth or reproduction of fungi in the laboratory often differ considerably from the temperatures at which diseases occur in plants. Similarly, disease-predictive models based on temperature are not likely to be entirely dependable if they have been developed under greenhouse conditions, where the full impact of normal interactions among the plant, the environment, and other microorganisms may not have been evident. Nevertheless, temperature is a primary factor governing disease processes, and the occurrence of diseases of particular turfgrass species is usually associated with certain seasons.

Humidity

Atmospheric water vapor in the turfgrass canopy is important in disease development. Free water on the leaf surfaces or nearly saturated air in the canopy is required for the germination of most fungal spores and the growth of mycelium. In many regions, humidity in the turfgrass canopy is often very high, especially when tiller density is high, the leaves are long, soil water content is high, and air movement is restricted. When humidity is high, the evaporation of water from leaf surfaces in the morning is reduced and the period of time favorable for pathogen growth is extended. Rapid drying aids disease control by dehydrating and killing active fungal structures that have not yet penetrated plants. For most pathogens, a period of more than 12 hours of leaf wetness is required for penetration. The duration of leaf wetness and the temperature during the infection process are the two most critical microclimatological factors determining disease incidence and severity.

The duration of leaf wetness can be reduced mechanically by moving a long, flexible pole, rope, or water hose across the grass surface in the morning to knock dew off the blades. This can also be accomplished with turf maintenance equipment, such as a mower or lightweight roller. Substantial reductions in dollar spot severity can be observed with daily lightweight rolling, most likely a result of dew removal and possibly the activation of microbes that are antagonistic to the dollar spot pathogen. Humidity in the canopy can be controlled to some degree by mowing at the suggested height, by planting seed at the recommended rate, by watering early in the morning and as infrequently as conditions allow, and by removing obstructions that reduce air movement across the turfgrass canopy. Periodic applications of soil surfactants or wetting agents can also help to reduce leaf wetness. Removal of obstructions may require changing the arrangement or spacing of shrubs or selectively removing or pruning shrubs and trees. Topographic and structural features should be considered before turf is planted in an enclosed area. When properly placed, high-powered fans may also increase air movement and reduce disease incidence where obstructions cannot be removed.

Humidity also has an important effect on plant and leaf temperature. The rate of transpiration of water from leaves decreases as the humidity of the air increases. Transpiration dissipates heat generated by incoming radiation, and therefore turf growing in humid sites with restricted air movement may become hotter and more susceptible to hot-weather pathogens than turf growing in less humid locations.

Water

Management of water in the turfgrass canopy and soil is of critical importance for the health of grasses. Soil water primarily influences the growth of plants and other organisms that live in the soil. Excess water favors succulent plant growth and increases humidity in the turfgrass canopy. Waterlogging can incite root dieback and favors the growth of pathogens that tolerate low oxygen concentrations, such as *Pythium* spp. It is important to ensure good soil drainage prior to planting, because the intensity and duration of rainfall cannot be regulated. The use of soil surfactants assists with infiltration and can reduce soil moisture accumulation in the thatch and at the soil surface. Thus, these materials can be an essential tool for disease management, especially on golf course putting greens.

Very dry soil also promotes diseases by injuring plants. Soil surfactants promote wetting of hydrophobic soils that may result from overly dry conditions or when fairy ring fungi deposit chemicals that promote hydrophobicity. Soil surfactants can help alleviate damage from fairy ring when rings become hydrophobic. The effect of dry soil on disease development is described in the introduction to this compendium (see Pathogens and Their Environment) and in Part II (see Abiotic Agents of Noninfectious Diseases).

Turfgrasses should be watered early in the morning or before dawn when dew is already present. This allows the turf to be irrigated without extending the leaf wetness period. Cooling temperatures at night are often conducive to the deposition of dew on leaf surfaces and the formation of guttation droplets on leaf blades. These wetting processes should be minimized to retard fungal growth. Midday watering, such as syringing in summer, can help to reduce the temperature of leaf surfaces for a short time during periods of radiation stress, but only if the atmospheric humidity is low. Routine watering, however, should not be performed during the middle of the day.

The amount of water that should be applied varies according to the type of soil, the topography, and temperature and rainfall patterns in the geographic region. Enough water should be applied to prevent wilting, and the actual amount may vary from area to area within a given turf. Turf should be watered thoroughly to wet the entire root zone and as infrequently as possible. The use of soil moisture sensors and probes can aid in determining irrigation amounts and schedules. Turfgrass managers can assess volumetric water content daily and can record where hot spots develop or identify areas that remain wet after irrigation or rainfall. These tools are extremely useful and may help in reducing disease severity if used appropriately. Each location displays different volumetric water contents, and tools should be calibrated on site for best results.

Light

The intensity and quality of light also influence diseases, although to a lesser degree than atmospheric humidity and temperature. Direct radiation can help to dry films of moisture on leaves and kill actively growing fungal structures. In spring, direct sunlight helps warm the soil and reduces the growth of pathogens that cause cool- or cold-weather diseases. In summer, however, high-intensity light promotes heat stress and thus favors the development of hot-weather diseases. Diseases such as powdery mildew, rust, and many leaf spots are more severe in heavily shaded areas than in sunny areas. The increased severity of these diseases is often associated with higher humidity and lower temperature and possibly with reduced air movement across the grass surface. Low light intensity can also reduce plant vigor, making turf more susceptible to infection. This is especially true for Bermudagrass stands. Many different diseases can affect Bermudagrass when its growth is reduced because of low light intensity. This typically occurs during the fall, winter, and spring but can occur during the summer as well under persistent cloudy conditions.

Fertilizers and Soil pH

A balance of available nutrients is required for optimal (not maximum) growth of turfgrasses. Nutrient imbalances often cause stresses in plants and increase their susceptibility to diseases.

The amount of fertilizer applied and the timing of applications can have a major impact on turfgrass diseases. Nitrogen is the element with the greatest impact on turfgrass growth and consequently influences the development of many turf diseases. Excessive nitrogen applications may increase the risk of severe outbreaks of diseases caused by species of *Pythium, Rhizoctonia, Drechslera, Pyricularia, Bipolaris, Typhula, Microdochium, Ustilago,* and other genera. In contrast, a deficiency of nitrogen can increase the susceptibility of grasses to diseases caused by *Laetisaria, Clarireedia, Colletotrichum,* and *Puccinia* spp. Evidence demonstrates that turfgrass growth rate may dictate disease severity more than rates of fertilizer. Fertilization affects turfgrass growth, and plants that are growing optimally continue to produce secondary metabolites that are required for plant defense against pathogen attack. Secondary metabolite production is lowest when plants are not growing well and is also reduced if plants are stimulated to grow excessively. Therefore, simply categorizing diseases as low- and high-nitrogen diseases may not be appropriate. Instead, turfgrass researchers are examining more-specific fertility recommendations for specific diseases. For example, research has demonstrated that anthracnose severity is diminished as nitrogen in the plant tissue reaches 3.5%. Achieving this concentration depends on many site-specific conditions including age, turf maturity, organic matter amount, irrigation source, and soil type.

Application of plant nutrients other than nitrogen should be based on periodic soil test results and local agronomic recommendations. Some diseases may become more severe when plant health is impaired by deficiencies of essential plant nutrients such as phosphorus, potassium, calcium, manganese, or magnesium. For example, sufficient manganese availability in the soil is important to limit take-all patch development in creeping bentgrass by facilitating plant resistance responses. Application of manganese sulfate combined with maintenance of soil pH at 5.5–6 (see below) can provide effective suppression of take-all patch.

Adjustments to soil pH with applications of lime or sulfur should be based on reliable soil test data and local recommendations. Soil pH governs the ability of plants to remove some nutrients from the soil and strongly influences soil microbial activity; pH of 6–7 in thatch and the underlying soil is considered ideal for nutrient uptake and microbial activity. Most organisms in thatch are saprophytic, and their activity helps to decompose the thatch and reduce the activity of pathogens with which they compete. Soil pH less than 6 is considered desirable only in a few situations, such as for the suppression of Microdochium patch, summer patch, dead spot, or take-all patch on golf or bowling greens.

While not considered an essential plant nutrient, silicon has been shown to suppress certain turfgrass diseases and improve tolerance to abiotic stresses such as traffic. Silicon helps to increase cell wall strength and plays a role in plant defense signaling. Quartz sands, highly weathered soils, and organic soils may be deficient and benefit from application of products containing calcium or potassium silicate. In these locations, silicon applications have suppressed several turfgrass diseases including large patch, brown patch, Pythium blight, dollar spot, and gray leaf spot.

Pesticides

Pesticides are often required to prevent damage from weeds, insects, and pathogens. These materials often have no adverse effect on turfgrass health, yet in some instances, they can indirectly influence pathogen activity. Herbicides that selectively control undesirable weeds in turf may reduce the vigor of the turfgrass by changing root and shoot metabolism and by altering its susceptibility to pathogens. Excessive applications may lead to the death of turfgrass plants. Synthetic auxin herbicides used to control broadleaf weeds in turf can also alter the metabolism of grass plants and increase susceptibility to leaf spot diseases.

Fungicides are applied to turfgrasses to suppress the activity of certain pathogens. Most fungicides, however, are not selective enough to affect only the pathogens they are intended to control. Some may suppress the activity of nontarget fungi and other microorganisms and thereby reduce the microbial competition that keeps unrelated pathogens at low levels of activity. For instance, fungicides of a specific chemical class can control one disease but favor another. These and other nontarget effects are more thoroughly described below in the sections Disease Management with Synthetic Chemistries and Nontarget Effects of Pesticides.

Mowing

Most diseases are favored by mowing. Mowers wound leaves, disperse infected clippings and spores, and remove photosynthetic tissues that generate the carbohydrates required for root and shoot growth. When leaf tips are severed, the wound may serve as a breach in the plant's defense system. If clippings are removed, fungicide residues are also removed from the turf stand.

Fungi that infect leaves may be transported by mowers. These fungi may exist as spores dislodged from leaves or as mycelium in infected tissue. The dissemination of pathogens is particularly facilitated by mowing grass when it is wet. The combination of leaf wetness, active fungal growth, and newly wounded tissue is highly favorable for leaf-blighting pathogens. A dramatic illustration can be seen when Pythium blight or gray leaf spot occurs in streaks coinciding with the direction of movement of a mower over an affected area. Such patterns are less likely to occur when mowing is delayed until leaves have dried, especially because dry clippings do not stick as readily to mowers as wet leaves do.

As mowing height is reduced, the plant partially compensates for the reduction in its carbohydrate-producing area by producing more tillers. The result is an increase in leaf density, which contributes to higher humidity in the leaf canopy. The vigor of each plant, however, is reduced. Low mowing can lead to an increase in susceptibility to many diseases. For instance, it may have an indirect effect on root diseases such as summer patch, Pythium root rot and root dysfunction, and take-all root rot. Low mowing reduces carbohydrate reserves and predisposes turf to abiotic stresses such as drought and/or heat stress, which makes root diseases more severe. Low mowing contributes to the development of a shallow root system, which is less capable of supporting plants if roots are also impaired by infection. Mowing also brings cut leaf tips more closely together, which favors "bridging" of mycelium from one leaf to the next. Low mowing height increases the severity of anthracnose (caused by *Colletotrichum cereale*) and many of the leaf spot diseases, as well as summer patch.

Increased frequency of mowing increases wounding and may enhance foliar diseases. However, when leaf blades are allowed to become excessively long before growth stops during the winter, they tend to become matted down, especially under snow or leaf litter. The matted leaves provide highly efficient food and moisture conduits that enable fungi to spread rapidly over large areas. Higher mowing heights and reduced mowing frequency increase gray leaf spot severity, for instance.

A dilemma often arises in weighing the advantages and disadvantages of removing clippings from the turf. Clippings do not significantly contribute to thatch accumulation, and their decomposition recycles nutrients and favors the growth of organisms that decompose thatch. Some fungi decompose thatch and can cause disease. In addition, clippings that are infested with certain pathogens may serve as a food base on which the pathogens sporulate and produce new inoculum. When all the benefits and detriments of returning or removing clippings are considered, it appears best to leave clippings on higher-mown turfs. If the clippings are not removed, adjustments in fertility and pH may be required, and the turf should be mowed frequently enough to prevent excessive matting of the clippings.

Thatch Control

Accumulation of thatch in a distinct and restrictive layer is a symptom of an imbalance between tissue production (primarily roots, stolons, and rhizomes) and decomposition in the ecosystem. In cool-season grasses, tissue production is most rapid in cool weather. Tissue production in warm-season grasses is most rapid from spring through summer months. In either case, thatch decomposes most rapidly during hot periods if adequate water and nitrogen are available and if the layer is well aerated. Hence, it is advisable to ensure that water, oxygen, and nitrogen levels are favorable during the summer to help decompose thatch. Regulation of these factors may also promote some diseases, however, because some pathogens are active decomposers of organic litter.

When thatch decomposition has been suppressed, mechanical removal of thatch by means of aerification and vertical mowers or power rakes may become necessary. Topdressing is particularly effective in managing thatch on golf and bowling greens. Thatch decomposition may be suppressed by (1) low soil temperatures; (2) conditions under which the turf is kept anaerobic; (3) excessive acidity (pH less than 5.5), as a result of failing to apply lime, or excessive alkalinity, as a result of applying large amounts of lime; (4) removal of clippings and allowing nitrogen to become depleted in thatch during periods of high temperature; (5) application of excessive amounts of nitrogen, especially in water-soluble and acid-reacting ammonium-based fertilizers; and (6) repeated applications of fungicides or certain insecticides, nematicides, and herbicides before or during summer when most thatch decomposition usually occurs.

Selected References

Beard, J. B. 2002. Turf Management for Golf Courses. 2nd ed. Ann Arbor Press, Chelsea, MI.

Couch, H. B. 1995. Diseases of Turfgrasses. 3rd ed. Krieger Publishing, Malabar, FL.

Giordano, P. R., Nikolai, T. A., Hammerschmidt, R., and Vargas, J. M. 2012. Timing and frequency effects of light weight rolling on dollar spot disease in creeping bentgrass putting greens. Crop Sci. 52:1371-1378.

Hempfling, J. W., Schmid, C. J., Wang, R., Clarke, B. B., and Murphy J. A. 2017. Best management practices effects on anthracnose disease of annual bluegrass. Crop Sci. 57:602-610.

Latin, R. 2021. A Practical Guide to Turfgrass Fungicides. 2nd ed. American Phytopathological Society, St. Paul, MN.

Turgeon, A. J., and Kaminski, J. E. 2019. Turfgrass Management. TurfPath, State College, PA.

Vargas, J. M., Jr. 2005. Management of Turfgrass Diseases. 3rd ed. John Wiley & Sons, Hoboken, NJ.

(Prepared by R. W. Smiley, P. H. Dernoeden, and B. B. Clarke; revised by J. P. Kerns)

Disease Management with Synthetic Chemistries

While turfgrass diseases can often be suppressed through use of resistant cultivars and proper cultural practices, synthetic chemistries are sometimes necessary to protect susceptible turfgrasses from infection or to control active infestations. Fungicides and nematicides are important tools for turfgrass managers in their endeavors to maintain high-quality turfgrasses for recreational or functional purposes.

Fungicides

Fungicides are synthetic chemistries used to prevent or control fungal growth in a variety of agricultural, industrial, and medical settings. While the term implies a lethal effect, many fungicides could be more accurately described as "fungistats" because they only temporarily inhibit the growth of target pathogens. Once the active ingredient degrades below concentrations that are inhibitory, the pathogen can resume growth and infection if environmental conditions are favorable. For this reason, fungicides are often reapplied at regular intervals to maintain consistent protection over time.

Fungicide classification. Fungicides can be categorized on the basis of several characteristics. They are often grouped into chemical classes based on their composition or molecular structure. Chemical class names are often used in the turfgrass industry in reference to a group of similar fungicide products. For example, triazole fungicides (e.g., propiconazole and tebuconazole) contain a five-membered heterocyclic ring typically with two carbon atoms and three nitrogen atoms. The carboxamide fungicides (e.g., boscalid and penthiopyrad) are amides of carboxylic acid with pyridine or pyrazole rings. Strobilurin fungicides (e.g., azoxystrobin and pyraclostrobin) are derived from naturally occurring antifungal compounds called strobilurins and share a similar chemical structure.

"Mode of action" describes the specific biochemical mechanism by which a fungicide inhibits fungal growth (Table 7). With modern scientific techniques, fungicide mode of action can be characterized in precise detail. For example, the QoI (quinone outside inhibitor) fungicides bind to the quinone outside site of the *bc*$_1$ complex in complex III of the electron transport chain, preventing the production of energy through respiration. Demethylation inhibitor (DMI) fungicides inhibit C_{14} demethylase in the microsomal membrane, an important step in the production of sterols that comprise cell membranes.

This is a more useful way to categorize fungicides since multiple chemical classes can share a similar mode of action. For example, triazole, imidazole, and pyrimidine fungicides all function as DMIs. Similarly, multiple chemistries can function as QoIs or succinate dehydrogenase inhibitors (SDHIs). Mode of action is a particularly important consideration in the effort to prevent fungicide resistance, discussed in more detail below.

Fungicides are also categorized on the basis of their "biokinetics" or "phytomobility," terms used to describe the physical movement of active ingredients on or in the plant after application. Active ingredients are broadly categorized as contacts, which remain on plant surfaces after application, or penetrants, which are absorbed into plant tissue after application. These distinctions have important practical implications for the selection and application of fungicides for turfgrass disease control.

Contact fungicides are often referred to as "protectants" because they form a prophylactic barrier on the plant surface to prevent spore germination or penetration of plant surfaces. They typically provide short-term protection (7–14 days) because the active ingredient is exposed to more degradative forces on the plant surface. Contact fungicides are best used as preventive treatments prior to pathogen infection. If applied on a curative basis, contact fungicides can prevent further disease spread by protecting adjacent, uninfected turfgrass plants.

Penetrant fungicides have both preventive and curative activity and can provide longer-lasting control than contact fungicides, generally 21–28 days. Upon application to plant surfaces, these active ingredients diffuse through the cuticle and epidermis into the underlying tissues (mesophyll cells in the foliage or cortical cells in roots). The degree and rate of absorption vary widely among active ingredients based on their chemical properties. For example, the QoI fungicide azoxystrobin (6 ppm water solubility; $\log K_{ow} = 2.5$) is absorbed into leaf tissue slowly over time; only up to 25% of the active ingredient is absorbed within 24 hours after application. In contrast, more lipophilic active ingredients like trifloxystrobin (0.6 ppm water solubility; $\log K_{ow} = 4.5$) or pyraclostrobin (1.9 ppm water solubility; $\log K_{ow} = 4$) are absorbed more rapidly, with at least 50% absorbed within 24 hours. In either case, a significant amount of active ingredient remains on the plant surface where it can serve as a protectant until it is absorbed, removed, or degraded.

Penetrant fungicides can be further categorized as local, acropetal, or systemic based on their movement once inside the plant. Local penetrants are absorbed into the mesophyll or cortical tissues but are not translocated from the application site in

TABLE 7. Modes of Action and Resistance Risk Classifications of Currently Available Turfgrass Fungicides[a]

Mode of Action	FRAC Code	Group Name	Target Site	Common Name	Resistance Risk
Nucleic acid metabolism	4	Phenylamides	RNA polymerase	mefenoxam	High
Cytoskeleton and motor protein	1	MBC (methyl benzimidazole carbamates)	β-Tubulin assembly in mitosis	thiophanate-methyl	High
	43	Benzamides	Delocalization of spectrin-like proteins	fluopicolide	Medium
Respiration	7	SDHI (succinate dehydrogenase inhibitors)	Complex II: Succinate dehydrogenase	flutolanil isofetamid fluopyram benzovindiflupyr fluindapyr penthiopyrad pydiflumetofen boscalid	Medium–high
	11	QoI (quinone outside inhibitors)	Complex III: Cytochrome bc_1 Qo site	azoxystrobin mandestrobin pyraclostrobin trifloxystrobin fluoxastrobin	High
	21	QiI (quinone inside inhibitors)	Complex III: Cytochrome bc_1 Qi site	cyazofamid	Medium–high
	29	2,6-Dinitroanilines	Uncoupler of oxidative phosphorylation	fluazinam	Low
Signal transduction	12	Phenylpyrroles	MAP/histidine kinase (*os-2, HOG1*)	fludioxonil	Low–medium
	2	Dicarboxamides	MAP/histidine kinase (*os-1, Daf1*)	iprodione	Medium–high
Lipid synthesis or transport	14	Aromatic hydrocarbons	Cell peroxidation	quintozene (PCNB)	Low–medium
	28	Carbamates	Cell membrane permeability	propamocarb	Low–medium
Sterol biosynthesis	3	DMI (demethylation inhibitors)	C_{14}-Demethylase	difenoconazole flutriafol mefentrifluconazole metconazole myclobutanil propiconazole prothioconazole tebuconazole triadimefon	Medium
Cell wall biosynthesis	19	Polyoxins	Chitin synthase	polyoxin D	Medium
Host plant defense induction	P1	Benzothiadiazole	Salicylate related	acibenzolar-*S*-methyl	Low
		Phosphonates	Unknown	fosetyl-Al potassium phosphite	Low
Multisite activity	M3	Dithiocarbamates	Multisite contact activity	mancozeb	Low
	M5	Chloronitriles	Multisite contact activity	chlorothalonil	Low

[a] Prepared by L. P. Tredway—© APS. Adapted from the Fungicide Resistance Action Committee (FRAC) Code List 2021 (Anonymous, 2021).

significant quantities. These active ingredients may move short distances via translaminar activity, in which active ingredient accumulates on the opposite leaf surface, or vapor phase activity, in which vapors of the active ingredient move short distances in the cuticular layer.

Acropetal penetrants enter xylem vessels and move with water and nutrients absorbed by the root system. Upward movement is driven by transpiration through the stomata on leaf surfaces. More-soluble active ingredients are typically translocated in the xylem more readily. To use the aforementioned QoI fungicides as examples, azoxystrobin is readily translocated from the application site to leaf tips, whereas the less-soluble trifloxystrobin and pyraclostrobin are not translocated and best classified as local penetrants (Fig. 333). Translocation rates can also be limited by low soil moisture, lack of root development, or low transpiration rates, such as during cool or wet weather.

Systemic penetrants, also sometimes called "true systemics," are translocated both upward in xylem vessels and downward in the phloem. The phosphonate fungicides fosetyl-Al and potassium phosphite are the only registered fungicides that function as true systemics. Translocation in the phloem, which carries the products of photosynthesis, moves these active ingredients to the growing point and newly formed leaf tissues. Phosphonates applied to the foliage can be translocated into the root system as well, but upward movement in the xylem is often more pronounced. When applied as soil drenches, the phosphonates are readily absorbed by root tissues and translocated acropetally. Soil drenches of these compounds are often employed in agricultural settings for control of Phytophthora root rot.

With modern analytical tools, translocation of active ingredients can be detected in very small quantities, but it is not always apparent whether these amounts are sufficient to provide disease control or suppression. An active ingredient may have translaminar or systemic activity against a pathogen that is particularly sensitive but not against other pathogens that are less sensitive. Furthermore, environmental conditions and plant physiological status can influence the uptake and translocation of active ingredients. Whether an active ingredient functions as a contact, penetrant, or systemic fungicide can vary depending on the target pathogen and the situation in which it is applied.

Fungicide resistance. Fungicide resistance occurs when a pathogen population is able to circumvent the activity of an active ingredient, leading to either reduced performance or control failure. Resistance is of increasing concern in the turfgrass industry and in agriculture as a whole. Many populations of the dollar spot (*Clarireedia* spp.) and anthracnose (*Colletotrichum* spp.) pathogens have developed resistance to multiple fungicide modes of action. In some locations, turfgrass managers have few options for managing these pathogens because of resistance issues.

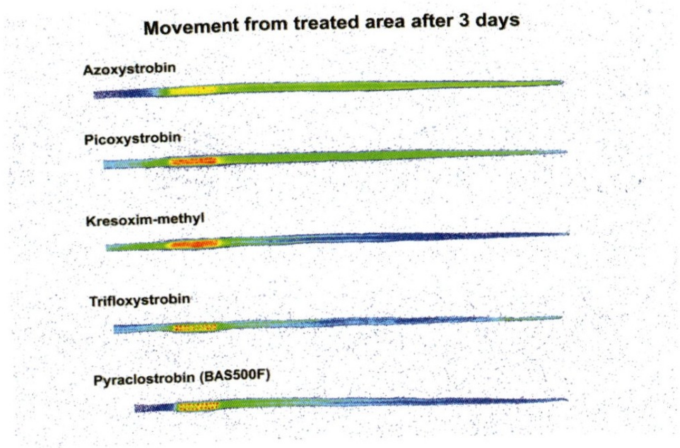

Fig. 333. Movement of QoI fungicides after application to the bases of wheat leaves. (Courtesy L. P. Tredway—© APS. Source: Syngenta Crop Protection)

Fungal pathogens are a population of individuals, which often vary in their sensitivity to a particular active ingredient. "Sensitivity" refers to the concentration of an active ingredient necessary to prevent growth of the target fungus. If a higher concentration of an active ingredient is needed to prevent growth, that individual is said to have reduced sensitivity. If the active ingredient fails to prevent growth even at very high concentrations, that individual is said to be insensitive or resistant. Resistance evolves when an active ingredient, or others that function through the same mode of action, are applied repeatedly over time, allowing individuals that are less sensitive to increase in the population through natural selection or "survival of the fittest." Once these strains become dominant in the population, control failure or a reduction in performance often ensues.

A variety of mechanisms are known to confer fungicide resistance. In most cases, a single point mutation in the fungal genome leads to a change in the target enzyme that reduces the chemical affinity between the fungicide and its binding site. The analogy of a "lock and key" is often used to describe this phenomenon, whereby if the shape of the lock (the binding site) changes, then the key (the active ingredient) no longer fits and does not control the pathogen.

Other mechanisms conferring fungicide resistance have been identified. Resistant individuals may overproduce the target enzyme to enable continued growth, even in presence of the active ingredient. These mechanisms are referred to as "target-site resistance," because they influence the direct interaction between the active ingredient and its intended target site. Target site resistance generally impacts all active ingredients sharing the same mode of action and may lead to a partial or complete loss of control.

"Non-target site resistance" mechanisms include toxin efflux pumps or detoxification systems that reduce the amount of active ingredient that reaches the target site. These mechanisms can lead to a partial or complete loss of control but, unlike target-site resistance, can also reduce the performance of other modes of action, albeit to a lesser degree.

The Fungicide Resistance Action Committee (FRAC) is a committee of crop protection scientists who develop standard recommendations for fungicide resistance management. As part of this effort, active ingredients are assigned to groups, designated by FRAC codes, based on their mode of action (Table 7). FRAC codes are prominently displayed on fungicide labels to assist users with resistance management programing.

FRAC groups are further categorized as having a low, medium, or high risk for fungicide resistance based on their modes of action (Table 7). Fungicides that inhibit a single biochemical reaction, called "single-site inhibitors," carry a medium or high risk for resistance because a single mutation in the target pathogen can confer resistance. Multisite inhibitors act at multiple, and often many, sites of action and generally carry a low risk for resistance. To date, all confirmed cases of fungicide resistance in turfgrass pathogens involve medium- or high-risk modes of action, with DMI, QoI, and SDHI resistance the most common.

Fungal species also vary in their propensity to develop resistance. Factors such as generation time, reproduction rate, and mode of reproduction all influence the likelihood that a pathogen can develop resistance to a particular mode of action. High-risk turfgrass pathogens include *Clarireedia, Colletotrichum,* and *Pyricularia* spp., which comprise the vast majority of documented cases of resistance in turfgrasses. *Pythium* spp. carry a moderate resistance risk, and resistance to metalaxyl or mefenoxam has been documented in *P. aphanidermatum.* Other turfgrass pathogens such as *Rhizoctonia* spp. or fairy ring fungi are at low risk, and to date, no cases of fungicide resistance have been confirmed in these pathogens.

In general, it is recommended to rotate and/or tank mix chemicals with different modes of action as part of a program approach to prevent or delay the development of resistance. Many fungicide labels now include resistance-management recommendations that limit the number of applications per season or the number of sequential applications before rotation to a product with a different mode of action. Additional measures may be necessary for high-risk pathogens when products with high-risk modes of action are used. When high-risk pathogens are targeted, fungicides with high-risk modes of action should not be applied more than once in succession and should be tank mixed with a low- or medium-risk chemistry to prevent the proliferation of resistant strains.

Fungicide application. Turfgrass fungicides are applied as liquid sprays over the turfgrass canopy or as granular carriers that are spread over the turf surface. Formulations designed for spray application include flowables (F), soluble concentrates (SCs), wettable powders (WP), water-dispersible granules (WDG), and various types of emulsions. These products are dissolved in water and applied through vehicle-mounted or handheld nozzles that deliver a known amount of spray solution per unit area. Spray volumes of 40–80 gallons of water per acre (375–750 L/ha) are recommended for most fungicide applications, but higher volumes, 120–160 gallons per acre (1,100–1,500 L/ha), are often used by lawn care operators when mowing heights are 7.5 cm or higher.

Uniform coverage of the turfgrass foliage is important to maximize control of foliar diseases. Nozzle type, droplet size, and spray volume are the most important factors influencing coverage uniformity. Flat fan nozzles that produce medium to coarse droplets generally provide the best foliar coverage. Larger droplets may not provide uniform coverage, especially at lower spray volumes, whereas smaller droplets are prone to drift. Many modern flat fan nozzles employ air induction technology to incorporate air into the spray solution as it passes through the nozzle chamber. This produces larger droplets that are less prone to drift but then shatter into many smaller droplets upon contact with the leaf surfaces. Air-induction nozzles that produce coarse to very coarse droplets generally provide adequate foliar coverage. Specification charts published by nozzle manufacturers provide recommended operating pressures to produce the desired droplet size and spray volume for each type of nozzle.

Most fungicide active ingredients have low solubility and high affinity for soil and organic matter; as a result, they are classified as immobile to moderately mobile in soil. Spray applications may fail to provide acceptable control of root-infecting pathogens unless the active ingredient is moved into the thatch and soil. Soil drenches (applications in 200 gallons of water or more per acre) provide effective root disease control but are impractical in many turfgrass settings. A foliar spray followed by overhead irrigation with 3–6 mm of water is an effective alternative to soil drenches and increases the amount of active ingredient to depths of 2.5–7.5 cm in the soil profile. Application of a wetting agent or soil surfactant may also facilitate movement of fungicide active ingredients into the turfgrass root zone.

Granular fungicides are produced from pellets of wood fibers, corn cobs, clay, limestone, or other minerals that are coated or impregnated with the fungicide active ingredient. These products are broadcast over the turf surface with a rotary spreader or similar implement. Because most of the active ingredient is placed on the thatch and soil surface, granular fungicides are typically most effective against root- and crown-infecting pathogens. Lightweight granular formulations are available that adhere to wet foliage and disperse in water and can provide effective control of foliar diseases.

Physiological effects of turfgrass fungicides. In addition to disease control, fungicides may have physiological effects, either positive or negative, on turfgrass plants. For example, many DMI fungicides exhibit growth-regulating effects on turfgrasses by inhibiting gibberellic acid production in plant cells. These growth-regulating properties may be beneficial or harmful depending on sensitivity of the grass species, presence of environmental stresses, and management factors. Certain DMIs such as difenoconazole, flutriafol, and mefentrifluconazole cause little to no growth regulation and demonstrate excellent safety to cool- and warm-season turfgrasses.

Several fungicide chemistries used for disease management also demonstrate positive effects on tolerance to abiotic stresses. Such "plant health" effects are of increasing interest in production agriculture to maximize yields or reduce inputs in marginally adapted growing environments. In turfgrasses, it is often difficult to differentiate disease control effects from improved stress tolerance because many turfgrass diseases are induced or enhanced by similar stresses. Regardless, chemistries with plant health benefits have become integral parts of disease management programs in many turfgrass settings.

Phosphonate fungicides such as fosetyl-Al and potassium phosphite are used widely to control *Pythium* spp. and certain other pathogens. These products contain various forms of the phosphite ion (PO_3^-), which inhibits the growth of certain pathogens by disrupting phosphate and glucose metabolism. Phosphite also triggers disease resistance responses in treated plants, such as accumulation of phenolic compounds and reactive oxygen species at infection sites. Furthermore, potassium phosphite has been shown to reduce the effects of ultraviolet B radiation in potato and increase expression of other genes involved in stress tolerance.

Some fungicide formulations include copper phthalocyanine pigments, large macrocyclic compounds that are green or blue. These pigments have very low solubility and remain on the leaf surfaces after application. While pigments provide no apparent disease control benefit, they improve the visual appearance of turfgrass immediately after application and repeat applications can improve the color, density, and overall quality of turfgrasses. Pigments have been shown to increase chlorophyll content and photochemical efficiency in treated turfgrasses. They are thought to reflect ultraviolet light and excess visual light energy to reduce production of damaging reactive oxygen species.

The beneficial effects of QoI fungicides on plant physiology have been studied extensively for their potential to increase crop yields. The QoIs inhibit fungal respiration by blocking electron transfer at the cytochrome bc_1 complex in the inner mitochondrial membrane. A similar mode of action in plant mitochondria reduces respiration rates and induces a range of other physiological responses, including increased production of nitric oxide, nitrate reductase, and antioxidants. Beneficial side effects such as reduced leaf senescence, increased chlorophyll content and photosynthetic efficiency, and increased nitrate assimilation have been documented in plants treated with QoI fungicides.

Acibenzolar-*S*-methyl (ASM) is a plant defense activator that is incorporated into certain fungicide formulations. This active ingredient controls or suppresses certain diseases through systemic acquired resistance, which triggers production in plants of pathogenesis-related proteins such as chitinase and glucanase that protect against fungal infection. Improved tolerance to abiotic stresses such as heat and drought has also been reported. ASM increases production of proteins such as rubisco, ATP synthase, heat shock proteins, and dehydrins, which may enhance turf tolerance to abiotic stresses. Turfgrasses treated with ASM also demonstrate reduced transpiration under high vapor pressure deficit conditions.

Nematicides

Fenamiphos, an organophosphate insecticide, was the only nematicide registered for use on turfgrasses for many years. It had broad-spectrum activity against plant-parasitic nematodes and long-term residual effects in the soil such that one application per year provided effective results. Use of this active ingredient in turfgrasses has been phased out because of environmental and human health concerns. While several alternative chemistries are available, they generally have more narrow spectra, are applied at lower rates, and have shorter residual times in the soil compared with fenamiphos. Multiple nematicide applications per season are now often necessary to manage nematodes. Product selection and timing should be guided by the nematode species present and their relative population levels based on laboratory assays.

1,3-Dichloropropene (1,3-D) is a soil fumigant historically used in production agriculture, but it can be applied to established turfgrasses under a Special Local Need (24c) label approved by certain states. Specialized equipment is used to inject a liquid formulation containing 1,3-D into the root zone and then immediately close the injection slits by rolling. Once in the soil, 1,3-D vaporizes and disperses through the root zone to control nematodes on contact. At least 13 mm of irrigation is applied immediately after application to minimize vapor loss and potential injury to the turfgrass foliage.

Other currently available nematicides are formulated for spray application to established turfgrasses. Abamectin is a fermentation byproduct of the soil actinomycete *Streptomyces avermitilis* with activity against nematodes and certain insects. Fluopyram is an SDHI fungicide (Table 7) that also has activity against certain plant-parasitic nematodes through a similar mode of action. Furfural is a byproduct of sugarcane processing with nematicidal activity through an unknown mode of action. These products are typically applied to the turfgrass foliage and followed immediately with 2.5–6 mm of irrigation to move the active ingredient into the root zone. Abamectin has a high affinity for thatch and organic matter, so it is recommended that it be tank mixed with a soil surfactant to improve penetration through the thatch layer.

Focused efforts in nematicide development by the crop protection industry promises to deliver more options for nematode management over the coming years.

Selected References

Anonymous. 2021. FRAC Code List 2021: Fungal control agents sorted by cross resistance pattern and mode of action. Fungicide Resistance Action Committee. https://www.frac.info/docs/default-source/publications/frac-code-list/frac-code-list-2021--final.pdf?sfvrsn=f7ec499a_2

Bartlett, D. W., Clough, J. M., Godwin, J. R., Hall, A. A., Hamer, M., and Parr-Dobrzanski, B. 2002. Review: The strobilurin fungicides. Pest Manage. Sci. 58:649-662.

Clarke, B. B., Vincelli, B., Koch, P., and Munshaw, G. 2020. Chemical Control of Turfgrass Diseases 2020. Univ. Ky. Ext. Publ. PPA-1. http://www2.ca.uky.edu/agcomm/pubs/ppa/ppa1/ppa1.pdf

Cortes-Barco, A., Hsiang, T., and Goodwin, P. 2010. Induced systemic resistance against three foliar diseases of *Agrostis stolonifera* by (2*R*,3*R*)-butanediol or an isoparaffin mixture. Ann. Appl. Biol. 157:179-189.

Crow, W. T. 2020. Nematode Management for Golf Courses in Florida. Univ. Fla. Inst. Food Agric. Sci. Ext. Bull. ENY-008. https://edis.ifas.ufl.edu/pdf/IN/IN12400.pdf

Daniel, R., and Guest, D. 2005. Defence responses induced by potassium phosphonate in *Phytophthora palmivora*-challenged *Arabidopsis thaliana*. Physiol. Mol. Plant Pathol. 67:194-201.

Hutchens, W. J., Gannon, T. W., Shew, H. D., Ahmed, K. A., and Kerns, J. P. 2020. Soil surfactants influence fungicide movement in United States Golf Association putting green soil. J. Environ. Qual. 49:450-459.

Hutchens, W. J., Gannon, T. W., Shew, H. D., and Kerns, J. P. 2019. Effect of post-application irrigation on fungicide movement and efficacy against *Magnaporthiopsis poae*. Crop Prot. 122:106-111.

Jespersen, D., and Huang, B. 2017. Effects of trinexapac-ethyl and Daconil Action (acibenzolar-*S*-methyl and chlorothalonil) on heat and drought tolerance of creeping bentgrass. Crop Sci. 57:S-138-S-146.

Latin, R. 2021. A Practical Guide to Turfgrass Fungicides. 2nd ed. American Phytopathological Society, St. Paul, MN.

Lyr, H., ed. 1995. Modern Selective Fungicides. 2nd ed. Gustav Fischer Verlag, Jena, Germany.

Oka, Y. 2020. From old-generation to next-generation nematicides. Agronomy 10:1387.

Shekoofa, A., Rosas-Anderson, P., Carley, D. S., Sinclair, T. R., and Rufty, T. W. 2016. Limited transpiration under high vapor pressure deficits of creeping bentgrass by application of Daconil Action. Planta 243:421-427.

Stevenson, K. L., McGrath, M. T., and Wyenandt, C. A., eds. 2019. Fungicide Resistance in North America. 2nd ed. American Phytopathological Society, St. Paul, MN.

Walters, D. R., Ratsep, J., and Havis, N. D. 2013. Controlling crop diseases using induced resistance: Challenges for the future. J. Exp. Bot. 64:1263-1280.

(Prepared by L. P. Tredway)

Nontarget Effects of Pesticides

Pesticides are frequently used where high-quality turfgrasses are desired. The effectiveness and target specificity of pesticides is well documented. The impact of a pesticide on any organism or process other than the target organism is called a "nontarget effect." Such effects can be unpredictable and even deleterious, but they are usually outweighed by the benefits provided. Most undesirable nontarget effects can be avoided by proper pesticide selection, proper application timing, and alternation of pesticides. Pesticide applicators should follow all label instructions and precautions to minimize the risk of nontarget effects.

Nontarget effects of herbicides applied to turfgrasses include root injury, which can have a considerable impact on turfgrass diseases, particularly those caused by root pathogens. Preemergence herbicides applied for annual grass control are more likely to injure turfgrass roots than many postemergent herbicides. Prodiamine has been reported to exacerbate the severity of leaf spot caused by *Drechslera dictyoides* and necrotic ring spot, and pendimethalin has been reported to increase the incidence of Drechslera leaf spot. In greenhouse experiments, benefin slightly promoted brown patch and dollar spot in Bermudagrass. Dithiopyr and bensulide have been shown to enhance anthracnose in *Poa annua*.

Root inhibition of turfgrasses by the broadleaf herbicides 2,4-D and dicamba has also been documented, and dicamba, MCPP, and 2,4-D have been reported to enhance leaf spot caused by *Bipolaris sorokiniana*. This fungus is more pathogenic to older, senescing leaves than to young leaves. Hence, the enhancement of the disease by these synthetic auxin herbicides has been attributed to their ability to accelerate leaf senescence.

Organophosphates and carbamates are broad-spectrum pesticides that are used for insect and nematode control in turfgrasses. They also impact nontarget organisms such as pollinators, beneficial insects, and soil microbes. For example, they can reduce populations of beneficial predators and parasites in the turfgrass ecosystem, which can then lead to a resurgence of insect pests. A single application of chlorpyrifos or isofenphos, for example, has been reported to reduce populations of predatory mites, spiders, and insects in turf by 60%. Application of materials toxic to earthworms, such as bendiocarb, carbaryl, and ethoprop, has been linked to thatch accumulation. In contrast, a nematicide application has been linked to a reduction of anthracnose, and the insecticide chlordane has been reported to suppress take-all patch. Modern insecticide and nematicide chemistries, such as diamides, avermectins, or SDHIs, are more specific toward their intended targets and are less likely to impact beneficial organisms. Use of these chemistries where appropriate can minimize the risk of nontarget effects in the management of insect pests or nematodes.

A great deal of information has been amassed on the nontarget effects of fungicides applied to turfgrasses. Among the documented deleterious effects associated with frequent use of certain fungicides are (1) selection of fungicide-resistant biotypes of pathogens (see Disease Management with Synthetic Chemistries), (2) promotion of nontarget diseases, (3) increased thatch buildup, (4) phytotoxicity, (5) disease resurgence, (6) enhanced microbial degradation of fungicides, reducing their effectiveness, and (7) enhanced growth of algae. Fungicides have also been reported to provide some nontarget benefits. For example, propiconazole and triadimefon increase posttransplant rooting and sod strength of *Poa pratensis*. Color enhancement elicited by fosetyl-Al and other fungicides improves turf quality in the absence of disease. Fungicides used to control brown patch, red thread, and stripe smut have been shown to indirectly reduce encroachment by crabgrass (*Digitaria* spp.) by promoting a dense stand of turf.

Fungicides applied to control one disease may increase the severity of other diseases. Over 100 examples of fungicide-induced increases in turfgrass diseases have been reported. Among the best-documented examples of fungicides that cause a significant increase in nontarget diseases are the following: azoxystrobin and flutolanil can increase dollar spot; chlorothalonil can increase summer patch and stripe smut in *Poa pratensis*; thiram can increase stripe smut; thiophanate-methyl may increase crown rust; and iprodione may enhance yellow tuft. The increase in disease is attributed to a disturbance of the delicate balance between antagonistic and pathogenic microorganisms in the turfgrass ecosystem.

Turfgrass managers have observed that some diseases recur more rapidly and severely in turf previously treated with fungicides than in adjacent nontreated areas. Resurgence of dollar spot and brown patch are probably the most common examples. Little is known about the nature and frequency of disease resurgence in turfgrasses.

Generally, phytotoxicity associated with the use of fungicides is not severe. Most phytotoxicity occurs when fungicides are applied to golf putting greens and bowling greens during periods of high-temperature stress. Fungicides formulated as emulsifiable concentrates are most likely to cause a foliar burn when applied in hot weather. A misapplication of ethazole or chloroneb at an excessive rate can be very injurious to greens. Copper-based products and quintozene (PCNB) may yellow turf when applied in warm weather. Quintozene in granular form that is washed away by a rainstorm and accumulates in low areas can also be highly phytotoxic. Furthermore, late winter or spring applications of quintozene for snow mold control can cause yellowing or light brown discoloration if temperatures rise unexpectedly. Chlorothalonil can severely injure some cultivars of creeping red fescue and other fine-leaf fescue species. Repeated applications of certain DMI fungicides, such as myclobutanil, propiconazole, or triadimefon, often elicit a blue green color in leaves and suppress foliar growth in most turfgrass species. High rates and frequent applications of these fungicides are particularly injurious to greens during periods of hot, humid weather. Other DMI fungicides, such as difenoconazole or prothioconazole, do not exert growth regulating effects and demonstrate excellent turf safety under these conditions.

When applied repeatedly, certain fungicides may cause thatch to accumulate. The accumulation of thatch is primarily the result of an increase in the rate of root and rhizome production in fungicide-treated turfgrasses. Thatch buildup has also been attributed to fungicide-induced acidity, which impairs the activity of thatch-degrading microorganisms. Fungicides that are directly toxic to earthworms and thatch-degrading microorganisms also contribute to thatch accumulation. In most reports of thatch accumulation associated with fungicides, the increase in thatch has not been agronomically significant.

The reduced effectiveness of iprodione for dollar spot control on golf courses in Italy has been attributed to enhanced degradation by soil microorganisms. Microorganisms are responsible for breaking down many pesticides in the environment. Some microbial populations can rapidly increase in response to continuous use of certain fungicides from the same chemical class. Microbes use the active ingredients of these fungicides as an energy source. These fungicides are thus degraded more rapidly so that their residual effectiveness diminishes over time. The loss of residual effectiveness of a fungicide may be an indicator that resistant biotypes are building up in the turf. In many cases, however, loss of residual effectiveness is caused by a buildup of populations of microbes that use the fungicide as an energy source.

DMI fungicides may promote the growth of blue-green filamentous algae on golf and bowling greens. The cause of this phenomenon is unknown. Open canopies or low turf density favor algal growth, in part by increasing the penetration of sunlight to the thatch or soil surface. The growth regulating effects of DMI fungicides may cause leaves to grow more upright, thus promoting sunlight penetration to the thatch layer. In contrast, chlorothalonil, copper hydroxide, and mancozeb have been shown to suppress algal growth on greens.

Harmful side effects usually are either isolated events or occur only after repeated application of fungicides belonging to the same chemical class over several years. The harmful effects of fungicides generally are much less important than the control of the diseases for which they were intended.

Selected References

Dernoeden, P. H., Krusberg, L. R., and Sardanelli, S. 1990. Fungicide effects on *Acremonium* endophyte, plant-parasitic nematodes, and thatch in Kentucky bluegrass and perennial ryegrass. Plant Dis. 74:879-881.

Dernoeden, P. H., and McIntosh, M. S. 1991. Disease enhancement and Kentucky bluegrass quality as influenced by fungicides. Agron. J. 83:322-326.

Dernoeden, P. H., Murray, J. J., and O'Neill, N. R. 1985. Non-target effects of fungicides on turfgrass growth and enhancement of red thread. Pages 579-593 in: Proc. Int. Turfgrass Res. Conf., 5th. F. Lemaire, ed. INRA Publications, Versailles, France.

Smiley, R. W. 1981. Nontarget effects of pesticides on turfgrasses. Plant Dis. 65:17-23.

Smiley, R. W., Craven Fowler, M., Kane, R. T., Petrovic, A. M., and White, R. A. 1985. Fungicide effects on thatch depth, thatch decomposition rate, and growth of Kentucky bluegrass. Agron. J. 77:597-602.

(Prepared by R. W. Smiley, P. H. Dernoeden, and B. B. Clarke; revised by L. P. Tredway)

Biological Control

Although pathogens cause disease, turfgrass health is heavily dependent on nonpathogenic microorganisms. In particular, beneficial soil microbes are responsible for increasing the availability of nutrients in the soil, forming symbiotic associations with roots, producing substances stimulatory to plant growth, and protecting plants against infection from pathogens. Biological control attempts to take advantage of these microbial attributes to limit the activity of plant pathogens. This may be achieved by the introduction of microbial antagonists into the soil and on plant parts or the manipulation of the microbial community that is already present in the turfgrass ecosystem. The goal of biocontrol is to reduce pathogen activity by suppressing inoculum production, protecting plant surfaces from infection, and/or inducing the plant's natural defense against disease. For example, the application of composts or other well-decomposed sources of organic matter to turf may reduce pathogen activity by introducing large populations of antagonistic microorganisms into the soil or by stimulating the activity of the native soil microbial community. Cultural practices may also indirectly affect disease severity by changing the environment so that it favors microbial antagonists rather than plant pathogens.

Some of the more commonly studied microbial antagonists include members of the fungal genera *Coniothyrium, Gliocladium,* and *Trichoderma* and the bacterial genera *Bacillus, Enterobacter, Paenibacillus, Pseudomonas, Stenotrophomonas,* and *Streptomyces.* These microorganisms interfere with pathogens through the production of antibiotics (antibiosis), parasitism, competition, or by inducing the plant's natural defense against infection (a process called "induced systemic resistance"). Microbial antagonists that produce antibiotics may destroy pathogen propagules, prevent the formation of inoculum, or displace the pathogen from the infection court. Some antagonists such as *Pseudomonas, Bacillus, Enterobacter,* and *Gliocladium* spp. produce antibiotics inhibitory to pathogen growth and development. Many soilborne and plant-associated microorganisms can colonize aboveground and belowground plant parts and effectively protect the plant from infection.

Antagonists such as *Trichoderma* and *Coniothyrium* spp. parasitize pathogen propagules and mycelium, whereas *Stenotrophomonas, Lysobacter,* and *Streptomyces* spp. produce chitinases and glucanases that break down the cell walls of susceptible pathogens. In addition, some species of *Pseudomonas* and *Enterobacter* are effective competitors with plant pathogens, thereby preventing them from colonizing and infecting the plant host. Other biocontrol agents, including *Bacillus subtilis* strains QST 713 and GB03, induce systemic resistance in turf to suppress dollar spot, brown patch, and summer patch when applied on a preventive basis.

Nearly all microorganisms that are antagonistic to turfgrass pathogens are capable of surviving on dead and decaying organic matter. Turf swards that are not intensively managed, such as turf on home lawns, parks, and cemeteries, normally support adequate populations of antagonists. However, intensively managed turfs such as on golf courses and athletic fields often utilize methods that discourage the development of a large and diverse microbial community. For example, the low organic matter content in sand-based root zones can reduce microbial populations. In addition, some studies suggest that the repeated use of broad-spectrum pesticides that persist in the environment may further reduce the population of antagonists. Other studies, however, have shown that pesticide applications do not adversely affect general microbial activity but may reduce the population of specific microorganisms to varying degrees in turfgrass stands.

To take full advantage of the benefits of antagonistic microorganisms and successfully manage turfgrass diseases with reduced fungicide inputs, an abundant and diverse microbial community must be maintained in turfgrass soils. Often this requires the addition of organic matter to the soil to support microbial growth. Some of the best sources of organic matter and antagonistic microbes can be found in composts and natural organic fertilizers. During composting, organic matter is broken down under controlled temperature and aeration to form a uniform and nutritive material after stabilization. Composting is fundamentally a microbial process involving the succession of microorganisms during various phases of organic matter decomposition. It is only after decomposition slows that a more stable and disease-suppressive microflora develops. Some studies have shown that, when applied as a topdressing, composts and organic fertilizers can suppress diseases such as brown patch, dollar spot, Pythium blight and root rot, and red thread, whereas others have found little or no evidence that natural organic fertilizers or composts reduce disease by improving general soil microbial levels. Indeed, some composted sewage sludges and manures have been shown to intensify diseases such as dollar spot. Such a wide range in response to disease should not be surprising, since composts of different origins can vary in disease suppressive properties and the spectrum of diseases controlled. This is primarily the result of the variability in microbial populations among different composts, which can vary in the quality of organic matter and the stage of decomposition.

Studies have shown that the application of specific antagonists can reduce the severity of turfgrass diseases. For example, applications of the fungal antagonist *Typhula phacorrhiza* have been reported to reduce the severity of gray snow mold by 74%, a level of control equivalent to that of the standard fungicide quintozene (PCNB). In another study, a reduction in brown patch severity of up to 70% was observed with the application of nonpathogenic, binucleate *Rhizoctonia*-like fungi. Moreover, strains of *Enterobacter cloacae* were found to effectively suppress dollar spot in the field and Pythium blight and brown patch in greenhouse trials. In *Agrostis* spp. turfs, fluorescent *Pseudomonas* spp., isolates of *Phialophora* spp., and *Gaeumannomyces graminis* have been reported to control take-all patch in greenhouse trials. Moreover, species of the fungal antagonist *Trichoderma* were found to effectively suppress brown patch but were ineffective against southern blight. In field studies, *Pseudomonas aureofaciens* strain Tx-1, *Bacillus licheniformis,* and strains of *Bacillus subtilis* have been shown to effectively control dollar spot when disease pressure is low, whereas *Lysobacter enzymogenes* (previously reported as *Stenotrophomonas maltophilia*) reduced brown patch severity as well as leaf spot caused by *Bipolaris sorokiniana.* In general, however, levels of disease suppression with microbial inoculants have not been consistent, and introduced microbial antagonists have often failed to reduce the severity of turf diseases from region to region. Such variation in the performance can be related to the turfgrass species and cultivars grown, the timing and frequency of applications, the formulation of the inoculants, environmental conditions, and other factors.

Several microbial-based products (biofungicides) are currently sold for the prevention of turfgrass diseases. While these products can often suppress turfgrass diseases alone under low to moderate pressure, they have been most effective when used in combination with reduced rates of compatible synthetic fungicides. If turfgrass managers are to successfully utilize biofungicides to prevent disease, scientists will need to improve their understanding of the biology and ecology of the biocontrol agents in these products. To be effective, the microbial antagonists must be able to be applied before disease develops, survive in turfgrass ecosystems, and remain at high population levels in the soil or on the plant. Additionally, biological control agents, including endophytic fungi (see Endophytic Fungi), must be compatible with management inputs. In particular, they must be tolerant of the fungicides and cultural practices currently used in turf management programs. Just as pathogens are influenced by environmental conditions, so too are biological control agents. Turfgrass managers, therefore, must learn not only how to manage the pathogen, but also how to manage and encourage the associated antagonistic microflora if biocontrol is to be effective.

Selected References

Coelho, L., Reis, M., Guerrero, C., and Dionísio, L. 2021. Biological control of turfgrass diseases with organic composts enriched with *Trichoderma atroviride.* Biol. Control 159. doi.org/10.1016/j.biocontrol.2021.104620

Cook, R. J., and Baker, K. F. 1983. The Nature and Practice of Biological Control of Plant Pathogens. American Phytopathological Society, St. Paul, MN.

Davis, J. G., and Dernoeden, P. H. 2002. Dollar spot severity, tissue nitrogen and soil microbial activity in bentgrass as influenced by nitrogen source. Crop Sci. 42:480-488.

Hoitink, H. A. J., and Fahy, P. C. 1986. Basis for the control of soilborne plant pathogens with composts. Annu. Rev. Phytopathol. 24:93-114.

Jeon, S. W., Naing, K. W., Lee, Y. S., Nguyen, X. H., Kim, S. J., and Kim, K. Y. 2015. Promotion of growth and biocontrol of brown patch disease by inoculation of *Paenibacillus ehimensis* KWN38 in bentgrass. Hortic., Environ. Biotechnol. 56:263-271.

Kobayashi, D. Y., and El-Barrad, N. 1996. Selection of bacterial antagonists using enrichment cultures for the control of summer patch disease of Kentucky bluegrass. Curr. Microbiol. 32:106-110.

Köhl, J., Kolnaar, R., and Ravensberg, W. J. 2019. Mode of action of microbial biological control agents against plant diseases: Relevance beyond efficacy. Front. Plant Sci. 10:845.

Lawton, M. B., and Burpee, L. L. 1990. Effect of rate and frequency of application of *Typhula phacorrhiza* on biological control of Typhula blight of creeping bentgrass. Phytopathology 80:70-73.

Liu, L. X., Hsiang, T., Carey, K., and Eggens, J. L. 1995. Microbial populations and suppression of dollar spot disease in creeping bentgrass with inorganic and organic amendments. Plant Dis. 79:144-147.

Lo, C.-T., Nelson, E. B., and Harman, G. E. 1997. Improved biocontrol efficacy of *Trichoderma harzianum* 1295-22 for foliar phases of turf diseases by use of spray applications. Plant Dis. 81:1132-1138.

Nelson, E. B., Burpee, L. L., and Lawton, M. B. 1994. Biological control of turfgrass diseases. Pages 409-428 in: Handbook of Integrated Pest Management for Turf and Ornamentals. A. Leslie, ed. CRC Press, Boca Raton, FL.

Nelson, E. B., and Craft, C. M. 1991. Introduction and establishment of strains of *Enterobacter cloacae* in golf course turf for the biological control of dollar spot. Plant Dis. 75:510-514.

Nelson, E. B., and Craft, C. M. 1992. Suppression of dollar spot on creeping bentgrass and annual bluegrass turf with compost-amended topdressings. Plant Dis. 76:954-958.

Palumbo, J. D., Yuen, G. Y., Jochum, C. C., Tatum, K., and Kobayashi, D. Y. 2005. Mutagenesis of β-1,3-glucanase genes in *Lysobacter enzymogenes* strain C3 results in reduced biological control activity toward Bipolaris leaf spot of tall fescue and Pythium damping off of sugar beet. Phytopathology 95:701-707.

Powell, J. F., Vargas, J. M., Jr., Nair, M. G., Detweiler, A. R., and Chandra, A. 2000. Management of dollar spot on creeping bentgrass with metabolites of *Pseudomonas aureofaciens* (TX-1). Plant Dis. 84:19-24.

Thompson, D. C., Kobayashi, D. Y., and Clarke, B. B. 1998. Isolation of rhizosphere competent bacteria for the suppression of summer patch disease on Kentucky bluegrass, and their establishment in field turfgrass. Soil Biol. Biochem. 30:257-263.

Wong, P. T. W., and Baker, R. 1984. Suppression of wheat take-all and Ophiobolus patch by fluorescent pseudomonads from a *Fusarium*-suppressive soil. Soil Biol. Biochem. 16:397-403.

(Prepared by R. W. Smiley, P. H. Dernoeden, and B. B. Clarke; revised by D. Kobayashi and B. B. Clarke)

Part IV. Disease Diagnosis

Many of the diseases that affect turfgrasses have similar signs and symptoms, often making it difficult to diagnose the causal agent. In addition, an incorrect diagnosis can lead to delays in managing the disease, potentially resulting in significant damage to the turfgrass sward and increased fungicide use. Diagnosis of turfgrass maladies requires knowledge of grass species and their responses to environmental conditions, knowledge of infectious and noninfectious agents, keen observation skills, and the ability to accurately interpret all accumulated evidence. Although the majority of turfgrass diseases are caused by fungi, bacteria, chromists (such as *Pythium, Labyrinthula,* and *Sclerophthora* spp.), viruses, and nematodes can also infect turf (see Diseases and Their Causal Agents in the introduction to this compendium). Moreover, abiotic disorders can mimic diseases caused by biotic agents and must be considered when a diagnosis is attempted. An effective diagnostic approach follows a logical sequence of steps, and considerable experience is necessary to achieve a high degree of competency in this process. This section provides guidelines for disease diagnosis. It must be emphasized, however, that combinations of pathogens are the rule rather than the exception and that disease symptoms can vary greatly in different grass species and under different environmental conditions.

Disease in turfgrasses, as in other plants, involves interaction among a susceptible grass species (i.e., a host), a pathogen, and an environment favorable for disease development. These contributing factors must be considered when one is attempting to make an accurate disease diagnosis.

The Art and Science of Disease Diagnosis: A Stepwise Approach

Step 1. Identify the Affected Grass Species

Many turfgrass pathogens preferentially attack certain species or cultivars of grass. Knowing the grass species can dramatically reduce the list of possibilities and enhance the likelihood of an accurate diagnosis. Therefore, it is important to identify the grass species being affected (see Grasses Managed as Turfs in the introduction). In a mixed stand, it is also important to note whether the problem affects all species present in the turf sward or is selectively affecting a particular grass. Moreover, it can be helpful to establish the identity of the affected cultivars by referring to records of the seed or sod mixture that was used to establish the turf.

Step 2. Observe the Symptoms

The appearance of disease symptoms on affected plants is an important source of diagnostic information. There are two approaches to assessing symptoms in diseased turfgrass areas: (1) observe symptoms in the overall stand of grass and (2) observe symptoms on individual plants.

Visible disease patterns may be present in the turfgrass stand. These symptoms can vary and are of great importance in disease diagnosis. For example, are symptomatic plants uniformly scattered across an area, or are they confined to specific locations such as low-lying wet areas or high and dry areas? Injury from disease may appear as small circular spots, irregular concentrated patches, or large rings or circles, or it may lack a defined pattern. Some diseases invariably cause rings or circles in a stand of turf, while others cause more diffuse symptoms. Occasionally, injury appears in bands, streaks, or other patterns that suggest the involvement of equipment or chemical agents. Thatch and soil should be inspected closely for potential insect pests, and records should be checked to see whether excessive rates of fertilizer, plant growth regulators, or pesticides have been applied.

Individual plant symptoms may include leaf spots, lesions, or blighting as well as wilt, yellowing, stunting, and stem or root rots. They are generally less variable than symptoms that appear over the entire stand. It is essential to carefully observe the appearance of symptoms on the affected part of individual plants. Leaf spots can be highly diagnostic for some diseases such as dollar spot and Bipolaris leaf spot, since they are often distinctive in color, size, and shape. Leaf blighting is distinct from leaf spots. Leaf-blighting pathogens typically produce a brownish necrosis or rot of leaf or sheath tissue without the defined border that is characteristic of leaf spots. The size, shape, and extent of the blight can provide valuable clues and may involve the entire leaf blade.

Step 3. Observe the Environmental Conditions

The cultural and environmental factors present during the onset of disease are an additional source of diagnostic information. Some diseases are more severe in wet areas, while others tend to develop more often in dry soils or when the plant is under drought stress. In addition, some diseases are more common in full sun, whereas others are more severe in shady locations where the leaf wetness period is extended. The temperature of the air and soil, light intensity and duration, and moisture conditions just prior to and during disease development can also provide critical diagnostic information. The condition of the affected site can further aid the diagnostic process. Air and water drainage, soil conditions (texture, pH, and moisture level), sun and shade, slope of the terrain, age of the stand, and proximity to other plantings or structures may all be significant factors in turf disease development. Prior chemical applications including plant growth regulators, pesticides, and fertilizers to the site may also provide important clues for disease diagnosis.

Management practices can have a significant impact on turfgrass diseases. Mowing practices including the height and frequency of cut can affect disease severity directly by wounding tissue or indirectly by influencing the duration of leaf wetness or plant health. Fertilization, topdressing, and irrigation practices can also affect disease development and should be taken into account when attempting to diagnose a turfgrass disease. Moreover, practices that stress or weaken turf commonly trigger or amplify stress-induced diseases like anthracnose, dollar spot, red thread, and certain rusts.

Step 4. Identify Visible or Microscopic Signs of the Pathogen

The appearance of the causal agent can be a critical and important aid to disease diagnosis. Spores, mycelium, and fruiting structures of some turfgrass pathogens may be visible without magnification. The fungi that cause smut, powdery mildew, and rust diseases produce spores that may accumulate to the point that they are visible on grass foliage. The presence of the pink, antler-like stromata of the fungus that causes red thread is another example of a significant diagnostic sign of disease. Removal of leaves and leaf sheaths associated with stems and stolons may reveal additional signs such as mycelium, infection mats, or sclerotia. When the pathogen is visible, its appearance and presence are often the most important clues in the diagnostic process.

A 10–20× hand lens may assist in identifying the cause of a disease. Mechanical injury and insect feeding can be distinguished from leaf damage caused by infectious agents when viewed with a hand lens. Fungal fruiting bodies, such as acervuli, sporodochia, pycnidia, or perithecia, may also be present and visible at low magnification.

Visualizing signs on leaf and stem tissue. When a pathogen is not visible with the naked eye, microscopic examination of tissues exhibiting symptoms may provide valuable diagnostic information. In Table 8, important morphological characteristics of fungal pathogens of turfgrasses are provided to aid in the diagnosis of diseases based on spores and hyphae. Although the identification of pathogens by use of these characteristics is by no means foolproof, the table should be helpful when used in conjunction with visible symptoms and recent environmental conditions. There is a lot more to accurately identifying plant pathogens than relying solely on microscopic observations. However, they can add to the overall body of diagnostic information and should increase the chances of accurately identifying turfgrass diseases. If microscopic examination does not reveal the presence of mycelium, spores, or fungal structures, tissues exhibiting symptoms should be incubated for 24–48 hours in a moist chamber. A high-humidity environment often induces sporulation or hyphal proliferation that includes not only the causal agent, but also saprophytes and inactive pathogens; therefore, it is important that tissue not be allowed to sit in

TABLE 8. Guide to Microscopic Identification of Selected Genera of Fungi and Fungus-like Organisms[a]

A. Organisms That Form Spores and Fruiting Structures on Leaves or Stems

	Spores Produced Freely on Condiophores		Spores Produced in:			
Spore Character	**Light-Colored Sporulation**	**Dark-Colored Sporulation**	**Pycnidium**	**Perithecium or Pseudoperithecium**	**Sporodochium**	**Acervulus**
Number of cells						
One	*Ramularia* *Blumeria*	*Nigrospora*	*Pseudoseptoria* *Phyllosticta*	*Phyllachora*	*Lectera*	*Colletotrichum*
Two	*Rhynchosporium*	*Cladosporium*	*Ascochyta* *Didymella*			
One to three	*Graminopassalora* *Pyricularia*					
Three to 10	*Fusarium* *Mastigosporium*	*Bipolaris* *Curvularia* *Pyrenophora* *Exserohilum*			*Fusarium*	
One to many	*Cercospora* *Spermospora* *Catenulocercospora*		*Septoria* *Stagonospora*		*Microdochium*	
Muriform cells (cross-walls in two or three axes)		*Alternaria*		*Gaeumannomyces* *Leptosphaerulina* *Magnaporthiopsis* *Ophiosphaerella*		

B. Organisms That Do Not Form Easily Identified Spores

				Hyphopodia	
Mycelium Color[b]	**Nonseptate Hyphae**	**Septate Hyphae with Clamp Connections[c]**	**Septate Hyphae without Clamp Connections**	**Simple**	**Lobed**
Colorless or white, cottony	*Pythium*				
Variable			All fungi in part A	*Gaeumannomyces*	*Candidacolonium* *Gaeumannomyces* *Magnaporthiopsis*
Brown			*Rhizoctonia* *Waitea*		
White, cobwebby			*Clarireedia* *Sclerotinia*		
Gray to white		*Typhula* *Coprinopsis* *Ceratorhiza* *Athelia*			
Pink		*Limonomyces*	*Laetisaria*		

[a] Prepared by R. W. Smiley, P. H. Dernoeden, and B. B. Clarke; revised by M. Tomaso-Peterson—© APS.

[b] Viewed en masse.

[c] Clamp connections are small "bridges" around each septum in the hyphae.

a moist chamber for extended periods and to look for emerging signs that may aid in the diagnosis.

Mycelium or spores produced on infected tissue can also be mounted on a glass slide for microscopic examination. Reproductive structures that are sometimes present in leaf or stem lesions include pycnidia, perithecia, or pseudothecia. When present, these structures should be removed from plant tissue, mounted intact on a slide, and gently crushed by applying pressure to the coverslip. The identity of the pathogen may also be determined on the basis of the color, size, shape, and number of septa of mature spores.

If fruiting bodies are present within the leaf or stem tissue, it may be necessary to cut thin cross-sections of these structures to observe their morphology and accompanying spores. To do this, place the spore-bearing surface of the plant facing upward and cut very thin slices with a razor blade using a forward sliding motion across the fruiting structures. Make multiple sections and place the thinnest on a slide for viewing. This technique is particularly useful for fungi that produce spores in acervuli (*Colletotrichum cereale*), pycnidia (*Septoria* and *Neoascochyta* spp.), and perithecia (*Gaeumannomyces, Leptosphaerulina, Magnaporthiopsis,* and *Ophiosphaerella* spp.). The color, size, and shape of a reproductive structure can be very helpful in the identification of the causal agent. If a bacterial disease is suspected, leaves or roots should be cut to see whether rapid streaming of bacteria emerges from the vascular tissue.

Visualizing signs on roots. The techniques described above are particularly useful when attempting to diagnose foliar diseases of turfgrasses. However, because turf soils and thatch abound with saprophytic fungi, it may be more difficult to determine the cause of disease on subterranean plant parts. Helpful diagnostic techniques include washing roots, rhizomes, or stolons and preparing thin sections of the affected tissue to view under a microscope. In the case of root diseases caused by *Pythium* spp., stems and roots are often stained and tissues microscopically inspected for the presence of coenocytic (nonseptate) mycelium and oospores. To determine the population of parasitic nematodes in a turf stand, soil must be collected from the root zone of symptomatic turf and sent to a diagnostic laboratory for analysis. It may also be informative to send soil from the root zone of a healthy stand of turf to compare populations of plant-parasitic nematodes. This can help determine whether the symptoms are associated with high nematode levels.

Other diagnostic techniques for root diseases include isolating pathogens on culture media, treating root or stem tissues with chemicals to render them transparent, applying stains to selectively color fungi, and growing susceptible plants in infested soil to "trap" a pathogen in emerging roots or shoots. In addition, molecular techniques and immunochemical assays are available for the identification of plant pathogens; however, these techniques require specialized equipment and considerable skill on the part of the diagnostician. Such expertise is available in many university and private diagnostic laboratories.

If pathogens are not observed on or in the affected tissue, it is possible that the observed symptoms result from abiotic causes such as heat or moisture stress, atmospheric pollution, mechanical injury, or phytotoxicity. See Distinguishing Abiotic and Biotic Issues in Turfgrasses for a review of the process for differentiating between abiotic and biotic causes of diseases.

Step 5. Verify a Diagnosis: Collection and Submission of Samples

It is often helpful, particularly for those who are inexperienced at identifying turfgrass diseases, to send a sample of the affected turf to a plant diagnostic laboratory to confirm the diagnosis. Submitted samples should come from active disease areas and be representative of the entire stand. Moreover, samples should be obtained prior to the application of fungicides or other products that might inhibit growth of the pathogen. Samples should always be collected from the edge of an infection center and include both healthy and infected plants. Samples should never be taken from the dead centers of affected patches or from old infection centers, because they are of little value to the diagnostician. Samples should be at least 10 cm in diameter and include soil to a depth of 6 cm. Smaller samples taken with a soil probe are generally inadequate and will likely yield inconclusive or inaccurate results.

After collection, the roots are covered with aluminum foil to prevent the soil from dislodging and contaminating the foliage. Samples should be packed tightly in a sturdy box to prevent movement and contamination of foliage with soil particles during shipment. *Note:* Do not put samples in watertight wrapping (plastic bags) or cover foliage with wet paper towels, because this leads to a rapid deterioration of samples and proliferation of saprophytes or inactive pathogens. Turf samples should be kept out of the sun and not allowed to dry out prior to shipment.

The diagnostician who receives turf samples will not have an opportunity to observe the disease in the field. Therefore, it is often helpful to send digital images or photos to provide additional clues. It is also essential that the following information accompanies the sample submission form: (1) identification of the grass species affected and a description of the extent of the damage; (2) a description of the cultural and environmental conditions of the affected area; (3) a complete and accurate description of the symptoms, including the date of first appearance; and (4) a list of all plant health care products, including fungicides, that have been applied recently with the dates of application.

Samples should reach a diagnostician in the same condition they were in in the field. Thus, rapid delivery of samples to the laboratory is essential. If personal delivery of a sample is impractical, it should be shipped by using the quickest carrier available. Priority mail or overnight delivery services are valuable for this purpose. Moreover, samples should not be sent at the end of the week (Friday) when they may remain undelivered over the weekend and deteriorate.

Advances in Disease Diagnostics

The traditional methods of disease diagnosis described above rely on visual symptoms and signs, incubation, culturing, and microscopy. While this approach can be used to effectively diagnose many turfgrass diseases, it has limitations. Symptoms can be nondescript, many diseases look alike in the field, and symptoms may vary with turf species and environment. Similarly, not all pathogens produce signs, some are not easily culturable, and others, like viruses, are too small to be seen with a light microscope.

Advances in molecular techniques and immunochemical assays provide diagnosticians with additional tools to help diagnose plant diseases. However, in most cases, such tools serve as complements to traditional diagnostic methods rather than alternative approaches. Most molecular and immunochemical assays are laboratory based and still require a trained specialist to narrow down the list of possible causal agents before an assay can be selected. Moreover, once a potential pathogen is identified, someone with extensive knowledge of pathogen biology is still required to determine whether the organism detected is the primary pathogen or just another member of the turf microbial community. Nevertheless, these assays have tremendous potential to enhance traditional disease diagnostic capabilities. A review of the molecular tools currently available to diagnose turfgrass diseases is provided by Stackhouse et al.

Serological techniques. Serological techniques such as enzyme-linked immunosorbent assays (ELISAs) and immunostrip testing have been used since the 1990s to identify turfgrass diseases caused by *Pythium, Rhizoctonia,* and *Clarireedia* spp. Many can be used in the field by practitioners as well as diagnosticians in the laboratory. They are also cost effective but typically are not as sensitive as many of the nucleic acid technologies, such as conventional polymerase chain reaction (PCR), multiplex PCR, quantitative PCR (qPCR), RAPD-PCR, and DNA sequencing.

Nucleic acid-based techniques. Conventional PCR is a common and effective method for identifying plant pathogens and has been used in turf since the early 1990s. Although PCR started primarily as a research tool, it is now frequently used in plant diagnostic laboratories. This method utilizes short sequences of DNA that are unique to a specific pathogen to make an identification. PCR has been used in turf for the detection of anthracnose, dead spot, selected fairy ring pathogens, gray leaf spot, necrotic ring spot, *Pythium* diseases, and take-all patch. Multiplex PCR is more convenient than conventional PCR and allows for several pathogens to be identified at one time in one reaction. There are not many multiplex PCR assays currently in use as diagnostic tools because they can be difficult to develop, but they have been used for the detection of *Pythium* spp. and *Gaeumannomyces graminis* in turf.

Real-time PCR (qPCR) is an extremely sensitive method that allows for quantitative analysis of PCR reactions in real time. It has been used to diagnose turfgrass diseases caused by *Clarireedia* spp., *Magnaporthiopsis poae, Rhizoctonia solani, Puccinia* spp., and *Acidovorax avenae,* one of the causal agents of bacterial etiolation. RAPD-PCR has been used in the past for turf disease diagnosis, but its use has declined with the development of other, more robust, molecular methods.

Genome sequencing is a powerful research tool that is very useful for the identification of emerging pathogens and organisms that cannot be cultured. Currently, most genome sequencing is generated with next-generation sequencing (NGS) technologies. NGS is just starting to be used for diagnosing turfgrass diseases such as dollar spot caused by *Clarireedia* spp.

Isothermal techniques. Isothermal diagnostic techniques came into use for turf disease diagnosis in the early 2000s. They include loop-mediated amplification (LAMP) and recombinase polymerase amplification (RPA). LAMP is one of the newest rapid disease diagnostic techniques in turf and can be used in the field. LAMP assays are more sensitive and rapid than conventional PCR and have been developed to detect bacterial wilt (*Xanthomonas translucens*), brown patch (*Rhizoctonia solani*), gray leaf spot (*Pyricularia oryzae*), necrotic ring spot (*Ophiosphaerella korrae*), summer patch (*Magnaporthiopsis poae*), and take-all patch (*Gaeumannomyces avenae*), as well as root-knot nematodes (*Meloidogyne* spp.) in turfgrasses. The RPA method has been used in turf for the identification of ectotrophic root-infecting pathogens such as *Gaeumannomyces avenae*, *Ophiosphaerella korrae,* and *Magnaporthiopsis poae.* They are extremely sensitive assays but have had issues with nonspecificity, especially with closely related species.

Future applications of other technologies. Several additional technologies are at the forefront of disease diagnostics and are likely to be of increasing importance in the future. They include portable high-throughput sequencing units, remote sensing, and smartphone applications for disease identification. The phone-size MinION genome sequencer is an example of how high-throughput sequencing can be used for rapid sequencing and pathogen identification in the laboratory or field. This technology can detect more than one pathogen in a sample and allows for testing without any prior information about the potential causal agent. High-throughput sequencing has been used for on-site disease detection in other plants but is only now being developed for use in turf.

Remote sensing is another technology that may prove useful for disease diagnosis in the field. In remote sensing, hyperspectral imaging is used to identify diseases and plants under stress by detecting differences in visible, near-infrared, and shortwave infrared light. Spectroradiometers can be handheld or mounted on drones for use in the field. The goal is to use remote sensing to predict diseases in the field before symptoms become apparent.

Finally, smartphone applications continue to be developed and refined for a range of applications in turfgrass science including assessing plant health and diagnosing and predicting turfgrass diseases. Many of these applications make use of disease images and weather data to help identify and predict diseases in the field. Other applications provide chemical control and cultural management information via online resources and GPS technology to help turf managers control diseases after a diagnosis is made. As these programs continue to evolve, they will likely incorporate new molecular technologies and artificial intelligence into smartphones to further enhance field-based, disease diagnostic, and predictive capabilities. However, as useful as these new technologies become, they will complement rather than replace traditional approaches to disease diagnostics that incorporate experience and sound judgment to accurately identify turfgrass pathogens.

Selected References

Agrios, G. N. 2005. Plant Pathology. 5th ed. Academic Press, San Diego, CA.

Barnett, H. L., and Hunter, B. B. 1998. Illustrated Genera of Imperfect Fungi. 4th ed. American Phytopathological Society, St. Paul, MN.

Couch, H. B. 1995. Diseases of Turfgrasses. 3rd ed. Krieger Publishing, Malabar, FL.

Kammerer, S. 2012. Turfgrass disease diagnosis. Green Sec. Rec. 50(4):1-4.

Riley, M. B., Williamson, M. R., and Maloy, O. 2002. Plant disease diagnosis. Plant Health Instr. doi:10.1094/PHI-I-2002-1021-01

Shurtleff, M. C., and Averre, C. W. 1997. The Plant Disease Clinic and Field Diagnosis of Abiotic Diseases. American Phytopathological Society, St. Paul, MN.

Smith, J. D., Jackson, N., and Woolhouse, A. R. 1989. Fungal Diseases of Amenity Turf Grasses. E. & F. N. Spon, London.

Stackhouse, T., Martinez-Espinoza, A. D., and Ali, E. E. 2020. Turfgrass disease diagnosis: Past, present, and future. Plants (Basel) 9(11):1544.

Stowell, L. D., and Gelernter, W. D. 2001. Diagnosis of turfgrass diseases. Annu. Rev. Phytopathol. 39:135-155.

(Prepared by R. W. Smiley, P. H. Dernoeden, and B. B. Clarke; revised by B. B. Clarke and J. Crouch)

Distinguishing Abiotic and Biotic Issues in Turfgrasses

Turfgrasses can be damaged by various abiotic and biotic issues. Distinguishing between the two can be challenging and without prior knowledge of turfgrass culture can lead to incorrect diagnoses of disease. There are, however, important factors turfgrass managers and diagnosticians should consider. In addition to following the diagnostic procedures outlined in the previous section, turfgrass managers and diagnosticians should take the following steps to determine whether or not they are dealing with abiotic damage.

Step 1. Observe Symptom Patterns

Many turfgrass diseases express stand symptoms in some form of a circular or ringlike pattern. Although spots are defined as being different from patches, circles, and rings, they are all essentially circles or rings of varying sizes. Great examples are diseases like spring dead spot, fairy ring, brown patch, and dollar spot. However, pathogens that disseminate spores tend to produce irregular symptom patterns as incidence and severity increase and may lead to confusion during the diagnostic process. Good examples of diseases that develop in this manner are rust, gray leaf spot, and Pythium blight.

Most turfgrass disease outbreaks are random in their distribution. If an entire property or every putting green is affected, the odds that it is caused by a biotic factor are very low. One of the most important things a turfgrass manager or diagnostician must do during the diagnostic process is to look at the overall symptom pattern and make careful observations as to whether the damage is linear or nonlinear. Disease diagnosis is far more accurate when photos accompany the samples so that the diagnostician can observe the pattern and distribution of the symptoms.

Linear damage. In most turfgrass systems, the majority of management practices, such as mowing, fertilization, application of chemicals, and cultivation procedures, are performed in straight lines. If damage or decline occurs in straight lines, it is apparent that the damage is a result of a mishap from a management practice (Fig. 334). This is not a firm rule, however, because diseases can occur in straight lines after aggressive cultural practices that have injured the turf and predisposed those plants to disease. For example, an outbreak of leaf spot in turfgrass that had recently been verticut during a stressful period may produce symptoms in straight lines (that is, along the verticutting channels in the turf canopy).

Nonlinear damage. If the damage is not in a straight-line pattern, there still could be some distinguishing factors that separate the damage from a disease. Examination of the periphery of the damaged area for gradual transitions can identify indicators of biotic problems. Most diseases display a gradual transition from healthy to dead turf. Typically, diseases do not have a distinct line of demarcation between healthy and dead turf. Exceptions to this are diseases like spring dead spot in Bermudagrass and zoysiagrass. Abiotic damage usually affects all the turfgrass plants in the affected area, and regrowth occurs only when the abiotic stress dissipates. In contrast, as diseases progress and pathogens colonize new, unaffected tissue, it is common to see regrowth in the center of stand symptoms. It is also important to make note of odd shapes with angles or repetitive patterns.

There are three key tips for distinguishing abiotic from biotic damage. (1) Biotic diseases progress over time. They start out small and gradually increase in size and severity. (2) In mixed stands of turfgrass species, biotic diseases usually attack only one species and not the other. (3) Weeds are resistant to most turfgrass diseases.

Step 2. The Disease and Diagnostic Triangles

Disease and diagnostic triangles (Fig. 335) can be used when a turf manager is trying to decide whether the damage observed is from something living or from something nonliving. For a disease to occur, all three sides of the disease triangle must occur in the same space and time. For a diagnosis to be correct, plant and stand symptoms must match signs of pathogen activity detected during the diagnostic process and recent management practices, recent weather, and any microclimate issues that may be present must be considered. Since turfgrass systems are typically perennial crops and most turfgrass pathogens are ubiquitous, the most important factor to consider is whether the environment has been favorable for the suspected disease during the time the damage occurred. Based on the host and environmental conditions, a list of potential diseases can be generated. Then through the process of elimination, some, if not all, possible diseases can be ruled out on the basis of the symptoms that are present in the affected stand of turfgrass.

Step 3. Submitting Samples

Solving perplexing turfgrass issues is often aided by facts from data collected by submitting various samples to help rule out possibilities. When troubleshooting any turfgrass issue, turfgrass managers should submit samples when appropriate if any factor may be a contributing problem. Samples can be submitted for soil nutrient analysis, tissue nutrient content, disease diagnosis, nematode quantification, analysis of water quality, or detection of herbicides as part of the diagnostic process. Each of these results can assist in distinguishing whether the damage is the result of an abiotic issue or a biotic issue. Occasionally, it is impossible to determine what caused decline, especially if turfgrass managers wait to engage their local turfgrass extension specialists or diagnostic services.

Fig. 334. Abiotic damage to a lawn. The straight lines indicate mower injury during heat and drought stress. (Courtesy E. L. Butler—© APS)

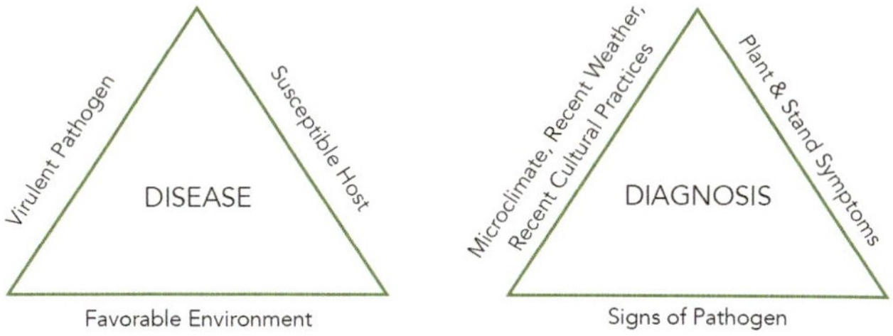

Fig. 335. Factors needed for disease to occur (disease triangle) and factors to consider when making a disease diagnosis (diagnosis triangle). (Courtesy E. L. Butler—© APS)

Selected References

Agrios, G. N. 2005. Plant Pathology. 5th ed. Academic Press, San Diego, CA.

Couch, H. B. 1995. Diseases of Turfgrasses. 3rd ed. Krieger Publishing, Malabar, FL.

Riley, M. B., Williamson, M. R., and Maloy, O. 2002. Plant disease diagnosis. Plant Health Instr. doi:10.1094/PHI-I-2002-1021-01

Shurtleff, M. C., and Averre, C. W. 1997. The Plant Disease Clinic and Field Diagnosis of Abiotic Diseases. American Phytopathological Society, St. Paul, MN.

Smith, J. D., Jackson, N., and Woolhouse, A. R. 1989. Fungal Diseases of Amenity Turf Grasses. E. & F. N. Spon, London.

Stackhouse, T., Martinez-Espinoza, A. D., and Ali, E. E. 2020. Turfgrass disease diagnosis: Past, present, and future. Plants (Basel) 9(11):1544.

Stowell, L. D., and Gelernter, W. D. 2001. Diagnosis of turfgrass diseases. Annu. Rev. Phytopathol. 39:135-155.

(Prepared by E. L. Butler and J. P. Kerns)

Appendix

Diseases of Turfgrasses

The following content is also available on the website of The American Phytopathological Society in the section "Common Names of Plant Diseases." Any updates that are subsequently made to the taxonomy and nomenclature will be reflected on the website.

ABBREVIATIONS OF NAMES OF TURFGRASS SPECIES

A = *Agrostis canina* L. (velvet bentgrass), *A. stolonifera* L. (creeping bentgrass), and *A. capillaris* L. (colonial bentgrass)
Ag = *Agropyron cristatum* (L.) Gaertn. (crested wheatgrass)
Ax = *Axonopus fissifolius* (Raddi) Kuhlm. (common carpetgrass)
B = *Buchloë dactyloides* (Nutt.) Engelm. (buffalograss)
C = *Cynodon dactylon* (L.) Pers. or *Cynodon* spp. hybrids (Bermudagrass)
C3 = Cool-season turfgrass genera [A, Ag, F, L, P, Ph]
C4 = Warm-season turfgrass genera [Ax, B, C, E, Pa, Pe, Pv, S, Z]
E = *Eremochloa ophiuroides* (Munro) Hack. (centipedegrass)
F = *Festuca arundinacea* Schreb. (tall fescue), *F. brevipila* R. Tracey (hard fescue), *F. ovina* L. (sheep fescue), *F. rubra* L. subsp. *fallax* (Thuill.) Nyman (chewings fescue), and *F. rubra* L. subsp. *rubra* (strong creeping red fescue)
L = *Lolium multiflorum* Lam. (Italian "annual" ryegrass) and *L. perenne* L. (perennial ryegrass)
P = *Poa annua* L. (annual bluegrass), *P. pratensis* L. (Kentucky bluegrass), and *P. trivialis* L. (rough bluegrass)
Pa = *Paspalum notatum* Flueggé (bahiagrass)
Pv = *Paspalum vaginatum* Sw. (seashore paspalum)
Ph = *Phleum nodosum* L. (turf timothy)
Pe = *Pennisetum clandestinum* Hochst. ex Chiov. (kikuyugrass)
S = *Stenotaphrum secundatum* (Walter) Kuntze (St. Augustine grass)
Z = *Zoysia matrella* (L.) Merr. (Manila grass), *Z. japonica* Steud. (Korean lawngrass), and *Z. tenuifolia* Willd. ex Thiele (Mascarene grass)

BACTERIAL DISEASES

Bacterial etiolation: A
Pantoea ananatis (Serrano)
Bacterial etiolation and decline: A
Acidovorax avenae Willems
Bacterial wilt: A, C, L, P,
Xanthomonas translucens Vauterin et al.
Bermuda stunting disease, C
Leifsonia xyli subsp. *cynodontis* (Davis et al.) comb. nov. (syn. *Clavibacter xyli* subsp. *cynodontis* Davis et al.)
Halo blight: F, L, P
Pseudomonas coronafaciens (Elliot) F. L. Stevens var. *atropurpurea* (Reddy & Godkin) Stapp

FUNGAL DISEASES

Acremonium snow mold
C3: *Nectria tuberculariformis* (Rehm ex Sacc.) G. Winter (syn. *Acremonium boreale* J. D. Sm. & J. G. N. Davidson)
Anthracnose
Ax: *Colletotrichum axonopodi* J. A. Crouch, B. B. Clarke, J. F. White & B. I. Hillman
E, Z: *Colletotrichum caudatum* (Sacc.) Peck (syns. *Ellisiella caudata* Sacc.; *Ellisiellina caudata* (Peck) Sousa da Câmara)
C3: *Colletotrichum cereale* Manns
E: *Colletotrichum eremochloae* J. A. Crouch & Tomaso-Peterson
Ax: *Colletotrichum hainanense* Wu Zhang & X. L. Niu
Pa: *Colletotrichum paspali* J. A. Crouch, B. B. Clarke, J. F. White & B. I. Hillman
Z: *Colletotrichum zoysiae* J. A. Crouch
A: *Lectera colletotrichoides* (J. E. Chilton) P. F. Cannon (syns. *Volutella colletotrichoides* J. E. Chilton; *Volutella colletotrichoides* var. *setosa* J. E. Chilton)
A, Ag, B, C, F, P: *Microdochium bolleyi* (R. Sprague) De Hoog & Herm.-Nijh. (syns. *Aureobasidium bolleyi* (R. Sprague) Arx; *Gloeosporium bolleyi* R. Sprague; *Idriella bolleyi* (R. Sprague) Arx)
Ascochyta leaf blight
L: *Ascochyta avenae* var. *avenae* (Petr.) R. Sprague & Aar. G. Johnson
F, L: *Ascochyta festucae-erectae* Henn.
C: *Ascochyta graminea* var. *graminea* (Sacc.) R. Sprague & Aar. G. Johnson
P: *Ascochyta hordei* var. *americana* Punith.
Ag, F: *Ascochyta hordei* var. *hordei* Hara
P: *Ascochyta hordeicola* Punith.
A, F, L, Ph: *Ascochyta leptospora* (Trail) Hara (syn. *Ascochyta graminicola* var. *leptospora* Trail)
L: *Ascochyta missouriensis* R. Sprague & Aar. G. Johnson
F, Ph, P: *Ascochyta phleina* R. Sprague
P, Ph, L: *Ascochyta phyllachoroides* f. *melicae* Fautrey
P: *Ascochyta poagena* Punith.
L: *Ascochyta rhodesii* Punith.
A, L: *Neoascochyta desmazieri* (Cavara) Qian Chen & L. Cai (syns. *Ascochyta agrostidis* Polozova; *Ascochyta desmazieri* Cavara)
L: *Neoascochyta graminicola* (Punith.) Qian Chen & L. Cai

Basal leaf blight: Pv
Waitea circinata var. *prodigus* S. J. Kammerer, Burpee & P. F. Harmon
(syns. *Chrysorhiza zeae* (Voorhees) T. F. Andersen & Stalpers; *Rhizoctonia zeae* Voorhees)
Bermudagrass decline and take-all root rot: C4
Candidacolonium cynodontis P. L. Vines & M. Tomaso-Peterson
Gaeumannomyces arxii M. Hern.-Restr. & Crous
Gaeumannomyces avenae (E. M. Turner) M. Hern.-Restr. & Crous
(syns. *Gaeumannomyces graminis* var. *avenae* (E. M. Turner) Dennis; *Ophiobolus graminis* var. *avenae* E. M. Turner)
Gaeumannomyces californicus M. Hern.-Restr. & Crous
Gaeumannomyces floridanus M. Hern.-Restr. & Crous
Gaeumannomyces graminicola M. Hern.-Restr. & Crous
Gaeumannomyces graminis (Sacc.) Arx & D. L. Olivier
(syns. *Gaeumannomyces graminis* var. *maydis* J. M. Yao, Yong C. Wang & Y. G. Zhu; *Ophiobolus graminis* (Sacc.) Sacc.)
Gaeumannomyces hyphopodioides M. Hern.-Restr. & Crous
Gaeumannomyces nanograminis P. L. Vines & M. Tomaso-Peterson
Gaeumannomyces oryzinus (Sacc.) Schrantz.
Gaeumannomyces tritici (J. Walker) M. Hern.-Restr. & Crous
Gaeumannomyces walkeri M. Hern.-Restr. & Crous
Gaeumannomyces wongoonoo P. Wong
Magnaporthiopsis cynodontis P. L. Vines & M. Tomaso-Peterson
Magnaporthiopsis incrustans (Landsch. & N. Jacks.) J. Luo & N. Zhang
(syn. *Gaeumannomyces incrustans* Landsch. & N. Jacks.)
Bipolaris and Exserohilum leaf spot and crown and root rot
Ax: *Bipolaris axonopicola* Y. P. Tan & R. G. Shivas
C: *Bipolaris coffeana* Sivan.
Ax,C, Pe: *Bipolaris cynodontis* (Marignoni) Shoemaker
(syns. *Cochliobolus cynodontis* R. R. Nelson; *Drechslera cynodontis* (Marignoni) Subram. & B. L. Jain; *Helminthosporium cynodontis* Marignoni)
A, C, P: *Bipolaris gigantea* (S. Ito) B. Lane, Stricker, M. E. Sm., S. L. Flory & Harmon
(syns. *Drechslera gigantea* S. Ito; *Helminthosporium giganteum* Heald & F. A. Wolf)
C, Z: *Bipolaris heveae* (Petch) Arx
(syns. *Cochliobolus heveicola* Tsukib. & W. H. Chung; *Drechslera heveae* (Petch) M. B. Ellis; *Helminthosporium heveae* Petch)
C: *Bipolaris maydis* (Y. Nisik. & Miyake) Shoemaker
(syns. *Cochliobolus heterostrophus* (Drechsler) Drechsler; *Drechslera maydis* (Y. Nisik. & Miyake) Subram. & B. L. Jain; *Helminthosporium maydis* Y. Nisik. & Miyake; *Ophiobolus heterostrophus* Drechsler)
Pe: *Bipolaris mediocris* (V. A. Putterill) Shoemaker
(syns. *Drechslera mediocris* (V. A. Putterill) Subram. & B. L. Jain; *Helminthosporium mediocre* V. A. Putterill)
C: *Bipolaris peregianensis* Alcorn
(syn. *Cochliobolus peregianensis* Alcorn)
P: *Bipolaris poae-pratensis* H. Deng & T. Y. Zhang
C, Pe: *Bipolaris sacchari* (E. J. Butler) Shoemaker
(syns. *Bipolaris sacchari* (Breda de Haan) Subram.; *Cercospora sacchari* Breda de Haan; *Drechslera sacchari* (E. J. Butler) Subram. & B. L. Jain; *Helminthosporium ocellum* Faris; *Helminthosporium sacchari* E. J. Butler)
Z: *Bipolaris simmondsii* Y. P. Tan & R. G. Shivas
C3, C4: *Bipolaris sorokiniana* (Sorokin) Shoemaker
(syns. *Cochliobolus sativus* (S. Ito & Kurib.) Drechsler ex Dastur; *Drechslera sorokiniana* (Sacc.) Subram. & B. L. Jain; *Helminthosporium acrothecioides* Lindf.; *Helminthosporium californicum* Mackie & G. E. Paxton; *Helminthosporium sativum* Pammel, C. M. King & Bakke; *Helminthosporium sorokinianum* Sorokin; *Ophiobolus sativus* S. Ito & Kurib.)
C: *Bipolaris stenospila* (Drechsler ex Faris) Shoemaker
(syns. *Drechslera stenospila* Subram. & B. L. Jain; *Helminthosporium stenospilum* Drechsler ex Faris)
Pe: *Bipolaris variabilis* Y. Marín, Y. P. Tan & Crous
C, Pe: *Bipolaris zeae* Sivan.
(syn. *Cochliobolus zeae* H. S. Chang)
C, Pe: *Bipolaris zeicola* (G. L. Stout) Shoemaker
(syns. *Cochliobolus carbonum* R. R. Nelson; *Drechslera carbonum* (Ullstrup) Sivan.; *Drechslera zeicola* (G. L. Stout) Subram. & B. L. Jain; *Helminthosporium carbonum* Ullstrup; *Helminthosporium zeicola* G. L. Stout)
Ag, C, Pe, S, Z: *Exserohilum rostratum* (Drechsler) K. J. Leonard & Suggs
(syns. *Bipolaris halodes* (Drechsler) Shoemaker; *Bipolaris rostrata* (Drechsler) Shoemaker; *Drechslera halodes* (Drechsler) Subram. & B. L. Jain; *Drechslera halodes* var. *elaeicola* Kovachich; *Drechslera rostrata* (Drechsler) M. J. Richardson & E. M. Fraser; *Exserohilum halodes* (Drechsler) K. J. Leonard & Suggs; *Helminthosporium halodes* Drechsler; *Helminthosporium halodes* var. *elaeicola* Kovachich; *Helminthosporium halodes* var. *tritici* Mitra; *Helminthosporium rostratum* Drechsler)
Blister smut: C3
Eballistra lineata (Cooke) R. Bauer, Begerow, A. Nagler & Oberw.
(syn. *Entyloma lineatum* (Cooke) Davis)
Entyloma speciosum J. Schröt & Henn.
Jamesdicksonia dactylidis (Pass.) R. Bauer, Begerow, A. Nagler & Oberw.
(syns. *Entyloma camusianum* Har.; *Entyloma dactylidis* (Pass.) Cif.)
Jamesdicksonia irregularis (Johanson) R. Bauer, Begerow, A. Nagler & Oberw.
(syn. *Entyloma irregulare* Johanson)
Ustilentyloma brefeldii (Krieg.) Vánky
(syns. *Entyloma brefeldii* Krieg.; *Entyloma poae* Liro)
Brown patch: C3
Rhizoctonia solani J. G. Kühn
(syn. *Thanatephorus cucumeris* (A. B. Frank) Donk)
Brown ring patch: A, L, P
Waitea circinata Warcup & P. H. B. Talbot
(syn. *Waitea nuda* Clémençon)
Brown stripe: C3, C4
Graminopassalora graminis (Fuckel) U. Braun, C. Nakash., Videira & Crous
(syns. *Cercosporidium graminis* (Fuckel) Deighton; *Passalora graminis* (Fuckel) Höhn.)
Cephalosporium stripe: L, P
Hymenula cerealis Ellis & Everh.
(syns. *Cephalosporium gramineum* Y. Nisik. & Ikata; *Phialophora cerealis* (Ellis & Everh.) Nirenberg & Dalchow)

Cercosporoid and cercosporoid-like diseases: C3, C4
A, B, S: *Catenulocercospora fusimaculans* (G. F. Atk.) C. Nakash., Videira & Crous
(syns. *Cercospora fusimaculans* G. F. Atk.; *Cercospora panici* Davis; *Cercospora panici-miliacei* Sawada; *Passalora fusimaculans* (G. F. Atk.) U. Braun & Crous; *Phaeoramularia fusimaculans* (G. F. Atk.) X. J. Liu & Y. L. Guo)
A: *Cercospora agrostidis* G. F. Atk.
C: *Cercospora cynodontis* Pavgi & R. A. Singh
F: *Cercospora festucae* Hardison
A: *Passalora agrostidicola* Phengs. & U. Braun
B, C: *Porocercospora seminalis* (Ellis & Everh.) Amaradasa, Amundsen, Madrid & Crous
(syns. *Cercospora seminalis* Ellis & Everh.; *Sporidesmium seminale* (Ellis & Everh.) U. Braun)
Cladosporium eyespot: C, F, P, Ph
Cladosporium phlei (C. T. Greg.) G. A. De Vries
(syn. *Heterosporium phlei* C. T. Greg.)
Copper spot: A, C
Microdochium sorghi (D. C. Bain & Edgerton ex Deighton) U. Braun
(syn. *Gloeocercospora sorghi* D. C. Bain & Edgerton ex Deighton)
Coprinus snow mold: A, Ag, P, F
Coprinopsis psychromorbida (Redhead & Traquair) Redhead, Vilgalys & Moncalvo
(syn. *Coprinus psychromorbidus* Redhead & Traquair)
Cream leaf blight: C, F
Limonomyces roseipellis Stalpers & Loer.
Crown rust: C3
Puccinia coronata Corda
(syn. *Solenodonta coronata* (Corda) Syd.)
Curvularia blight, ink spot (dog's footprint), leaf spot, stem and crown rot: C3, C4
F, Z: *Curvularia affinis* Boedijn
Pe: *Curvularia alcornii* Manamgoda, L. Cai & K. D. Hyde
C, Pe: *Curvularia australiensis* (Bugnic. ex M. B. Ellis) Manamgoda, L. Cai & K. D. Hyde
(syns. *Bipolaris australiensis* (Bugnic. ex M. B. Ellis) Tsuda & Ueyama; *Drechslera australiensis* Bugnic. ex M. B. Ellis)
C: *Curvularia brachyspora* Boedijn
B: *Curvularia buchloës* (Lefebvre & Aar. G. Johnson) Manamgoda et al.
(syns. *Bipolaris buchloës* (Lefebvre & Aar. G. Johnson) Shoemaker; *Drechslera buchloës* (Lefebvre & Aar. G. Johnson) Subram. & B. L. Jain; *Helminthosporium buchloës* Lefebvre & Aar. G. Johnson)
C: *Curvularia eragrostidis* (Henn.) J. A. Mey.
(syns. *Brachysporium eragrostidis* Henn.; *Cochliobolus eragrostidis* (Tsuda & Ueyama) Sivan.; *Curvularia maculans* (C. K. Bancr.) Boedijn; *Spondylocladium maculans* C. K. Bancr.)
C, Pa, Pe: *Curvularia geniculata* (Tracy & Earle) Boedijn
(syns. *Cochliobolus geniculatus* R. R. Nelson; *Helminthosporium geniculatum* Tracy & Earle)
C: *Curvularia hawaiiensis* (Bugnic. ex M. B. Ellis) Manamgoda, L. Cai & K. D. Hyde
(syns. *Bipolaris hawaiiensis* (M. B. Ellis) J. Y. Uchida & Aragaki; *Cochliobolus hawaiiensis* Alcorn; *Drechslera hawaiiensis* Bugnic. ex M. B. Ellis; *Pseudocochliobolus hawaiiensis* (Alcorn) Tsuda & Ueyama)
C: *Curvularia heteropogonicola* (Sivan.) Alcorn
(syn. *Exserohilum heteropogonicola* Sivan.)
B, F, P, S: *Curvularia inaequalis* (Shear) Boedijn
(syns. *Acrothecium arenarium* Moreau & Mme. Moreau; *Helminthosporium inaequale* Shear)
C: *Curvularia intermedia* Boedijn
(syn. *Cochliobolus intermedius* R. R. Nelson)
A, Ax, C, F, L, P, Pa, Pe, S, Z: *Curvularia lunata* (Wakker) Boedijn
(syns. *Cochliobolus lunatus* R. R. Nelson & F. A. Haasis; *Helminthosporium caryopsidum* Sacc.; *Pseudocochliobolus lunatus* (R. R. Nelson & F. A. Haasis) Tsuda et al.)
C, Z: *Curvularia malina* Tomaso-Peterson & Jo
C: *Curvularia mebaldsii* Y. P. Tan & R. G. Shivas
Ax, P: *Curvularia micropus* (Drechsler) Hern.-Restr., Y. P. Tan & Crous
(syns. *Bipolaris micropus* (Drechsler) Shoemaker; *Drechslera micropus* (Drechsler) Subram. & B. L. Jain; *Exserohilum paspali* J. J. Muchovej & Nesio; *Helminthosporium micropus* Drechsler)
C: *Curvularia pallescens* Boedijn
(syns. *Curvularia leonensis* M. B. Ellis; *Pseudocochliobolus pallescens* Tsuda & Ueyama [syn. *Cochliobolus pallescens* (Tsuda & Ueyama) Sivan.])
Pe: *Curvularia penniseti* (Mitra) Boedijn
(syn. *Acrothecium penniseti* Mitra)
Ph: *Curvularia protuberata* R. R. Nelson & Hodges
A, F, L, Pe, S: *Curvularia ramosa* (Bainier) Boedijn
(syn. *Brachycladium ramosum* Bainier)
C: *Curvularia senegalensis* (Speg.) Subram.
(syns. *Acrothecium falcatum* Tehon; *Brachysporium senegalense* Speg.; *Curvularia falcata* (Tehon) Boedijn)
A, B, C, F, L, P, Z: *Curvularia spicifera* (Bainier) Boedijn
(syns. *Bipolaris spicifera* (Bainier) Subram.; *Drechslera spicifera* (Bainier) Arx; *Helminthosporium spiciferum* (Bainier) Nicot)
A, C, L, P, Pe: *Curvularia trifolii* (Kauffm.) Boedijn
(syn. *Brachysporium trifolii* Kauffm.)
B, C, Z: *Curvularia verruculosa* M. B. Ellis
(syns. *Cochliobolus verruculosus* (Tsuda & Ueyama) Sivan.; *Pseudocochliobolus verruculosus* Tsuda & Ueyama)
Dead spot: A, C
Ophiosphaerella agrostidis Dern., M. P. S. Câmara, N. R. O'Neill, Berkum & M. E. Palm
Dollar spot: C3, C4
C3: *Clarireedia bennettii* C. Salgado, L. A. Beirn, B. B. Clarke & J. A. Crouch
Pv: *Clarireedia hainanense* H. Zhang & J. Hu
F: *Clarireedia homoeocarpa* (F. T. Benn.) L. A. Beirn, B. B. Clarke, C. Salgado & J. A. Crouch
(syn. *Sclerotinia homoeocarpa* F. T. Benn.)
C3, Pv: *Clarireedia jacksonii* C. Salgado, L. A. Beirn, B. B. Clarke & J. A. Crouch
C4: *Clarireedia monteithiana* C. Salgado, L. A. Beirn, B. B. Clarke & J. A. Crouch
Pv: *Clarireedia paspali* Jian Hu & Lamour
Downy mildew (*See* Yellow tuft)
Drechslera leaf spot (*See* Pyrenophora leaf spots; melting-out; crown, root, and rhizome rot): C3, C4
P: *Drechslera andersenii* A. Lam
Endophytic fungi: C3
Epichloë coenophiala (Morgan-Jones & W. Gams) C. W. Bacon & Schardl
(syns. *Acremonium coenophialum* Morgan-Jones & W. Gams; *Neotyphodium coenophialum* (Morgan-Jones & W. Gams) Glenn, C. W. Bacon & Hanlin)

Epichloë festucae var. lolii (Latch, M. J. Chr. & Samuels) C. W. Bacon & Schardl
(syns. *Acremonium lolii* Latch, M. J. Chr. & Samuels; *Neotyphodium lolii* (Latch, M. J. Chr. & Samuels) Glenn, C. W. Bacon & Hanlin)
Epichloë typhina (Pers.) Tul. & C. Tul.
(syns. *Acremonium typhinum* Morgan-Jones & W. Gams; *Claviceps typhinum* (Pers.) F. M. Bailey; *Cordyceps typhina* (Pers.) Fr.; *Dothidea typhina* (Pers.) Fr.; *Neotyphodium typhinum* (Morgan-Jones & W. Gams) Glenn, C. W. Bacon & Hanlin)

Fairway patch: C, Pe
Phialocephala bamuru P. T. W. Wong & C. Dong

Fairy ring: C3, C4
Species of Agaricales and gastromycetes, mostly in the genera *Agaricus, Agrocybe, Arachnion, Bovista, Calvatia, Chlorophyllum, Clitocybe, Coprinus, Hygrocybe, Lepiota, Lycoperdon, Marasmius, Panaeolus, Scleroderma, Tricholoma,* and *Vascellum*

Flag smut: C3
Urocystis agropyri (Preuss) A. A. Fisch. Waldh.

Frost scorch: A, F, P
Ceratorhiza rhizodes (Auersw.) Z. Xu, T. C. Harr., M. L. Gleason & Batzer
(syn. *Sclerotium rhizodes* Auersw.)
Rhizophydium graminis Ledingham

Fusarium leaf spot, crown and root rot: P
Fusarium acuminatum Ellis & Everh.
(syn. *Gibberella acuminata* Wollenw.)
Fusarium avenaceum (Fr.:Fr.) Sacc.
Fusarium cerealis (Cooke) Sacc.
(syn. *Fusarium crookwellense* L. W. Burgess, P. E. Nelson & Toussoun)
Fusarium culmorum (Wm. G. Sm.) Sacc.
Fusarium equiseti (Corda) Sacc.
Fusarium graminearum Schwabe
Fusarium heterosporum Nees & T. Nees
Fusarium incarnatum (Desm.) Sacc.
Fusarium poae (Peck) Wollenw.
Fusarium pseudograminearum O'Donnell & T. Aoki
(syn. *Gibberella coronicola* T. Aoki & O'Donnell)

Fusarium patch (*See* Microdochium patch)

Gray leaf spot
C3: *Pyricularia oryzae* Cavara
(syn. *Magnaporthe oryzae* B. C. Couch)
C4: *Pyricularia grisea* Cooke ex Sacc.
(syn. *Magnaporthe grisea* (T. T. Hebert) M. E. Barr)

Gray snow mold: C3
Typhula incarnata Lasch ex Fr.
(syns. *Clavaria elegantula* P. Karst.; *Sclerotium fulvum* Fr.; *Typhula itoana* S. Imai)

Large patch: C4
Rhizoctonia solani J. G. Kühn
(syn. *Thanatephorus cucumeris* (A. B. Frank) Donk)

Leaf blotch: C3
Rhynchobrunnera orthospora (Caldwell) B. A. McDonald, U. Braun & Crous
(syn. *Rhynchosporium orthosporum* Caldwell)
Rhynchosporium secalis (Oudem.) Davis

Leaf rust
C3: *Puccinia brachypodii* G. H. Otth
C: *Puccinia cynodontis* Lacr. ex Desm.
C3: *Puccinia recondita* Roberge ex Desm.
S: *Puccinia stenotaphricola* J. Walker
Z: *Puccinia zoysiae* Dietel
(syn. *Dicaeoma zoysiae* (Dietel) Syd.)
F, P: *Uromyces dactylidis* G. H. Otth

Leaf and sheath spot: C3, C4
Waitea oryzae J. A. Crouch & Cubeta
(Replacement for invalid names *Rhizoctonia oryzae* and *Waitea circinata* var. *oryzae*)
Waitea prodiga J. A. Crouch & Cubeta
(Replacement for invalid name *Waitea circinata* var. *prodigus*)
Waitea zeae (Voorhees) J. A. Crouch & Cubeta
(syns. *Chrysorhiza zeae* (Voorhees) T. F. Andersen & Stalpers; *Moniliopsis zeae* (Voorhees) R. T. Moore; *Rhizoctonia endophytica* var. *filicata* H. K. Saksena & Vaartaja; *Rhizoctonia zeae* Voorhees; *Waitea circinata* var. *zeae* Toda, Mushika, T. Hayak., Tanaka, Tani & Hyakum.)

Leptosphaerulina leaf blight: A, F, L, P
Leptosphaerulina australis McAlpine
(syns. *Leptosphaerulina gaeumannii* (E. Müll.) Wehm.; *Pleospora gaeumannii* E. Müll.; *Pseudoplea gaeumannii* (E. Müll.) Wehm.)
Leptosphaerulina oryzae (I. Miyake) Phook., J. K. Liu & K. D. Hyde
(syn. *Pleosphaerulina oryzae* I. Miyake)

Mastigosporium leaf spot (leaf fleck): A, F, L, P, Ph
Mastigosporium album Riess
Mastigosporium cylindricum R. Sprague
Mastigosporium kitzebergense U. Schlöss.
Mastigosporium rubricosum (Dearn. & Barth.) Nannf.

Microdochium patch: C3: A, Ag, F, L, P; C4: C, Pv
Microdochium paspali W. Zhang, Z. B. Nan & M. J. Hu
Microdochium poae J. Liang & L. Cai
Monographella nivalis (Schaffnit) E. Müll.
(syns. *Calonectria nivalis* Schaffnit [syn. *Gerlachia nivalis* (Ces. ex Berl. & Voglino) W. Gams & E. Müll.]; *Fusarium nivale* Ces. ex Berl. & Voglino Voglino; *Griphosphaeria nivalis* (Schaffnit) E. Müll. & Arx; *Lanosa nivalis* Fr. [syns. *Fusarium nivale* (Fr.) Sorauer; *Microdochium nivale* (Fr.:Fr.) Samuels & I. C. Hallett]; *Micronectriella nivalis* (Schaffnit) C. Booth)

Necrotic ring spot: C3
Ophiosphaerella korrae (J. Walker & A. M. Sm. bis) Shoemaker & C. E. Babc.
(syn. *Leptosphaeria korrae* J. Walker & A. M. Sm. bis)

Net blotch (*See* Pyrenophora leaf spots)

Nigrospora blight: F, L, P, S
Nigrospora oryzae (Berk. & Broome) Petch
Nigrospora osmanthi Mei Wang & L. Cai
Nigrospora sphaerica (Sacc.) E. W. Mason

Physoderma leaf spot and leaf streak: C3, C4
Physoderma graminis (Büsgen) De Wild.
(syn. *Cladochytrium graminis* Büsgen)
Physoderma paspali J. A. Stev.

Pink patch: C, C3, Pv
Limonomyces roseipellis Stalpers & Loer.

Pink snow mold (*See* Microdochium patch)

Powdery mildew: C3
Blumeria graminis (DC.) Speer
(syn. *Erysiphe graminis* DC.)

Pseudoseptoria leaf spot: C3
Pseudoseptoria donacis (Pass.) B. Sutton
Pseudoseptoria everhartii (Sacc. & P. Syd.) B. Sutton
Pseudoseptoria stomaticola (Bäumler) B. Sutton

Pyrenophora leaf spots; melting-out; crown, root, and rhizome rot: C3, C4
C, F, L: *Pyrenophora avenae* Ito & Kurib.
(syns. *Drechslera avenae* (Eidam) Scharif; *Helminthosporium* avenae Eidam)

L: *Pyrenophora biseptata* (Sacc. & Roum.) Crous
(syns. *Drechslera biseptata* (Sacc. & Roum.) M. J. Richardson & E. M. Fraser; *Helminthosporium biseptatum* Sacc. & Roum.)
A, F, L: *Pyrenophora catenaria* (Drechsler) Rossman & K. D. Hyde
(syns. *Bipolaris catenaria* (Drechsler) Somal; *Drechslera catenaria* (Drechsler) Ito; *Helminthosporium catenarium* Drechsler)
A, C, F, L: *Pyrenophora dematioidea* (Bubák & Wróbl.) Rossman & K. D. Hyde
(syns. *Helminthosporium dematioideum* Bubák & Wróbl.; *Marielliottia dematioidea* (Bubák & Wróbl.) Shoemaker)
F, L, P, Z: *Pyrenophora dictyoides* A. R. Paul & Parbery
(syns. *Drechslera dictyoides* (Drechsler) Shoemaker; *Helminthosporium dictyoides* Drechsler)
A, L: *Pyrenophora erythrospila* A. R. Paul
(syns. *Drechslera erythrospila* (Drechsler) Shoemaker; *Helminthosporium erythrospilum* Drechsler)
A: *Pyrenophora fugax* (Wallr.) Rossman & K. D. Hyde
(syns. *Drechslera fugax* (Wallr.) Shoemaker; *Drechslera stenacra* (Drechsler) Shoemaker; *Helminthorporium fugax* Wallr.; *Helminthosporium stenacrum* Drechsler)
A, L: *Pyrenophora grahamii* Rossman & K. D. Hyde
(syns. *Drechslera phlei* (J. H. Graham) Shoemaker; *Helminthosporium dictyoides* var. *phlei* J. H. Graham; *Helminthosporium phlei* (J. H. Graham) Scharif)
F, L, P: *Pyrenophora lolii* Dovaston
(syns. *Drechslera siccans* (Drechsler) Shoemaker; *Helminthosporium siccans* Drechsler)
L: *Pyrenophora nobleae* (McKenzie & D. Matthews) Rossman & K. D. Hyde
(syn. *Drechslera nobleae* McKenzie & D. Matthews)
L, P: *Pyrenophora poae* (Baudyš) Y. Marin & Crous
(syns. *Drechslera poae* (Baudyš) Shoemaker; *Drechslera vagans* (Drechsler) Shoemaker; *Helminthosporium poae* Baudyš)
L, P, Pe: *Pyrenophora seminiperda* (Brittleb. & D. B. Adam) Shoemaker
(syns. *Bipolaris cyclops* (Drechsler) R. Sprague; *Drechslera campanulata* (Lév.) B. Sutton; *Helminthosporium cyclops* Drechsler; *Pleosphaeria seminiperda* Brittleb. & D. B. Adam)
L, P: *Pyrenophora teres* Drechsler
(syns. *Helminthosporium fragosoi* Bubák; *Pyrenophora graminea* S. Ito & Kurib.)
F, P: *Pyrenophora trichostoma* (Fr.:Fr.) Fuckel
(syns. *Pleospora trichostoma* (Fr.:Fr.) Ces. & De Not.; *Sphaeria trichostoma* Fr.:Fr.)
F, P, S: *Pyrenophora tritici-repentis* (Died.) Drechsler
(syns. *Drechslera tritici-repentis* (Died.) Shoemaker; *Helminthosporium tritici-repentis* Died.; *Pleospora trichostoma* f. sp. *tritici-repentis* Died.; *Pleospora tritici-repentis* Died.)
A, F, L, P: *Pyrenophora triseptata* (Drechsler) Rossman & K. D. Hyde
(syns. *Drechslera triseptata* (Drechsler) Subram. & B. L. Jain; *Helminthosporium triseptatum* Drechsler)

Pythium foliar blight and root rot: C3, C4
Globisporangium ultimum (Trow) Uzuhashi, Tojo & Kakish.
(syns. *Globisporangium ultimum* var. *sporangiiferum* (Drechsler) Uzuhashi, Tojo & Kakish.; *Pythium ultimum* Trow; *Pythium ultimum* var. *sporangiiferum* Drechsler)
Pythium aphanidermatum (Edson) Fitzp.
(syns. *Nematosporangium aphanidermatum* (Edson) Fitzp.; *Rheosporangium aphanidermatum* Edson)
Pythium graminicola Subram.
(syn. *Pythium graminicola* var. *stagni* Höhnk)
Pythium torulosum Coker & P. Patt.
Pythium vanterpoolii V. Kouyeas & H. Kouyeas

Pythium root and crown rot: C3, C4
Globisporangium carolinianum (V. D. Matthews) Uzuhashi, Tojo & Kakish.
(syn. *Pythium carolinianum* V. D. Matthews)
Globisporangium debaryanum (R. Hesse) Uzuhashi, Tojo & Kakish.
(syns. *Eupythium debaryanum* (R. Hesse) Nieuwl.; *Pythium debaryanum* R. Hesse)
Globisporangium irregulare (Buisman) Uzuhashi, Tojo & Kakish.
(syns. *Pythium irregulare* Buisman; *Pythium irregulare* var. *hawaiiense* Sideris)
Globisporangium rostratum (E. J. Butler) Uzuhashi, Tojo & Kakish.
(syn. *Pythium rostratum* E. J. Butler)
Phytopythium vexans (de Bary) Abad, de Cock, Bala, Robideau, Lodhi & Lévesque
(syns. *Pythium vexans* de Bary; *Ovatisporangium vexans* (de Bary) Uzuhashi, Tojo & Kakish.)
Pythium aristosporum Vanterp.
Pythium arrhenomanes Drechsler
(syn. *Nematosporangium arrhenomanes* (Drechsler) Sideris)
Pythium catenulatum V. D. Matthews
Pythium graminicola Subram.
Pythium myriotylum Drechsler
Pythium periplocum Drechsler
(syn. *Pythium periplocum* var. *coimbatorense* M. S. Balakr.)
Pythium tardicrescens Vanterp.
Pythium volutum Vanterp. & Truscott

Pythium root dysfunction: A
Pythium aristosporum Vanterp.
Pythium arrhenomanes Drechsler
Pythium volutum Vanterp. & Truscott

Pythium snow blight: C3
Globisporangium iwayamae (S. Ito) Uzuhashi, Tojo & Kakish.
(syn. *Pythium iwayamae* S. Ito)

Ramularia leaf spot: A, Ag, L, P
Ramularia holci-lanati (Cavara) Deighton
Ramularia pusilla Unger

Rapid blight: A, P
Labyrinthula terrestris D. M. Bigelow, M. W. Olsen & Gilb.

Red leaf spot (*See* Pyrenophora leaf spots; melting-out; crown, root, and rhizome rot)

Red thread: C, C3, Pv, Z
Laetisaria fuciformis (McAlpine) Burds.
(syns. *Athelia fuciformis* (Berk.) Burds.; *Corticium fuciforme* (Berk.) Wakef.; *Isaria fuciformis* Berk.)

Rusts (*See* Crown rust, Leaf rust, and Stripe rust)

Seedling diseases: C3, C4
Species of *Fusarium, Microdochium, Pythium,* and *Rhizoctonia;* less often, species of *Bipolaris, Colletotrichum, Curvularia, Drechslera,* and others.

Septoria leaf spot: C3
P: *Septoria macropoda* Pass.
Septoria macropoda var. *grandis* R. Sprague
Septoria macropoda var. *septulata* (Gonz. Frag.) R. Sprague
Septoria oudemansii Sacc.

L: *Septoria loligena* R. Sprague
Septoria tritici var. *lolicola* R. Sprague & Aar. G. Johnson
F: *Septoria tenella* Cooke & Ellis
A: *Septoria calamagrostidis* (Lib.) Sacc.
Septoria secalis Prill. & Delacr.
Septoria secalis var. *stipae* R. Sprague
Septoria triseti Speg.
C: *Septoria cynodontis* Fuckel
Ag: *Septoria agropyrina* Lobik
Septoria elymi Ellis & Everh.
Smuts (*See* Blister smut, Flag smut, and Stripe smut)
Snow scald: C3
Sclerotinia borealis Bubák & Vleugel
(syns. *Myriosclerotinia borealis* (Bubák & Vleugel) L. M. Kohn; *Sclerotinia graminearum* Elenev ex Solkina)
Southern blight: C3, C4
Athelia rolfsii (Curzi) C. C. Tu & Kimbr.
(syns. *Corticium rolfsii* Curzi; *Sclerotium rolfsii* Sacc.)
Speckled snow mold: C3
Typhula ishikariensis var. *canadensis* J. D. Sm. & Årsvoll
Typhula ishikariensis var. *idahoensis* (Remsberg) Årsvoll & J. D. Sm.
Typhula ishikariensis var. *ishikariensis* S. Imai
Spermospora leaf spot: C3
Spermospora ciliata (R. Sprague) Deighton
Spermospora lolii MacGarvie & O'Rourke
Spermospora poagena (R. Sprague) MacGarvie & O'Rourke
Spring dead spot: C, Z
Ophiosphaerella herpotricha (Fr.) J. C. Walker
(syn. *Ophiobolus herpotrichus* (Fr.) Sacc.)
Ophiosphaerella korrae (J. C. Walker & A. M. Sm. bis) Shoemaker & C. E. Babc.
(syn. *Leptosphaeria korrae* J. Walker & A. M. Sm. bis)
Ophiosphaerella narmari (J. Walker & A. M. Sm. bis) H. C. Wetzel, Hulbert & Tisserat
(syn. *Leptosphaeria narmari* J. Walker & A. M. Sm. bis)
Stagonospora leaf spot
Pa: *Neottiosporina paspali* (G. F. Atk.) B. Sutton & Alcorn
(syn. *Stagonospora paspali* G. F. Atk.)
Ag, A, F, Ph: *Parastagonospora avenae* (A. B. Frank) Quaedvlieg, Verkley & Crous
(syns. *Septoria avenae* A. B. Frank; *Stagonospora avenae* (A. B. Frank) Bissett)
Ag, A, F, Ph, P: *Parastagonospora nodorum* (Berk.) Quaedvlieg, Verkley & Crous
(syns. *Septoria nodorum* (Berk.) Berk.; *Stagonospora nodorum* (Berk.) E. Castell. & Germano)
Ag, F, L, P: *Stagonospora arenaria* (Sacc.) Sacc.
Stem rust: C3, C4
Puccinia graminis Pers.:Pers.
Stripe rust: P
Puccinia striiformis Westend.
Stripe smut: C3
Ustilago salweyi Berk. & Broome
(syns. *Tilletia striiformis* (Westend.) Sacc.; *Uredo striiformis* Westend.; *Ustilago poae-annuae* W. H. Davis; *Ustilago poae-pratensis* W. H. Davis; *Ustilago striiformis* (Westend.) Niessl; *Ustilago striiformis* f. *agrostis-palustris* W. H. Davis; *Ustilago striiformis* f. *phlei-pratensis* W. H. Davis)
Summer patch: A, F, P
Magnaporthiopsis agrostidis P. Wong, Khemmuk & R. G. Shivas
Magnaporthiopsis meyeri-festucae J. Luo & N. Zhang
Magnaporthiopsis poae (Landsch. & N. Jacks.) J. Luo & N. Zhang
(syn. *Magnaporthe poae* Landsch. & N. Jacks.)
Superficial fairy ring: C3, C4
Coprinopsis kubickae (Pilát & Svrček) Redhead, Vilgalys & Moncalvo
(syn. *Coprinus kubickae* Pilát & Svrček)
Trechispora cohaerens (Schwein.) Jülich & Stalpers
(syns. *Corticium byssinum* var. *microsporum* Bres.; *Sporotrichum cohaerens* Schwein.)
Trechispora farinacea (Pers.:Fr.) Liberta
Take-all patch: A
Gaeumannomyces avenae (E. M. Turner) Hern.-Restr. & Crous
(syns. *Gaeumannomyces graminis* var. *avenae* (E. M. Turner) Dennis; *Ophiobolus graminis* var. *avenae* E. M. Turner)
Take-all root rot (*See* Bermudagrass decline and take-all root rot)
Tar spot: C3, C4
Phyllachora acuminata Starbäck
Phyllachora bulbosa D. G. Parbery
Phyllachora cynodontis (Sacc.) Niessl
Phyllachora dactylidis Delacr.
Phyllachora eragrostidis Chardón
Phyllachora fuscescens (Speg.) Sacc.
Phyllachora graminis (Pers.:Fr.) Fuckel
Phyllachora helvetica Fuckel
Phyllachora minutissima (Welw. & Curr.) A. L. Sm.
Phyllachora panici (Schwein.) Sacc.
(Replacement for invalid name *Phyllachora puncta*)
Phyllachora paspalicola Henn.
Phyllachora quadraspora Tehon
Phyllachora sylvatica Sacc. & Speg.
Thatch collapse: C3, C4
Sphaerobolus stellatus Tode
White blight: C3
Melanotus phillipsii (Berk. & Broome) Singer
(syn. *Agaricus phillipsii* Berk. & Broome)
Yellow patch: C3
Ceratobasidium cereale D. Murray & L. L. Burpee
(syns. *Ceratorhiza cerealis* (E. P. Hoeven) R. T. Moore; *Rhizoctonia cerealis* E. P. Hoeven)
Yellow ring: P
Trechispora alnicola (Bourdot & Galzin) Liberta
Yellow tuft (downy mildew): A, P, S
Sclerophthora macrospora (Sacc.) Thirum., C. G. Shaw & Naras.
Zonate leaf spot (See Pyrenophora leaf spots)
Zoysia patch (*See* Brown patch; Large patch)

MISCELLANEOUS DISEASES AND DISORDERS

Black layer: A
Anaerobic soil plus blue-green algae and/or sulfate-reducing bacteria
Fungus-like Organisms: C3, C4
Lagenocystis radicicola (Vanterp. & Ledingham) H. Copel.
Olpidium brassicae (Woronin) P. A. Dang.
Polymyxa graminis Ledingham
Slime molds (superficial, not pathogenic): C3, C4
Didymium squamulosum (Alb. & Schwein.) Fr.
Fuligo spp.
Mucilago crustacea A. Wigg.
Physarum cinereum (Batsch) Pers.
Physarum spp.

MOLLICUTE DISEASES

Aster yellows: Aster yellows phytoplasma
White leaf: Bermudagrass white leaf phytoplasma
Yellow dwarf: Unknown mollicute

NEMATODES, PARASITIC: C3, C4

Awl: *Dolichodorus* spp. Cobb
Cyst: *Heterodera* spp. Schmidt
Dagger: *Xiphinema* spp. Cobb
Lance: *Hoplolaimus* spp. von Daday
Lesion: *Pratylenchus* spp. Filipjev
Needle: *Longidorus* spp. (Micoletzky) Thorne & Swanger
Pin: *Paratylenchus* spp.
Pseudo-root-knot: *Hypsoperine* spp. Sledge & Golden
Ring: *Criconemella, Criconemoides, Macroposthonia,* and *Mesocriconema* spp.
Root-gall: *Subanguina* spp.
Root-knot: *Meloidogyne* spp. Goeldi
Sheath: *Hemicycliophora* spp.
Sheathoid: *Hemicriconemoides* spp.
Spiral: *Helicotylenchus* spp. Steiner
Stem gall: *Anguina pacificae* Cid del Prado Vera & Maggenti
Sting: *Belonolaimus* spp. Steiner
Stubby-root: *Paratrichodorus* and *Trichodorus* spp.
Stylet or stunt: *Tylenchorhynchus* spp. Cobb

VIRUS DISEASES

Centipedegrass mosaic: E
Panicum mosaic virus (PMV), genus *Panicovirus*
Mosaic: S
Sugarcane mosaic virus (SCMV), genus *Potyvirus*
St. Augustine decline: S
Panicum mosaic virus (PMV), genus *Panicovirus*

Glossary

a.i.—active ingredient
C—Celsius or centigrade
cm—centimeter (1 cm = 0.01 m = 0.39 in.)
F—Fahrenheit
g—gram (1 g = 0.035 oz.)
gal—gallon (1 gal liquid [U.S.] = 3.785 L)
in.—inch (1 in. = 2.54 cm)
kg—kilogram (1 kg = 2.2 lb)
L—liter (1 L = 1.06 quarts liquid [U.S.])
m—meter (1 m = 39.37 in.)
mg—milligram (1 mg = 0.001 g)
ml—milliliter (1 ml = 0.001 L)
mm—millimeter (1 mm = 0.001 m = 0.04 in.)
μm—micrometer (1 μm = 10^{-6} m)
nm—nanometer (1 nm = 10^{-9} m)
oz—ounce (1 oz = 28.35 g); fluid ounce (1 fl oz [U.S.] = 29.57 ml)

abiotic—nonliving; pertaining to nonliving agents that cause diseases or to the diseases caused by such agents
acervulus (pl. **acervuli**)—erumpent, saucer-shaped, cushionlike fruiting body bearing conidiophores, conidia, and sometimes setae
acid—having a pH less than 7
acropetal—upward from the base to the apex of a shoot of a plant (*see* **basipetal**)
aeciospore—dikaryotic spore produced in the aecium of a rust fungus; in heteroecious rust fungi, a spore stage that infects the alternate host
aecium (pl. **aecia**; adj. **aecial**)—the fruiting body in which the first dikaryotic spores (aeciospores) of a rust fungus are produced
aerial—occurring in air
agar—jellylike material derived from algae and used to solidify liquid culture media; a medium containing agar
alkaline—having basic (nonacidic) properties; having a pH greater than 7
alternate host—one of two kinds of plant on which a parasitic fungus (e.g., a rust fungus) must develop to complete its life cycle
alternative host—a plant that can be colonized by a parasite, other than its main host, but is not required by the parasite for completion of its developmental cycle
anamorph (adj. **anamorphic**)—the asexual form in the life cycle of a fungus, producing asexual spores (such as conidia) or no spores (syn. **imperfect state**; *see* **teleomorph**)
anastomosis (pl. **anastomoses**)—fusion between branches of the same or different structures (e.g., hyphae) to make a network
annual—a plant that completes its life cycle and dies within 1 year (*see* **perennial**)
antagonist—an organism or substance that counteracts or limits the action of another
anterior—toward the front or head (*see* **posterior**)
anthracnose—disease caused by acervuli-forming fungi (order Melanconiales) and characterized by sunken lesions and necrosis
antibiotic—a chemical compound produced by a microorganism that kills or inhibits the growth of other living organisms
aplerotic—pertaining to an oospore that does not fill the oogonium
apothecium (pl. **apothecia**)—open, cuplike or saucerlike, ascus-bearing fungal fruiting body (ascocarp), often supported on a stalk
appressed (**adpressed**)—closely flattened down or pressed against a surface
appressorium (pl. **appressoria**)—swollen, flattened portion of a fungal filament that adheres to the surface of a higher plant, providing anchorage for invasion by the fungus
arbuscule (adj. **arbuscular**)—branched haustorial structure formed by certain endomycorrhizal fungi within living cells of the root
arthroconidium (pl. **arthroconidia**)—jointed conidium composed of more than one cell and capable of becoming separated
ascocarp—sexual fruiting body (e.g., apothecium, ascostroma, cleistothecium, perithecium, or pseudothecium) producing asci and ascospores of ascomycetous fungi (syn. **ascoma**)
ascomycetes (adj. **ascomycetous**)—a group of fungi in the phylum Ascomycota (kingdom Fungi) characterized by the production of sexual spores (ascopores) within an ascus
Ascomycetes (adj. **ascomycetous**)—a group of fungi, also called the Ascomycotina, characterized by the production of sexual spores (ascospores) within an ascus
ascospore—sexual spore borne in an ascus
ascostroma (pl. **ascostromata**)—a fruiting body containing bitunicate (double-walled) asci in locules (cavities), usually dark with multiple locules, but sometimes single (*see* **pseudothecium**)
ascus (pl. **asci**)—saclike structure containing ascospores (typically eight) and usually borne in a fungal fruiting body
aseptate—lacking septa (cross-walls); nonseptate; coenocytic
asexual—vegetative; lacking sex organs, gametes, or sexual spores; pertaining to the imperfect or anamorphic stage of a fungus
asexual reproduction—any type of reproduction not involving the union of gametes and meiosis
auricle—a short, claw-shaped, rounded, or pointed attachment at the base of a grass leaf blade
avirulent—nonpathogenic; unable to cause disease (*see* **virulent**)
azygospore—haploid spore produced by apomictic development, characteristic of some mycorrhizal fungi

bacilliform—shaped like short rods with rounded ends
bacterium (pl. **bacteria**)—any of various prokaryotic, microscopic, single-celled organism with a cell wall that increases by binary fission
bar—unit of pressure used to express water potential (1 bar = 0.987 atmosphere)
basidiocarp—sexual fruiting body of a basidiomycetous fungus (syn. **basidioma**)
basidiomycetes (adj. **basidiomycetous**)—a group of fungi in the phylum Basidiomycota (kingdom Fungi) characterized by the formation of external basidiospores on basidia
basidiospore—haploid (1N) sexual spore produced on a basidium
basidium (pl. **basidia**; adj. **basidial**)—a structure on which haploid basidiospores are produced externally
basipetal—downward from the apex toward the base of a shoot; developing in the direction of the base, so that the apical part is oldest (*see* **acropetal**)
binucleate—having two nuclei per cell (*see* **multinucleate**)
biological control (**biocontrol**)—exploitation of the natural competition, parasitism, or antagonism of organisms for management of pests and pathogens
biotic—living; pertaining to diseases caused by living organisms
biotroph—organism that can live and multiply only on another living organism (syn. **obligate parasite**; *see* **necrotroph**)
biotype—a subdivision of a species, subspecies, or race based on some identifiable physiological trait, such as a specific virulence pattern

bitunicate—having two walls
blade—the flat portion of a grass leaf above the sheath
blastospore—asexual spore developed by budding or extrusion from an existing spore or cell
bleached—white to straw-colored, with reference to necrotic tissue
blend—a combination of different cultivars of the same turfgrass species
blight—sudden, severe, and extensive spotting, discoloration, wilting, or destruction of leaves, flowers, stems, or entire plants
blotch—irregularly shaped, usually superficial spot or blot
broadcast application—spray application of a fertilizer, pesticide, or other management input
bulbil—sclerotium-like structure lacking a distinct rind layer
bursa—extension or flap of cuticle at the side of the male nematode sex organ, used for orienting during mating

calcareous—rich in calcium carbonate (lime)
canopy—the expanded leafy top of a plant or plants
catenulate—formed in chains or rows
causal agent—organism or agent that produces a disease
cereal—grass grown for its edible seed
chasmothecium—the ascocarp of a powdery mildew fungus; it has no natural opening at maturity, but opens by the rupturing of its wall (*see also* **cleistothecium**)
chlamydospore—thick-walled or double-walled asexual resting spore formed from hyphal cells (terminal or intercalary) or by transformation of conidial cells and functioning as an overwintering stage
chloridoid—pertaining to grasses in the subfamily Chloridoideae (Eragrostoideae), adapted mostly to warm temperate to tropical climates
chlorophyll (adj. **chlorophyllous**)—one of a group of green pigments found in chloroplasts and important in photosynthesis
chloroplast—disklike structure containing chlorophyll, in which photosynthesis occurs in green plants
chlorosis (adj. **chlorotic**)—failure of chlorophyll development, caused by disease or a nutritional disturbance; fading of green plant color to light green, yellow, or white
chronic—pertaining to slow-developing, persistent, or recurring symptoms
chytrid—any fungus in the order Chytridiales
ciliate—having protoplasmic filaments (cilia) that are similar to flagella but are shorter and more numerous on the cell
circulative—pertaining to viruses that pass through the gut and circulate in the body of an insect vector before being transmitted to a host
clamp connection—a bridge- or buckle-like hyphal protrusion in basidiomycetous fungi, formed at cell division and connecting the newly divided cells
clavate—claviform, club-shaped
cleistothecium (pl. **cleistothecia**)—a spherical ascocarp that is closed at maturity (*see also* **chasmothecium**)
clypeus (pl. **clypei**)—shieldlike stromatic growth over one or more perithecia
coalesce—to grow together into one body (used with reference to lesions or patches that merge together as they enlarge)
coenocytic—having multiple nuclei embedded in cytoplasm without cross-walls; nonseptate
coleoptile—ephemeral tissue sheathing the first true leaf of a grass seedling
collar—the portion of a seedling or plant near the surface of the soil
colonization—establishment and ramification of a pathogen within a host plant
colonize—to infect and ramify through plant tissue during the growth of a pathogen
conidiogenous—producing and bearing conidia
conidiophore—simple or branched hypha on which conidia are produced
conidium (pl. **conidia**)—an asexual, nonmotile fungal spore that develops externally or is liberated from the cell that formed it
corpus—the narrow, cylindrical part of the esophagus of a nematode
cortex (adj. **cortical**)—the region of parenchymal tissue between the epidermis and the phloem in stems and roots; the region beneath the rind of a sclerotium
crown—compacted series of nodes from which tillers, culms, and roots arise
culm—the stem of grasses, cereals, and bamboos
cultivar (abbr. **cv.**)—a cultivated variety of a plant species, resulting from deliberate genetic manipulation and having defined characteristics such as color, growth habit, or disease resistance (*see* **variety**)
cultural practices—horticultural management practices used to promote the health of plants, such as application of nutrients, irrigation, and cultivation
cuticle (adj. **cuticular**)—noncellular outer layer of an insect or a nematode; water-repellent, waxy layer of epidermal cells of plant parts, such as leaves, stems, and fruit
cylindrical—shaped like a cylinder
cyst—with reference to fungi, a resting structure in a protective membrane or shell-like enclosure; with reference to nematodes, the egg-laden carcass of a female nematode; with reference to bacteria, a specialized type of bacterial cell enclosed in a thick wall, often dormant and resistant to environmental conditions
cytoplasm—the living protoplasm in a cell, except the nucleus

damping-off—death of a seedling before or shortly after emergence caused by decomposition of the root or lower stem (it is common to distinguish between preemergence and postemergence damping-off)
desiccate—to dry out
deuteromycete—any member of a class of fungi with no known sexual stages
diagnostic (n. **diagnosis**)—pertaining to a distinguishing characteristic important for the identification of a disease or other condition
diclinous—having the oogonium and antheridium on different hyphae
dikaryon (adj. **dikaryotic**)—a cell having two sexually compatible haploid nuclei per cell, which divide simultaneously (this phase is called the dikaryophase)
diploid—having two complete sets of chromosomes (2N chromosomes) (*see* **haploid**)
disease—abnormal functioning of an organism
disease complex—disease resulting from combined or sequential actions of two or more biotic or abiotic agents
disease cycle—the succession of all events and interactions among a host, a parasite, and the environment in the process of disease, from initial infection of the host by a causal agent, through pathogenesis, to overseasoning, until another infection occurs
dissemination—dispersal; spread of infectious material (inoculum) from diseased plants to healthy plants
DNA—deoxyribonucleic acid, the double-stranded helical molecule that contains genetic code information, consisting of repeating units, or nucleotides, each of which is composed of deoxyribose (a sugar), a phosphate group, and a purine (adenine or guanine) or a pyrimidine (thymine or cytosine) base
dolipore septum—a type of cross-wall in fungi in the group basidiomycetes, characterized by distinctive swellings and membranes in association with the septal pore
dormancy (adj. **dormant**)—a condition of suspended growth and reduced metabolism of an organism, generally induced by internal factors or environmental conditions as a mechanism of survival
dormant—resting; living in a state of reduced physiological activity

echinulate—having small spines projecting from the cell wall
ecology—the study of the interactions between individual organisms, between groups of organisms, and between organisms and their environment
ectomycorrhiza—mycorrhizal association in which the fungus does not penetrate the root (*see* **mycorrhiza**)
ectoparasite—parasite that feeds from the exterior of its host (*see* **endoparasite**)
ectotrophic—pertaining to fungi that develop primarily on the surface of roots
edaphic—of or pertaining to soil
ELISA—enzyme-linked immunosorbent assay, a serological test in which the sensitivity of the reaction is increased by attaching an enzyme that creates a colored product in the presence of one of the reactants
emergence—growth of a seedling shoot through the surface of the soil
endodermis—the layer of cells between the vascular tissue and the cortex of roots
endogenous—arising from inside

endomycorrhiza—mycorrhizal association in which the fungus penetrates the root (*see* **mycorrhiza**)
endoparasite—parasite that lives and feeds inside its host (*see* **ectoparasite**)
endophyte—plant developing inside another organism; any of various endoparasitic fungi associated with grass species
enzyme—protein that catalyzes a specific biochemical reaction
epidemic—an increase of disease in a population; a general and severe outbreak of disease (*see* **epiphytotic**)
epidemiology (adj. **epidemiological**)—the study of factors influencing the initiation, development, and spread of infectious disease; the study of disease in populations of plants
epidermis (adj. **epidermal**)—surface layer of cells of leaves and other plant parts
epiphytotic—pertaining to the widespread and destructive outbreak of a plant disease
erumpent—bursting or erupting through the substrate surface
etiolation—elongation of stems caused by reduced intensity of light
etiology—the study of the causes of diseases
eukaryote—any of various organisms consisting of cells containing a membrane-bound nucleus and other organelles, including all higher plants, animals, fungi, and protists (*see* **prokaryote**)
extracellular—not occurring or located within a cell
exudate—liquid excreted or discharged from diseased tissues, from roots and leaves, or by fungi

f. sp.—*see* ***forma specialis***
facultative—capable of changing lifestyle (e.g., from saprophytic to parasitic or the reverse)
facultative parasite—organism that is normally saprophytic but is capable of being parasitic
facultative saprophyte—organism that is normally parasitic but is capable of being saprophytic
fallow—pertaining to cultivated land kept free from a crop or weeds during the normal growing season
filament (adj. **filamentous**)—a thin, flexible, threadlike structure
filamentous—filiform, threadlike
filiform—long, needlelike
flagellum (pl. **flagella**)—hairlike, whiplike, or tinsel-like appendage of a motile cell, bacterium, or zoospore, providing locomotion
fleck—minute discolored spot in green tissue
flexuous—having turns or windings; capable of bending
flock—cottonlike tuft
focus (pl. **foci**)—small area of diseased plants within a population
foliar—pertaining to leaves
foot cell—basal cell, as of a spore or conidiophore
forma specialis (pl. ***formae speciales***)—a taxonomic group within a pathogenic species, defined by its host range (members of different *formae speciales* infect different groups of plants), abbr. **f. sp.** (pl. **ff. spp.**)
frass—excrement of an insect
friable—easily crumbled into small aggregates, as with soil
fruiting body—any of various complex, spore-bearing fungal structures
fumigant (v. **fumigate**)—a gas or volatile substance used to kill or inhibit the growth of microorganisms or other pests
fungicide (adj. **fungicidal**)—chemical or physical agent that kills or inhibits the growth of fungi
fungistat (adj. **fungistatic**)—chemical agent that inhibits fungal growth or sporulation but does not kill fungi
fusiform—spindle-shaped, tapering at the end

gametangium (pl. **gametangia**)—cell containing gametes or nuclei that act as gametes
gamete—sex cell
gene—a unit of the chromosome in the cells of an organism, controlling heritable characteristics of the organism
genus (pl. **genera**)—a taxonomic category that consists of a group of closely related (structurally or phylogenetically) species (the genus or generic name is the first name in a Latin binomial)
germ tube—hypha resulting from an outgrowth of the spore wall and cytoplasm after germination
germinate (n. **germination**)—to begin growth, as of a seed, spore, sclerotium, or other reproductive body
giant cell—enlarged, multinucleate cell formed in roots by repeated nuclear division without cell division, induced by secretions of certain sedentary plant-parasitic nematodes
girdle—to circle and cut through a stem, disrupting the phloem and xylem
globose—nearly spherical
Gramineae—taxonomic family containing the grasses
graminicolous—living upon grass
Gram-negative—with reference to bacteria, staining red or pink in the Gram staining procedure after treatment with Gram's stain
Gram-positive—with reference to bacteria, staining violet or purple in the Gram staining procedure after treatment with Gram's stain
gravid—containing an egg or eggs; capable of depositing eggs
gubernaculum—guiding piece for the spicule of a male nematode
guttation—exudation of watery, sticky liquid from hydathodes, especially along leaf margins

haploid—having a single complete set of chromosomes (*see* **diploid**)
haustorium (pl. **haustoria**)—specialized branch of a parasite formed inside host cells to absorb nutrients
hemibiotroph—an organism that is parasitic in living tissue for some time and then continues to live in dead tissue
herbicide—chemical agent used for killing or inhibiting the growth of plants (e.g., a weed or grass killer)
heterokaryon (adj. **heterokaryotic**)—a cell with genetically different nuclei
heterothallism (adj. **heterothallic**)—the condition in which sexual reproduction can occur only in the presence of genetically different mycelia (*see* **homothallism**)
heterozygous—having different forms (alleles) of a gene on homologous chromosomes (*see* **homozygous**)
hilum—the scar on a spore at the point of attachment to a conidiophore or sterigma
homothallism (adj. **homothallic**)—the condition in which sexual reproduction occurs in a single, self-fertile thallus (*see* **heterothallism**)
homozygous—having the same form (allele) of a gene on homologous chromosomes (*see* **heterozygous**)
honeydew—sugary ooze or exudate, often from aphids
host plant—living plant attacked by or harboring a parasite or pathogen and from which the invader obtains part or all of its nourishment
host range—the range of plants on which an organism, particularly a parasite, feeds
hyaline—transparent or nearly so; translucent; colorless
hybrid (v. **hybridize**)—offspring of two individuals of different genotypes
hydathode—epidermal leaf structure specialized for secretion or exudation of water; opening at the terminus of a leaf vein
hymenium—the continuous, spore-bearing layer of a fungal fruiting body
hypha (pl. **hyphae**; adj. **hyphal**)—tubular filament of a fungal thallus or mycelium, the basic structural unit of a fungus
hyphopodium (pl. **hyphopodia**)—specialized, usually flattened hypha of an epiphytic fungus, for attachment to or penetration of a host
hypogenous—situated below the base of an oogonium

immune—not capable of being infected by a given pathogen
imperfect state—the asexual form in the life cycle of a fungus, producing asexual spores (such as conidia) or no spores (syn. **anamorph**; *see* **perfect state**)
in situ—in the original place or environment
in vitro—in glass, on artificial media, or in an artificial environment; outside a host
in vivo—within a living organism
infection—the process in which an organism enters, invades, or penetrates and establishes a parasitic relationship with a host plant
infection court—a site in or on a host plant where infection can occur
infection cushion—organized mass of hyphae formed on the surface of a plant and producing numerous infective hyphae
infection peg—specialized, narrow hyphal strand produced on the underside of an appressorium and serving to penetrate host cells (syn. **penetration peg**)
infectious—with reference to disease, capable of spreading from plant to plant

infective—able to attack a host and cause infection; with reference to vectors, carrying or containing a pathogen and able to transfer it to a host plant
infest (n. **infestation**)—to attack as a pest (used especially of insects and nematodes); to contaminate, as with microorganisms; to be present in large numbers
inflorescence—a flower or flower cluster
injury—damage caused by transitory interaction with an agent such as an insect, a chemical, or an unfavorable environmental condition
inoculate (n. **inoculation**)—to place inoculum in an infection court; to insert a pathogen into healthy tissue
inoculum (pl. **inocula**)—pathogen or its parts, capable of causing infection when transferred to a favorable location
inoculum density—a measure of the number of propagules of a pathogenic organism per unit of area or volume
intercalary—inserted within (e.g., located along a hypha, as opposed to located at the end of a hypha)
intercellular—between or among cells
internode (adj. **internodal**)—the portion of a stem between two successive nodes
interveinal—between veins (of a leaf)
intracellular—occurring or located within a cell or through cells
isolate—(n.) a culture or subpopulation of a microorganism separated from its parent population and maintained under controlled conditions; (v.) to remove from soil or host material for growth in pure culture
isometric—with reference to virus particles, icosahedral in structure, appearing approximately round

juvenile—an immature form that appears similar to but usually smaller than the adult and is not sexually mature (e.g., insects with gradual metamorphosis, nematodes)

knot—a gall; a localized abnormal swelling

larva (pl. **larvae**)—the immature, wormlike or caterpillar stage in the life cycle of an insect that undergoes complete metamorphosis
latent—present but not manifested or visible, as a symptomless infection by a pathogen
latent infection—infection unaccompanied by visible symptoms
latent period—the period between infection and the production of new inoculum; the time after a vector has acquired a pathogen and before it can be transmitted
leaf axis—the axis parallel to the venation of a grass leaf
leaf bud—primordial of a leaf
leaf spot—a plant disease lesion typically restricted in development in the leaf after reaching a characteristic size
lesion—a localized diseased area or wound
life cycle—the cyclical stages in the growth and development of an organism
lignin—a complex organic substance or group of substances that impregnate the cell walls of xylem vessels and certain other plant cells, constituting one of the main components of woody tissue
ligule—in grass leaves, an outgrowth from the upper and inner side of the leaf blade where it joins the sheath
lobulate—lobed
lunate—crescent-shaped, like a new moon

macerate—to cause disintegration of tissues by separation of cells; to soften by soaking
macroconidium (pl. **macroconidia**)—the larger of two kinds of conidia formed by certain fungi (*see* **microconidium**)
manual transmission—the spread or introduction of inoculum to an infection court by hand manipulation
mating types—compatible strains (usually designated + and – or A and B) necessary for sexual reproduction in heterothallic fungi
mechanical injury—injury of a plant by abrasion, mutilation, or wounding
mechanical transmission—the spread or introduction of inoculum to an infection court by human manipulation, accompanied by physical disruption of host tissues (wounding)
median bulb—*see* **metacorpus**
medium (pl. **media**)—a mixture of organic or inorganic chemical compounds and water, providing the nutrients needed for the growth of a microorganism in vivo; with reference to higher plants, a mixture of fertilizers and other components in which a plant is growing
medulla—the central part of an organ or tissue; the inner part of a sclerotium enclosed in a rind
meiosis—the process of nuclear division in which the number of chromosomes per nucleus is halved (i.e., converting the diploid state to the haploid state) (*see* **mitosis**)
melanin (adj. **melanoid**)—a dark brown to black pigment; common in sclerotia and other survival structures
meristem (adj. **meristematic**)—plant tissue characterized by frequent cell division, producing cells that become differentiated into specialized tissues
mesophyll—the central, internal, nonvascular tissue of leaves
metabolism—the transformation of nutrients into energy and byproducts within the cell
metabolite—any chemical participating in metabolism; a nutrient
metacorpus—the swollen posterior portion of a nematode corpus, also called the median bulb
microclimate—weather conditions on a small scale, such as at the surface of a plant or within a planting
microconidium (pl. **microconidia**)—the smaller of two kinds of conidia formed by certain fungi (*see* **macroconidium**)
microflora—the combination of all microorganisms in a particular environment
microorganism—an organism small enough that as an individual it can be seen only with the aid of microscope; microbe
midrib—the central, thickened vein of a leaf
migratory—migrating from place to place on a plant or from plant to plant when feeding (*see* **sedentary**)
mildew—thin coating of mycelial growth and spores on the surfaces of infected plant parts
mitosis—the process of nuclear division in which the chromosome number remains the same (*see* **meiosis**)
mixture—a combination of two or more turfgrass species
MLO—*see* **phytoplasma**
mold—any microfungus (especially an economically important saprobe) with conspicuous, profuse, or woolly superficial growth (mycelium or spore masses) on various substrates, commonly growing on damp or decaying matter and on the surface of plant tissues
mollicute—any of a group of prokaryotic organisms bounded by flexuous membranes and lacking cell walls (*see* **phytoplasma**; **spiroplasma**)
molt—to shed a cuticle or body encasement during a phase of growth
monilioid—having swellings at regular intervals, like a string of beads
monogenic—determined by a single gene (*see* **polygenic**)
monoclinous—having the gametangium and oogonium originating from the same hypha
morphology (adj. **morphological**)—the study of the form of organisms; the form and structure of organisms
mosaic—disease symptom characterized by nonuniform coloration of plant tissue, with intermingled normal and light green or yellowish patches, usually caused by a virus (often used interchangeably with **mottle**)
motile—capable of self-propulsion by means of flagella, cilia, or amoeboid movement
mottle—disease symptom characterized by light and dark areas in plant tissue, appearing in an irregular pattern, usually caused by a virus (often used interchangeably with **mosaic**)
mucilaginous—viscous, slimy
mulch—a layer of material, such as organic matter or plastic, applied to the surface of the soil for purposes such as retention of water and inhibition of weeds
multicellular—consisting of more than one cell
multinucleate—having more than one nucleus per cell (*see* **binucleate**)
multiseptate—having many septa (cross-walls)
muriform—having transverse and longitudinal septa
mutation (n. **mutant**)—an abrupt heritable or genetic change in a gene or an individual as a result of an alternation in a gene or chromosome or an increase in chromosome number
mycelium (pl. **mycelia**; adj. **mycelial**)—mass of hyphae constituting the body (thallus) of a fungus
mycoparasite—a fungus that attacks another fungus
mycoplasmalike organism—*see* **phytoplasma**

mycorrhiza (pl. **mycorrhizae**; adj. **mycorrhizal**)—a symbiotic association between a nonpathogenic or weakly pathogenic fungus and the roots of a plant

myxomycete—organism characterized by a multinucleate, motile mass of protoplasm

necrosis (adj. **necrotic**)—death of cells or tissue, usually accompanied by blackening or browning

necrotroph—parasite that typically kills host cells and obtains its energy from dead host cells (*see* **biotroph**)

nematicide—agent, usually a chemical, that kills nematodes

nematode—any of various nonsegmented roundworms parasitic on plants or animals or free living in soil or water

node (adj. **nodal**)—the enlarged portion of a shoot at which a leaf or bud arises

noninfectious disease—disease caused by an abiotic agent and not transmissible from one plant to another

nonpersistent transmission—a type of virus transmission in which a virus is acquired and transmitted by a vector after a short feeding time and is retained by the vector for only a short time (syn. **stylet-borne transmission**)

nonseptate—lacking cross-walls; coenocytic

obclavate—shaped like an upside-down club

obligate parasite—organism that can grow only as a parasite in association with its host plant and cannot be grown in artificial culture media (syn. **biotroph**)

obligate saprophyte—organism that is limited to a saprophytic mode of life

obovoid—shaped like an upside-down egg

obtuse—rounded or blunt; pertaining to an angle greater than 90° but less than 180°

odontostyle—the protrusible hollow spear or grooved tooth in nematodes of the class Adenophorea (often called a spear)

oogonium (pl. **oogonia**)—female gametangium of fungi in the group oomycetes, containing one or more gametes

oomycetes (adj. **oomycetous**)—a group of funguslike organisms in the phylum Oomycota (kingdom Chromista); typically has nonseptate mycelium, asexual sporangia and zoospores, and sexual oospores

oospore—the thick-walled, sexually derived resting spore of organisms in the group oomycetes

ooze—mass of bacterial cells mixed with host fluids

organelle—any of various membrane-bound structures (e.g., mitochondria, chloroplasts) having a specialized function within a cell

ostiole (adj. **ostiolate**)—pore; opening in the papilla or neck of a perithecium, pseudothecium, or pycnidium, through which spores are released

oversummer—to survive or persist through the summer

overwinter—to survive or persist through the winter

ovate—oval

ovoid—egg-shaped

panicoid—pertaining to grasses in the subfamily Panicoideae, adapted mostly to subtropical and tropical climates

papilla (pl. **papillae**; adj. **papillate**)—a small, blunt projection

paraphysis (pl. **paraphyses**)—a hairlike cell within a fungal fruiting structure

parasite (adj. **parasitic**)—organism that lives in intimate association with another organism, on which it depends for its nutrition, but not necessarily as a pathogen

parthenogenesis (adj. **parthenogenetic**)—reproduction by the development of an unfertilized egg

pasteurization—the process by which a material, usually a liquid, is treated with heat to eliminate selected harmful microorganisms

patch—in grasses, a distinctly delimited, somewhat circular area in which most or all plants are affected by disease

pathogen (adj. **pathogenic**)—a disease-producing organism or agent

pathogenesis—production and development of disease

pathogenicity—ability to cause disease

pathology—the study of diseases

pathotype—a subdivision of a pathogenic species characterized by its pattern of virulence or avirulence to differential host varieties

pathovar (abbr. **pv.**)—a subdivision of a plant-pathogenic bacterial species, defined by host range (a pathovar of a bacterial species is equivalent to a *forma specialis* of a fungal species)

penetration—initial invasion of a host by a pathogen

penetration peg—specialized, narrow hyphal strand produced on the underside of an appressorium and serving to penetrate host cells (syn. **infection peg**)

perennial—a plant that survives for several to many years (*see* **annual**)

perfect—sexual; capable of sexual reproduction

perfect state—the sexual form in the life cycle of a fungus (syn. **teleomorph**; *see* **anamorph**)

perithecium (pl. **perithecia**)—the flask-shaped or subglobose, thin-walled fruiting body (ascocarp) of an ascomycetous fungus, containing asci and ascospores, and having a pore (ostiole) at its apex, through which ascospores are expelled or released

persistent transmission—a type of virus transmission in which a virus is acquired and transmitted by a vector after a relatively long feeding time and remains transmissible for a prolonged period while in association with the vector (syn. **circulative transmission**)

pest—any organism that damages plants or plant products

pesticide—chemical agent used to control pests

pH—the negative logarithm of the effective hydrogen ion concentration, used as a measure of acidity (pH 7 is neutral; values less than pH 7 are acidic; values greater than pH 7 are alkaline)

phasmid—posterior sensory pore of certain nematodes

phenotype—the composite of observable physical qualities of an organism produced by the interaction of its genotype with the environment

phialide—end cell of a conidiophore with one or more open ends through which a basipetal succession of conidia develops

phialospore—conidium produced on a phialide

phloem—food-conducting, food-storing tissue in the vascular system of roots, stems, and leaves

photosynthate—a product of photosynthesis

photosynthesis—manufacture of carbohydrates from carbon dioxide and water in the presence of chlorophyll, using light energy and releasing oxygen

physiologic race—a group of individuals belonging to the same species and differing from other members of the species in their behavior or other characteristics but not in morphology; physiologic form

phytoalexin—any of numerous antibiotics produced by plants under attack by a pathogen or parasite, in a phytochemical defense mechanism analogous to the production of antibodies in animals

phytopathology—the study of plant diseases (syn. **plant pathology**)

phytoplasma—any of various plant-parasitic pleomorphic mollicutes (prokaryotes lacking cell walls) found in phloem tissue (previously called mycoplasmalike organisms, or MLOs)

phytotoxic—harmful to plants (usually with reference to chemicals)

pigment—a colored compound, such as chlorophyll, in the cells of plants or fungi

plant pathology—the study of plant diseases (syn. **phytopathology**)

plasmodium (pl. **plasmodia**)—naked multinucleate mass of protoplasm moving and feeding in amoeboid fashion

plate mycelium—mass of usually dark-pigmented mycelium and haustorium-like cells on plant tissues

pleomorphic—able to assume various shapes (and perhaps sizes); characterized by the occurrence of more than one distinct form in the life cycle of an organism

plerotic—pertaining to an oospore that fills the oogonium, as in *Pythium* spp.

polar—at one end (pole) of the cell

polygenic—pertaining to, or governed by, many genes (*see* **monogenic**)

polyhedral—having many sides or faces

pooid—pertaining to grasses in the subfamily Pooideae (Festucoideae), adapted mostly to temperate and subarctic climates

posterior—toward the back or tail (*see* **anterior**)

predispose (n. **predisposition**)—to make prone to infection and disease

primary inoculum—inoculum (usually from an overwintering source) that initiates disease in the field, as opposed to inoculum that spreads disease during the season (syn. **initial inoculum**)

procorpus—cylindrical portion of the corpus (esophagus) anterior to the metacorpus in nematodes

prokaryote—any of various organisms lacking internal membrane-bound organelles and a distinct nucleus, including bacteria and mollicutes (*see* **eukaryotic**)
promycelium (pl. **promycelia**)—in rust and smut fungi, a germ tube issuing from a teliospore and bearing basidiospores
propagule—any part of an organism capable of independent growth
protectant—an agent, usually a chemical, applied to a plant surface in advance of exposure to a pathogen to prevent infection
protein—any of numerous nitrogen-containing organic compounds composed of amino acids
protist (pl. **protists**)—any member of the kingdom Protoctista, comprising a diverse group of unicellular and some simple multinuclear and multicellular eukaryotic organisms, including some algae, protozoans, and water molds, many of which possess flagella
protoplasm—the living contents of a cell
protoplast—the living contents of a cell exclusive of the cell wall
pseudomonad—any bacterium of the genus *Pseudomonas*
pseudothecium (pl. **pseudothecia**)—perithecium-like fruiting body containing asci and ascospores that are dispersed rather than borne in an organized hymenium; an ascostroma having a single locule or cavity and containing bitunicate asci
pustule—a small, blisterlike elevation of plant epidermis, formed as spores of a pathogen emerge
pv.—pathovar
pycnidiospore—spore (conidium) produced in a pycnidium
pycnidium (pl. **pycnidia**)—asexual, globose or flask-shaped fruiting body of certain imperfect fungi producing conidia
pycniospore—haploid, sexually derived spore formed in the pycnium of a rust fungus
pycnium (pl. **pycnia**)—globose or flask-shaped, haploid fruiting body of rust fungi, bearing receptive hyphae and pycniospores
pyriform—pear-shaped

quarantine—an enforced isolation imposed to prevent the spread of disease; pertaining to plant disease, legislative control of the transport of plants or plant parts to prevent the spread of pests or pathogens
quiescent—dormant or inactive

race—subgroup or biotype within a species or variety, distinguished from other races by virulence, symptom expression, or host range, but not by morphology
receptive hypha—the part of the pycnium (spermagonium) of a rust fungus that receives the nucleus of a pycniospore (spermatium)
reniform—kidney-shaped
resistant (n. **resistance**)—possessing properties that prevent or impede disease development (*see* **susceptible**)
respiration—series of chemical reactions that make energy available through oxidation of carbohydrates and fat
resting spore—spore, often thick-walled, that can remain alive in a dormant state for some time, later germinating and capable of initiating infection
reticulate—having netlike markings
rhizomycelium—filamentous, haustorium-like fungal thallus that resembles mycelium
rhizosphere—the microenvironment in the soil immediately around plant roots
rind—an outer layer
RNA—ribonucleic acid, any of several nucleic acids composed of repeating units of ribose (a sugar), a phosphate group, and a purine (adenine or guanine) or a pyrimidine (uracil or cytosine) base, transcribed from DNA, and involved in translation to proteins
root hair—threadlike, single-celled outgrowth from a root epidermal cell
rot—softening, discoloration, and often disintegration of plant tissue as a result of fungal or bacterial infection
runner hyphae—thickened hyphal strands
rust—any of several diseases caused by specialized fungi in the group basidiomycetes, some of the spores of which are a rusty color

sanitation—destruction or removal of infected and infested plants or plant parts; decontamination of tools, equipment, containers, work space, hands, etc.
saprobe—*see* **saprophyte**
saprophyte (adj. **saprophytic**)—organism that obtains nourishment from nonliving organic matter (syn. **saprobe**)
satellite virus—a virus that accompanies another virus and depends on it for its multiplication
scald—a necrotic condition in which tissue is usually bleached and has the appearance of having been exposed to high temperatures
sclerotium (pl. **sclerotia**)—a vegetative resting body of a fungus, composed of a compact mass of hyphae with or without included host tissue, and usually having a darkened rind
scorch—any symptom that suggests the action of flame or fire on the affected tissue, often at the margins of leaves
scutellum (adj. **scutellar**)—cotyledon of a grass embryo
secondary inoculum—inoculum produced by an infection established previously during the same growing season
secondary organism—organism that multiplies in already diseased tissue but is not the primary pathogen
sedentary—remaining in a fixed location (*see* **migratory**)
seed—ripened ovule consisting of an embryo and stored food enclosed by a seed coat
seed treatment—application of a biological agent, chemical substance, or physical treatment to seed, to protect the seed or plant from pathogens or to stimulate germination or plant growth
seedborne—carried on or in seed
selective medium—a culture medium containing substances that inhibit or prevent the growth of certain species of microorganisms
senesce (adj. **senescent**; n. **senescence**)—to decline, as with maturation, age, or disease stress
septate—having one or more septa (cross-walls)
septum (pl. **septa**; adj. **septate**)—a dividing wall; in fungi, a cross-wall
sessile—with reference to leaves, leaflets, flowers, florets, fruit, ascocarps, basidiocarps, etc., lacking a stalk, petiole, pedicel, stipe or stem; with reference to nematodes, permanently attached; not capable of moving about
seta (pl. **setae**)—bristle-like or hairlike structure, usually deep yellow or brown and thick-walled
sexual spore—spore produced during the sexual cycle of a fungus
sheath—the lower part of a grass leaf, clasping the culm; a membranous cover
sieve element—a tube-shaped living cell in the phloem, functioning in the transport of dissolved organic substances in the plant (syn. **sieve tube element**)
sign—an indication of disease from direct observation of a pathogen or its parts on a host plant (*see* **symptom**)
sinuous—having many curves, bends, or turns
slime mold—any of various saprophytic organisms in the class Myxomycetes, forming vegetative amoeboid plasmodia and spores
smut—any of several diseases caused by fungi in the group basidiomycetes that typically release masses of black, dusty teliospores at maturity
soil drench—application of a solution or suspension of a chemical to the soil, especially a pesticide to control soilborne pathogens
soilborne—carried on or beneath the soil surface
sorus (pl. **sori**)—compact fruiting structure, especially a spore mass of rust or smut fungi
sp. (pl. **spp.**)—species (used only after a genus name; the singular form is used to refer to an undetermined species; the plural form is used to refer to two or more species without naming them individually)
spear—an odontostyle, as opposed to a stomatostyle
spermagonium (pl. **spermagonia**)—structure in which male reproductive cells are produced; in rust fungi, a globose or flask-shaped, haploid fruiting body, also called a pycnium, composed of receptive hyphae and spermatia (pycniospores)
spermatium (pl. **spermatia**)—a male sex cell; a nonmotile male gamete; a haploid male gamete; in rust fungi, a pycniospore
spicule—the copulatory organ of male nematodes
spiroplasma—a spiral-shaped plant-pathogenic mollicute (prokaryote lacking a cell wall)
sporangiophore—sporangium-bearing body of a fungus
sporangiospore—nonmotile, asexual spore borne in a sporangium
sporangium (pl. **sporangia**)—saclike fungal structure in which the entire contents are converted into an indefinite number of asexual spores
spore—a reproductive structure of fungi and some other organisms, containing one or more cells; a bacterial cell modified to survive an adverse environment

sporidium (pl. **sporidia**)—the basidiospore of rust fungi, smut fungi, and other members of the group basidiomycetes
sporocarp—spore-bearing fruiting body
sporodochium (pl. **sporodochia**)—superficial, cushion-shaped asexual fruiting body consisting of a cluster of conidiophores
sporulate—to produce spores
sporulating—producing and (often) liberating spores
spot—a symptom of disease characterized by a limited necrotic area, as on leaves, flowers, and stems
stele—central cylinder of vascular tissue, especially in roots
sterigma (pl. **sterigmata**)—a small, usually pointed projection that supports a spore
sterile—unable to reproduce sexually; free of living microorganisms
sterile fungus—a fungus that is not known to produce any kind of spores
sterilization (adj. **sterilized**)—the total destruction of living organisms by various means, including heat, chemicals, or irradiation
stigma—the portion of a flower that receives pollen and on which pollen germinates
stipe—stalk
stippling—a pattern of small dots or speckles on leaves where chlorophyll is absent
stolon—a slender, horizontal stem that grows close to the soil surface, also called a runner
stoma (pl. **stomata**; adj. **stomatal**)—a structure composed of two guard cells and the opening between them in the epidermis of a leaf or stem, functioning in gas exchange, also called a stomate
stomatostyle—the protrusible, needle-like feeding tube (often called a stylet) in nematodes of the class Secernentea
strain—a distinct form of an organism or virus within a species, differing from other forms of the species biologically, physically, or chemically
streak—necrosis along vascular bundles in leaves or stems of grasses
striate (n. **striations**)—to mark with delicate lines, grooves, or ridges
stripe—necrosis of tissue between vascular bundles in leaves or stems of grasses
stroma (pl. **stromata**)—compact mass of mycelium (with or without host tissue) that supports fruiting bodies or in which fruiting bodies are embedded
stunting—reduction in height of a vertical axis resulting from a progressive reduction in the length of successive internodes or a decrease in their number
stylet—the stiff, slender, hollow feeding organ of plant-parasitic nematodes or sap-sucking insects, such as aphids or leafhoppers; with reference to nematodes, a stomatostyle, as opposed to an odontostyle
stylet knobs—the basal protuberances, usually three, of the stylet of a nematode
subepidermal—located or occurring just below the epidermis
subglobose—not uniformly spherical
subspherical—not uniformly spherical
substrate—the substance on which an organism lives or from which it obtains nutrients; a chemical substance acted upon, often by an enzyme
sunscald—injury of plant tissues burned or scorched by direct sunlight
suscept—a susceptible plant
susceptible (n. **susceptibility**)—prone to develop disease when infected by a pathogen (*see* **resistant**)
suspensor—hypha supporting a gamete, gametangium, or especially a zygospore
sward—a lawn composed of grasses or other ground covers such as clovers
swarm spore—zoospore, as in *Pythium* spp.
symbiosis (adj. **symbiotic**)—mutually beneficial association of two different kinds of organisms
symptom—an indication of disease from the reaction of a host (e.g., canker, leaf spot, wilt) (*see* **sign**)
symptomatology—the set of symptoms characteristic of a particular disease
symptomless carrier—a plant that, although infected with a pathogen (usually a virus), produces no obvious symptoms
syn.—synonym
syncytium (pl. **syncytia**)—a multinucleate structure in root tissue formed by the dissolution of common cell walls induced by secretions of certain sedentary plant-parasitic nematodes (e.g., cyst nematodes)
syndrome—the total effects of a disease on a plant, whether produced all at once or successively
synergism (adj. **synergistic**)—interaction in which the total effect of interacting factors is greater than the additive effects of the factors acting individually
synnema (pl. **synnemata**)—compact or fused, generally upright conidiophores, with branches and spores forming a headlike cluster (syn. **coremium**)
systemic—pertaining to a disease in which the pathogen (or a single infection) spreads generally throughout a plant; pertaining to chemicals that spread internally through a plant

taxonomy (adj. **taxonomic**)—the science of naming and classifying organisms
teleomorph—the sexual form in the life cycle of a fungus (syn. **perfect state**; *see* **anamorph**)
teliospore—thick-walled resting spore produced by some fungi, notably rust and smut fungi, from which a basidium is produced
telium (pl. **telia**)—the fruiting body (sorus) in which the teliospores of a rust fungus are produced
thallus—the vegetative body of a fungus
thatch—the layer of plant litter from long-term accumulation of dead plant roots, crowns, rhizomes, and stolons
tiller—a lateral shoot, culm, or stalk arising from a crown bud, common in grasses
tissue—a group of cells, usually of similar structure, that perform the same or related functions
tolerance (adj. **tolerant**)—the ability of a plant to endure disease, adverse conditions, or chemical injury without serious damage or yield loss
toxicity—the capability of a substance to interfere with the vital processes of an organism
toxin—poisonous substance of biological origin
tracheid—xylem cell that conducts water and solutes from roots to shoots
translocation—movement of water, nutrients, chemicals, or food materials within a plant
translucent—sufficiently clear to allow the passage of light
transmit (n. **transmission**)—to spread or transfer, as in spreading an infectious pathogen from plant to plant or from one generation of plants to another
transpiration—water loss by evaporation from leaf surfaces and through stomata
truncate—to end abruptly, as though an end has been cut off
turgid—swollen; inflated; plump or swollen as a result of internal water pressure
turgidity—the state of being rigid or swollen as a result of internal water pressure

ubiquitous—constantly encountered; widespread
unicellular—one-celled (*see* **multicellular**)
uniflagellate—having one flagellum
urediniospore—asexual, dikaryotic, often rusty-colored spore produced in the uredinium of a rust fungus (syns. **urediospore**, **uredospore**); the "repeating stage" of a heteroecious rust fungus (i.e., the spore stage capable of infecting the host plant on which it is produced)
uredinium (pl. **uredinia**)—fruiting body (sorus) producing urediniospores of a rust fungus (syn. **uredium**)

vacuole—generally spherical organelle within a plant cell, bound by a membrane and containing dissolved materials such as metabolic precursors, storage materials, or waste products
variety (adj. **varietal**)—a plant type within a species, resulting from deliberate manipulation and having recognizable characteristics such as color, shape of flowers, fruits, seeds, and form (*see* **cultivar**)
vascular—pertaining to fluid-conducting tissues (xylem and phloem) in plants
vascular bundle—strand of conductive tissue, usually composed of xylem and phloem (in leaves, small bundles are called veins)
vascular cylinder—the cylinder of vascular tissue in stems and roots (syn. **stele**)

vascular wilt disease—xylem disease that disrupts normal uptake of water and minerals, resulting in wilting and yellowing of foliage
vector—a living organism (e.g., insect, mite, bird, higher animal, nematode, parasitic plant, human) able to carry and transmit a pathogen and disseminate disease; in genetic engineering, a cloning vehicle, or self-replicating DNA molecule, such as a plasmid or virus, used to introduce a fragment of foreign DNA into a host cell
vegetative—pertaining to the somatic or asexual parts of a plant, which are not involved in sexual reproduction
vein—a small vascular bundle in a leaf
vermiform—worm-shaped
verrucose—having small, rounded processes or warts
vesicle—the thin sac in which zoospores are differentiated and from which they are released; the bulbous head terminating the conidiophores of *Aspergillus* spp.; a structure formed by an endomycorrhizal fungus within living cells of the root
vessel—water-conducting structure of xylem tissue with pit openings in end walls
viable (n. **viability**)—in a living state; able to germinate, as seeds, fungus spores, sclerotia, etc.; capable of growth
virion—complete virus particle
viroid—an infectious, nonencapsidated (naked) circular, single-stranded RNA
virulence—degree or measure of pathogenicity; relative capacity to cause disease
virulent—highly pathogenic; capable of causing severe disease (*see* **avirulent**)
viruliferous—virus-laden, usually with reference to insect or nematode vectors
virus—any of numerous submicroscopic, intracellular obligate parasites consisting of a core of infectious nucleic acid (either RNA or DNA) usually surrounded by a protein coat

water-soaked—pertaining to plant tissues or lesions that appear wet, dark, and usually sunken and translucent as a disease symptom
whorl—circular arrangement of like parts
wilt—drooping of leaves and stems from lack of water caused by inadequate water supply or excessive transpiration; vascular disease that interrupts normal water uptake
witches'-broom—disease symptom characterized by an abnormal, massed, brushlike development of many weak shoots arising at or close to the same point

xylem—water- and mineral-conducting, food-storing, supporting tissue of a plant

yellows—plant disease characterized by chlorosis and stunting of the plant

zonate—targetlike; pertaining to a symptom appearing in concentric rings
zoosporangium—sporangium, or spore case, that bears zoospores
zoospore—fungal spore with flagella, capable of locomotion in water
zygomycetes (adj. **zygomycetous**)—a group of fungi in the phylum Zygomycota (kingdom Fungi) that produce resistant, spherical sexual spores following fusion of hyphal gametangia
zygospore—the sexual resting spore formed from the union of gametangia of fungi in the group zygomycetes

Selected References

Agrios, G. N. 2005. Plant Pathology. 5th ed. Academic Press, San Diego, CA.

D'Arcy, C. J., Eastburn, D. M., and Schumann, G. L. 2001. Illustrated Glossary of Plant Pathology. Plant Health Instructor. doi:10.1094/PHI-I-2001-0219-01

Holliday, P. 2001. A Dictionary of Plant Pathology. 2nd ed. Cambridge University Press, Cambridge.

Kirk, P. M., Cannon, P. F., David, J. C., and Stalpers, J. A., eds. 2001. Ainsworth & Bisby's Dictionary of the Fungi. 9th ed. CABI Publishing, Wallingford, U.K.

Stearn, W. T. 1995. Botanical Latin: History, Syntax, Terminology, and Vocabulary. 4th ed. Timber Press, Portland, OR.

Index